CRC Handbook of Animal Models of Renal Failure

Editors

Stephen R. Ash, M.D., F. A. C. P.
Director of Research
Ash Medical Systems, Inc.
Professor of Comparative Medicine
School of Veterinary Medicine
Purdue University
West Lafayette, Indiana

and

Jerry A. Thornhill, D. V. M.
Director
Veterinary Internal Medicine and
Dialysis Center Ltd.
Veterinary Specialty Clinic Building
Riverwoods, Illinois
Research Associate, Section of Nephrology
Department of Medicine
Hines V. A. Hospital
Hines, Illinois

CRC Press
Taylor & Francis Group
Boca Raton London New York

CRC Press is an imprint of the
Taylor & Francis Group, an informa business

First published 1985 by CRC Press
Taylor & Francis Group
6000 Broken Sound Parkway NW, Suite 300
Boca Raton, FL 33487-2742

Reissued 2018 by CRC Press

Library of Congress Cataloging in Publication Data

Main entry under title: CRC Handbook of animal models of renal failure.

Bibliography: p.
Includes index.
1. Renal insufficiency—Animal models. 2. Dogs—
Diseases. I. Ash, Stephen R., 1945-
II. Thornhill, Jerry, A., 1941- . III. Title:
C.R.C. handbook of animal models of renal failure.
[DNLM: 1. Kidney failure, Chronic. 2. Kidney Failure,
Acute. 3. Disease Models, Animal. WJ 342 C911]
RC918.R4C73 1985 616.6'1 84-14223
ISBN 0-8493-2975-2

A Library of Congress record exists under LC control number: 84014223

Publisher's Note
The publisher has gone to great lengths to ensure the quality of this reprint but points out that some imperfections in the original copies may be apparent.

Disclaimer
The publisher has made every effort to trace copyright holders and welcomes correspondence from those they have been unable to contact.

ISBN 13: 978-1-315-89391-4 (hbk)
ISBN 13: 978-1-351-07301-1 (ebk)

Visit the Taylor & Francis Web site at http://www.taylorandfrancis.com and the
CRC Press Web site at http://www.crcpress.com

FOREWORD

Failure of the kidneys, or renal failure, is a serious condition in which toxic waste substances continually accumulate in the blood. Untreated, renal failure will result in malfunction of almost every organ system in the body (uremia). Death of the patient will occur within days, unless the dialysis procedure is instituted.

Dialysis is a chemical process in which concentration gradients cause substances to diffuse across semipermeable membranes. It may be performed clinically, by removal of blood from the body, and perfusion of cellulosic membrane packages (hemodialysis), or by placement of salt solution in the abdomen with transfer of substances across the peritoneal membrane (peritoneal dialysis). Both of these dialysis procedures are, in theory, quite simple, but in practice difficult. With intermittent dialysis procedures, patients may be successfully maintained for several weeks (in presence of acute renal failure), or for many years (in end-stage renal disease, or ESRD). With the exception of the artificial heart, dialysis is the only artificial organ therapy successfully able to support patients for long periods of time. Dialysis is always instituted before transplantation is considered.

The unparalleled medical success of dialysis in supporting patients has resulted in unparalleled problems for society, the medical community, and families of patients. Both hemodialysis and peritoneal dialysis are costly, risky, and require extensive patient training and/or assistance. In addition, hemodialysis frequently causes nausea, headache, and low blood pressure during the therapy, thus necessitating the presence of a partner, nurse, or technician during the procedure.

Disposable items are quite expensive, for hemo- or peritoneal dialysis, as are the expenses of maintaining home support and training programs. The overall cost for support of each patient with ESRD ranges from \$16,000/year for continuous ambulatory peritoneal dialysis (at home), to \$25,000/year for in-center hemodialysis.

Since Medicare began coverage of ESRD therapy in 1973, the number of dialysis patients in the U.S. has increased from 20,000 to 70,000. The toal expenditure in the U.S. for ESRD therapy is \$1.5 × 10-9. Hospitals and insurance companies are now beginning to limit coverage of ESRD therapy. "Uniform reimbursemnt" is designed to encourage the less expensive, home dialysis techinques, as the therapy for the majority of patients. However, the population of patients now on dialysis are older and more frequently have disease of several organ systems, than the population of the 1970s. Therefore, training of these patients, and rehabilitation, will be very difficult. Also, long term complications of dialysis are more evident and damaging.

It would thus appear, that we have "painted ourselves into a corner", in this area of medical practice. It has been necessary to implement a rather "imperfect" therapy in support of patients, since the "conservative" therapy is eventually ineffective, in most patients, and strict "patient selection" is untenable in a caring and informed society. The resulting strains on medical and economic resources is a problem to be faced directly and solved, since similar strains will certainly develop after introduction of other artificial organ therapies, and will soon develop in provision of medical care in general (10% of our gross national product now goes to health care).

The solution to the dialysis "problem" lies in three areas of research: (1) improvements in techniques of dialysis, with greater simplicity, safety, and cost effectiveness, (2) understanding of the diseases which initiate renal failure, and the mechanisms of kidney damage, and (3) analysis and dissection of the toxic effects of uremia on the various organ systems, and the contribution of failure in other organs to the final renal "shutdown". Research in these areas have been severely limited by the lack of suitable animal models of renal failure. While a variety of kidney dieases such as glomerulonephritis, pyelonephritis, interstitial nephritis, etc. have been created in laboratory animals, these diseases do not create kidney damage fast enough or serious enough to create uremia and diminish the exretory efficiency of the kidney to nearly zero in a reasonable time. In this text, we present a variety of methods of creation of renal failure, by author's experience in the study and support of laboratory animal models of renal failure.

Ironically, any long-term study of uremia is also a study of dialysis. If severe renal failure is instituted in animals such as dogs, uremic complications in the gastrointestinal, cardiovascular and endocrine organs are even more severe than in humans. As such, the dog is an excellent model of uremic complications, but without dialysis, the animal will merely be studied in the rapid, downhill course leading to death in 2 to 3 days. For this reason, considerable effort has been expended at several centers, to adapt human hemo- and peritoneal dialysis techniques to support of uremic dogs. These efforts have not only allowed study of the uremic condition, but also allowed identification and analysis of the etiology of various complications of dialysis.

Also presented in this text are three studies of the mechanisms of renal damage and renal failure in animal models. The etiologic agents of these studies are all "clinically relevant" and result in the renal failure of a portion of the ESRD or acute dialysis population: (1) diabetes mellitus, (2) cardiovascular shock and toxin exposure, and (3) aminoglycoside antibiotics and the septic conditions for which they are used. Particularly in the case of diabetes mellitus, these disease conditions are resulting in an increasing number of patients on ESRD therapy. Of the 1.2 million diabetics on insulin in the U.S., 14,000 have already developed ESRD, and now make up 20% of the ESRD population.

A wide variety of other kidney diseases also can result in renal failure, including glomerulonephritis, interstitial nephritis, hypertension and collagen vascular diseases. For a review of each of these diseases, a number of excellent textbooks exist. Also, for study of uremic toxins, studies of human serum by chromatography have effectively identified enough biochemically distinct compounds to occupy biochemistry for many years. These studies are reviewed elsewhere. Similar investigations of the toxic effect of nitrogenous compounds are now being performed in the controlled conditions of dialytic support of animal models of renal failure.

Stephen R. Ash, M.D., F.A.C.P.
Jerry A. Thornhill, D.V.M.

CONTRIBUTORS

Leslee Angell
Veterinary Assistant
Acorn Research Laboratory
Austin, Texas

Stephen R. Ash, M. D., F.A.C.P.
Director of Research
Ash Medical Systems, Inc.
Associate Professor of Comparative Medicine
School of Veterinary Medicine
Purdue University
West Lafayette, Indiana

John T. Daugirdas, M. D.
Assistant Section Chief, Renal and Hypertension Section
Hines V.A. Hospital
Hines, Illinois
Assistant Professor of Medicine
Loyola University Stritch School of Medicine
Maywood, Illinois

Alison Booth Harter
Chief Research Technician
Acorn Research Laboratory
Austin, Texas

Mary Hiatt
Department of Biomedical and Chemical Engineering
University of Texas at Austin
Austin, Texas

Todd S. Ing, M.D.
Chief, Renal and Hypertension Section
Hines V.A. Hospital
Hines, Illinois
Professor of Medicine
Loyola University Stritch School of Medicine
Maywood, Illinois

Elsa Janle-Swain, Ph.D.
Visiting Instructor
Department of Biology
Purdue University
Research Consultant
Ash Medical Systems, Inc.
West Lafayette, Indiana

Jack W. Moncrief, M.D.
President
Acorn Research Laboratory
Co-Director, Dialysis and Transplantation
Austin Diagnostic Clinic
Austin, Texas

Carol V. Mullins-Blackson, R.N.
Research Coordinator
Acorn Research Laboratory
Austin, Texas

Robert P. Popovich, Ph.D.
Vice President
Acorn Research Laboratory
Professor
Department of Biomedical and Chemical Engineering
University of Texas at Austin
Austin, Texas

W. K. Pyle, Ph.D.
Department of Biomedical Engineering
University of Texas at Austin
Austin, Texas

J. Edmond Riviere, D.V.M., Ph.D.
Associate Professor of Pharmacology/Toxicology
Laboratory of Toxicokinetics
School of Veterinary Medicine and Interdepartmental Toxicology Program
North Carolina State University
Raleigh, North Carolina

Everett E. Simmons, D.V.M.
Acorn Research Laboratory and
Burnett Road Animal Hospital
Austin, Texas

A. J. Sorrels, R.N.
Director of Clinical Research
Acorn Research Laboratory
Austin, Texas

George A. Tanner, Ph.D.
Professor of Physiology
Department of Physiology and
Biophysics
Indiana University School of Medicine
Indianapolis, Indiana

Jerry A. Thornhill, D.V.M.
Director
Veterinary Internal Medicine and
Dialysis Center Ltd.
Veterinary Specialty Clinic Building
Riverwoods, Illinois
Research Associate
Section of Nephrology, Department of
Medicine
Hines V.A. Hospital
Hines, Illinois

TABLE OF CONTENTS

CURRENT STATUS OF HEMODIALYSIS IN VETERINARY MEDICINE AND THE DEVELOPMENT OF THE CANINE ANIMAL MODEL FOR HEMODIALYSIS-RELATED DISORDERS IN MAN

Jerry A. Thornhill and Stephen R. Ash

INTRODUCTION

Hemodialysis is a process by which blood is removed from the body, circulated through an artificial kidney composed of semipermeable membranes bathed in a physiological solution, and returned to the body. Dialysis, or the diffusion of substances across semipermeable membranes due to chemical gradients, results in mass transfer during the procedure. Due to their increased diffusibility and to favorable membrane permeability, transfer of small molecules is favored.

In 1913, it was first demonstrated that dialysis of blood could remove small molecules which are normally excreted in the urine.[1] In 1944, Dr. Willem Kolff[2] of the Netherlands developed a hemodialysis machine, a rotating drum device that was to serve as a prototype for the present-day artificial kidney, with large enough surface area to successfully treat human patients with renal failure. With the development of cellophane for use as a semipermeable membrane and the employment of heparin as an anticoagulant, the technical feasibility of Kolff's work was made possible. Before the development of the rotating drum, Kolff found that an artificial sausage skin made of cellulose acetate (cellophane) filled with 50 cc of blood to which had been added 400 mg % urea would lose its urea within 15 min if the bag were suspended in saline and moved up and down. As Kolff learned and later applied to the rotating drum, the secret of the effectiveness of removal of small molecules from the blood is the intense movement of both blood and dialyzing fluid.[3]

Following its introduction in the Netherlands, dialysis spread to other parts of Europe and to North America by the late 1940s. Many design changes and refinements of hemodialyzers occurred during the next 3 decades, and with the development of blood access devices by the middle of this period in the 1960s, long-term maintenance dialysis then became practiced on patients with end-stage kidney disease.[4-8] Currently, some 55,000 patients in the U.S. are being supported by chronic maintenance dialysis.[9]

In human medicine, the artificial kidney has become the most successful artificial internal organ, being able to support large numbers of patients over long periods of time, and all aspects of its performance have been extensively reviewed in the literature. In veterinary medicine, however, hemodialysis has played an insignificant role in the management of renal failure due primarily to technical difficulties, complications of the procedure, and cost involved. Therefore, only a limited number of reviews have been written on its application as a therapeutic modality for animals.[10-20]

In this chapter the theory of hemodialysis as well as the equipment needed and technique involved for veterinary application will be discussed. Special emphasis will be placed on the development of hemodialysis in the research dog, followed by an overview of hemodialysis support of clinical canine patients presented to Purdue University (West Lafayette, Ind.) with oliguric acute renal failure. As will be appreciated, in the laboratory-induced, or spontaneously occurring dialysis-dependent uremic dog, many of the hemodialysis related disorders in man can be modeled.

DIALYSIS EQUIPMENT

Basic Principles

The basic principles of hemodialysis have remained unchanged since the early days

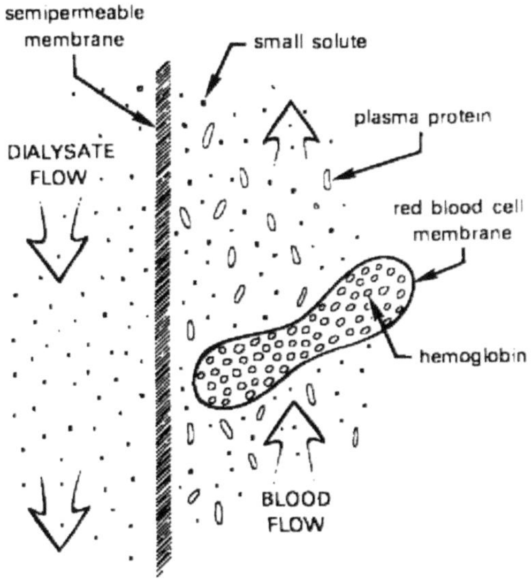

FIGURE 1. Semipermeable membrane of a hemodialyzer allowing blood to interface with dialysate (pumped countercurrent to blood) creating concentration gradients for small solute (urea, creatinine, sodium, potassium, etc.) removal while preventing protein and blood cell loss.

of the Kolff rotating-drum artificial kidney. Anticoagulated blood is circulated through one of the many types of hemodialyzers (i.e., coil, hollow fiber, or plate dialyzer) by a dialysis machine, where it interfaces with a balanced salt solution (dialysate) across a semipermeable membrane (Figure 1). Solute moves from blood to dialysate through the membrane by diffusing down a concentration gradient. Its removal from blood is proportional to the membrane surface area of the dialyzer and to the flow rates of blood and dialysate through the dialyzer which are generally countercurrent to establish steeper gradients. Clearance of solute from blood to dialysate is nonspecific and is primarily a function of molecular size, with smaller molecules such as urea and creatinine passing more efficiently than larger molecules such as peptide hormones.

Water and accompanying salt are removed from the patient by ultrafiltration. The hydrostatic pressure gradient developed between blood and dialysate across semipermeable membranes within the dialyzer allow these ingredients of the plasma to be extracted. The hydrostatic pressure gradient can be increased by increasing the negative pressure (vacuum) on the dialysate side of the dialyzer.

Hemodialysis Machines

The hemodialysis machine, such as shown in Figures 2 and 3, is the so-called artificial kidney and has a twofold responsibility: (1) pumping of patient's blood through the membrane packets of a dialyzer and (2) storage, heating, and circulation of dialysate around the membrane packets of a dialyzer. With the above hemodialysis machine, the operator adds concentrated dialysate, containing sodium, potassium, calcium, magnesium, chloride, and either bicarbonate or acetate to 100 to 200 ℓ of deionized water in the storage tank, to provide a "batch" recipe solution. Dialysis is performed until the solution is used up or becomes saturated with nitrogenous wastes.[21]

The batch system is bulky and requires operator time to prepare the dialysate (i.e., loading the storage tank with 100 to 200 ℓ of deionized water and thoroughly mixing

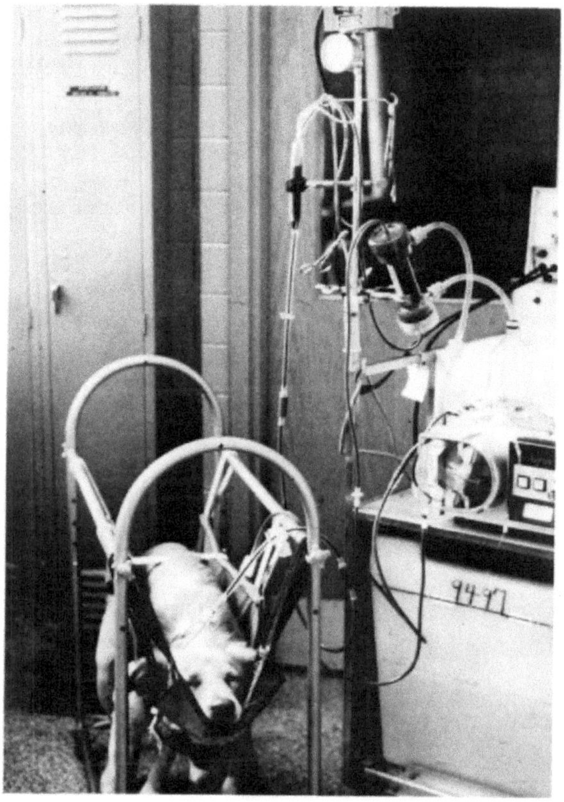

FIGURE 2. Travenol® RSP, Heme-Alert® II hemodialysis machine. Designed to be used with a coil dialyzer, but used at Purdue with a hollow fiber dialyzer. The dialysate warmer dial (A), single pass converter (B) by Cordis Dow®, and dialysate reservoir (C) are all essential parts of the dialysate delivery system.

with concentrated dialysate), but has worked well with the Purdue University team. An important step is the final check of the dialysate batch for salt content, which assures that concentrate was added to water in the tank. This can be done with a conductivity meter (Figure 4). Failure to add salt to the dialysate would result in dialysis of blood against a hypotonic solution, causing massive hemolysis.

In more updated dialysis machines, "proportioning" dialysate mixing systems have simplified dialysate preparation. Proportioning machines dilute concentrated dialysate from a bottle with treated (deionized) tap water to a ratio of approximately 34:1. No reservoir tank for batching dialysate is needed in the proportioning system, since dialysate is continuously prepared, used, and discarded down a drain. Uninterrupted monitoring of the electrical conductivity by the machine of diluted dialysate solution affords control over the accuracy of its ionic concentration.

Flow of dialysate solution from the dialysis machine through the dialyzer may be recirculating, single pass, or recirculating-single pass (RSP). In the recirculating delivery system, dialysate which interfaces with blood in the dialyzer begins with zero concentration of uremic wastes. However, as dialysis proceeds, these wastes gradually increase as dialysate continually recirculates from and to the batch through the dialyzer, reaching saturation usually within 3 to 5 hr. In the single pass delivery system,

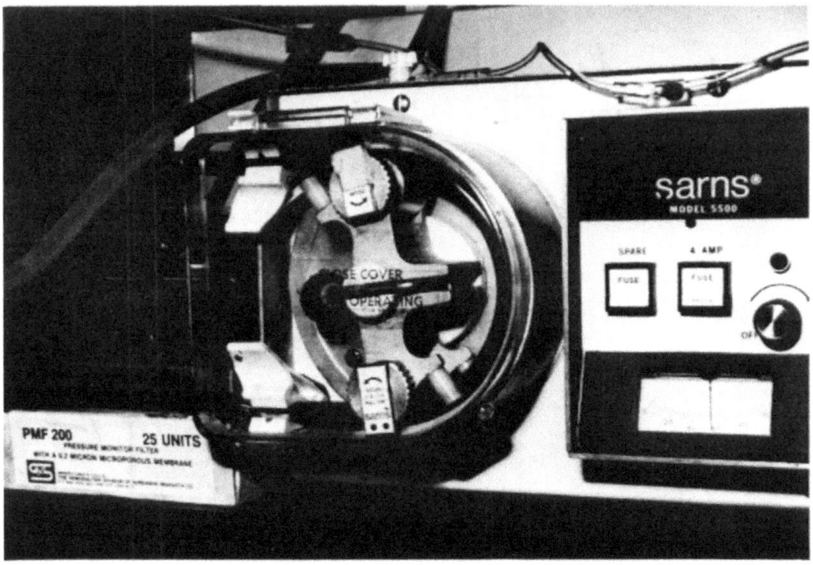

FIGURE 3. A Sarns® roller pump (Sarns, Inc., Ann Arbor, Mich.) regulates blood flow to the dialyzer.

FIGURE 4. Travenol® conductivity meter which measures the electrical conductivity of dissolved ionizable substances (primarily NaCl) in the dialysate bath. (Photo courtesy of Travenol Laboratories, Inc., Deerfield, Ill.)

fresh dialysate constantly interfaces with blood enhancing the efficiency of solute removal by maintaining a high gradient from blood to dialysate (Figure 2). In the third type, RPS, as the name implies, a portion of the dialysate is recirculated and a portion is fresh.

Dialysate may also be regenerated, thus preventing its saturation with uremic wastes.

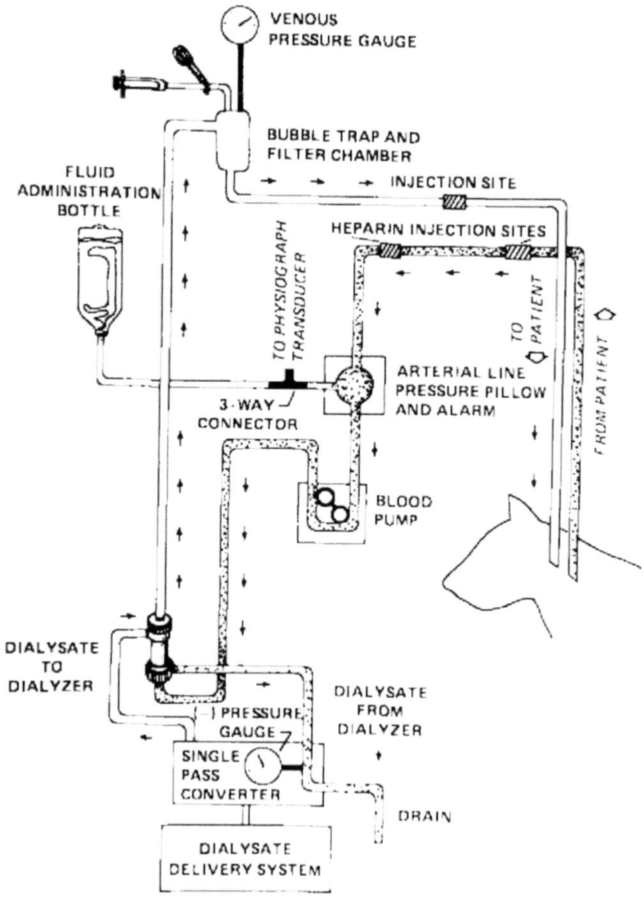

FIGURE 5. Illustrated scheme of hemodialysis. Shaded blood line (a) represents arterial blood flow to the dialyzer. Clear line (v) represents venous return from the dialyzer. Dialysate flow to and from the dialyzer by the SPC is represented by clear and shaded dialysate lines, respectively.

In the Redy® system, a small volume of dialysate is regenerated each cycle by running it through a sorbent cartridge containing zirconium phosphate, activated charcoal, hydrated zirconium oxide, alumina, and urease. Urea and other waste products are either directly removed, or metabolically converted to other compounds and removed, as dialysate passes through the cartridge. Calcium, magnesium, and potassium are also removed by the cartridge obligating careful monitoring of their blood levels to determine concentrations of these elements to be reinfused.[22-24]

Dialysis machines are equipped with monitoring devices in the delivery system to warn of technical problems associated with the hemodialysis procedure. These devices give either audio or visual alarms requiring manual correction or shutdown of equipment by the operator or fail-safe alarms which automatically turn off the equipment.

The following monitors are present in most dialysis machines:

1. Blood leak detector — This is usually a photocell located in the dialysate delivery and return system which observes discolorations in the dialysate, indicating dialyzer membrane leak, sounding an alarm. The photocell monitor is very sensitive and may give false alarms with air bubbles in the line.
2. Arterial line negative pressure alarm (Figure 5) — This monitor is situated on the

arterial blood line before the blood pump. Incorporated into the arterial blood line is a small sac known as a "mouse". The sac has a normal distensibility by the patient's blood pressure. If the sac collapses because of occlusion of the arterial line before it reaches the pump or because the pump pulls blood faster than the artery can supply it, an alarm sounds or the roller pump is automatically shut off.

3. Venous bubble trap and pressure monitor (Figure 5) — In this system, the drip chamber collects blood exiting the dialyzer, preventing air bubble escape back to the patient. The pressure gauge monitors changes in blood pressure also in the venous extracorporeal circuit. Increased resistance in this line will increase the ultrafiltration fraction in the dialyzer, and complete occlusion will cause blood to back up into and completely fill the drip chamber, creating exceeding pressure on the gauge triggering the alarm. The venous pressure monitor is used in conjunction with the arterial pressure monitor to calculate ultrafiltration rates.
4. Arterial pressure monitor — This monitor warns of pressure changes between the blood pump and dialyzer. The monitor is used in conjunction with the venous pressure monitor to calculate expected ultrafiltration rates.

Other components of the dialysis machine and system include the blood roller pump which can be set at variable speeds to influence the rate of blood flow through the dialyzer, a heater to regulate dialysate temperature (which must be at body temperature or slightly above to prevent significant loss of heat from the patient during extracorporeal circulation), and disposable arterial and venous blood lines with numerous injection sites. A negative pressure gauge on the single pass converter (SPC) regulates the amount of negative pressure applied to the dialyzer from the dialysate side, which can greatly influence ultrafiltration. With the Cordis Dow® SPC*, negative pressure from −1 to −500 mmHg can be applied to the dialyzer, greatly augmenting fluid removal from the patient. Other alarm monitoring devices include heater alarms, conductivity alarms on proportioning machines, foam detectors for blood, and air-fluid indicators.

Hemodialyzers

In general, three basic types of disposable hemodialyzers are available: coil, parallel plate, and hollow fiber. In the coil dialyzer, blood enters a spiral membrane which is wrapped around a porous screen over a plastic core (Figures 6A and 6B). Due to high internal resistance, blood must be pumped and courses from inside the coil to the outside, interfacing with dialysate, which is pumped through the porous screen between the layers of the coil. A high dialysate flow rate (10 ℓ/min) is required to assure adequate mixing of dialysate near the membranes. With this high flow, recirculation of dialysate is necessary in order to conserve fluid.[25] A distinct disadvantage for use of this system in veterinary medicine is the large priming volume of blood needed, which can approach 1 ℓ, as membranes are compliant and increase their fill capacity as transmembrane pressure increases.

In the parallel plate dialyzer, due to low resistance, blood pressure supplied through an arteriovenous shunt is adequate to perfuse the membranes, although a blood pump can be used to better regulate flow. The flow device consists of plastic boards, machined on their mating surfaces to carry dialysate fluid by means of longitudinal grooves. Paired membrane sheets, with molded plastic ports inserted at their inflow and outflow ends, are inserted between each board forming blood compartments (Figures 7A and 7B). Dialysate is pumped countercurrent to blood flow, optimizing solute removal, and a dialysate flow rate of approximately 500 mℓ/min allows the use of a

* Cordis Dow Corporation, 999 Brickell Avenue, Miami, Fla. 33145.

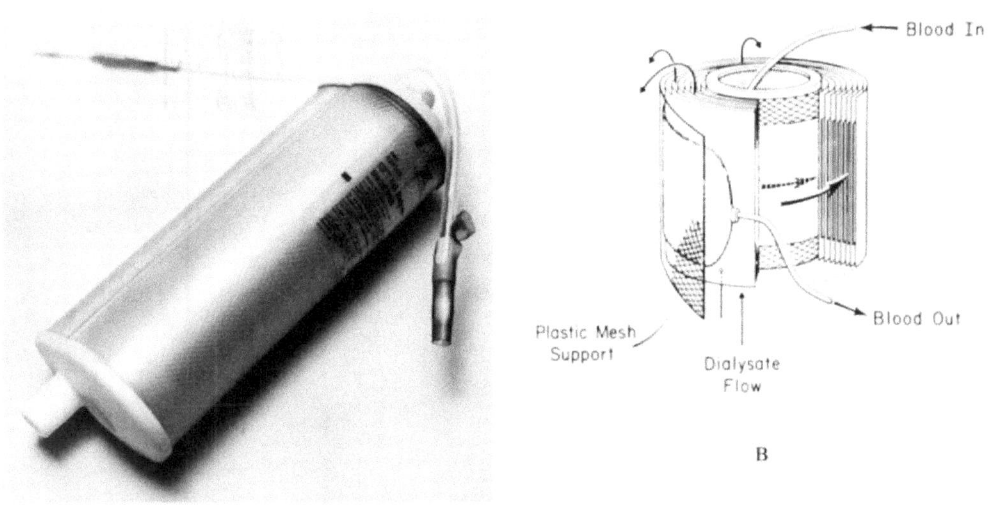

B

A

FIGURE 6. (A) Travenol® CD coil dialyzer. (B) Drawing of a coil dialyzer tubular membrane. Dialysate flow is from bottom to top of dialyzer around the membrane. (From Colton, C. K. and Lowrie, E. G., in *The Kidney*, Vol. 2, 2nd ed., Brenner, B. M. and Rector F. C., Eds., W. B. Saunders, Philadelphia, 1981. With permission.)

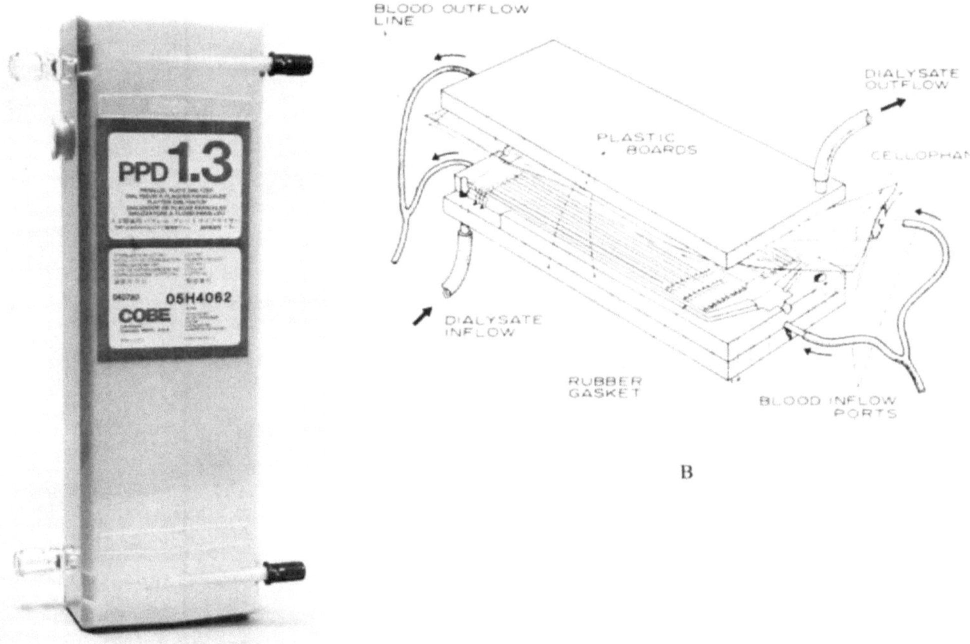

B

A

FIGURE 7. (A) Cobe® (Cobe Laboratories, Inc., Lakewood, Colo.) PPD parallel-plate dialyzer. (B) Drawing of a parallel-plate dialyzer membrane packet. Dialysate flow is countercurrent to blood flow. (From Colton, C. K. and Lowrie, E. G., in *The Kidney*, Vol. 2, 2nd Ed., Brenner, B. M. and Rector, F. D., Eds., W. B. Saunders, Philadelphia, 1981. With permission.)

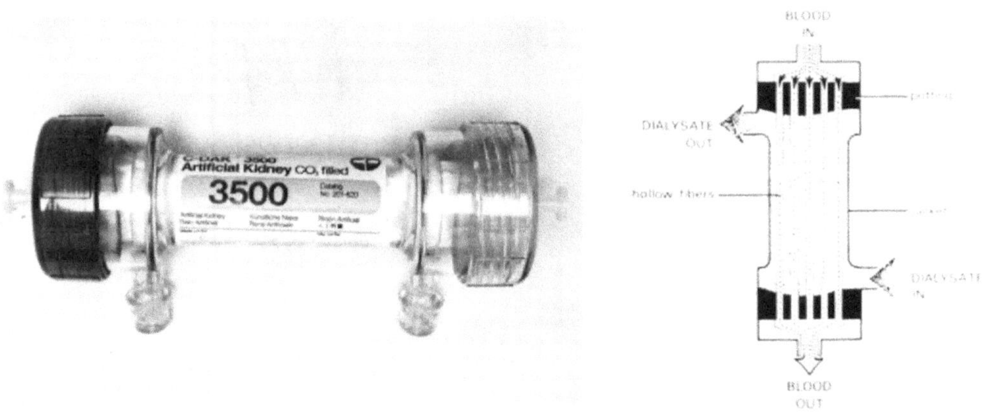

A B

FIGURE 8. (A) Cordis Dow® C-Dak ™ 3500 hollow fiber dialyzer. (B) Drawing of hollow fiber membrane tubes. Dialysate flow is countercurrent to blood flow.

SPC system. Parallel plate dialyzers have not received the attention given to hollow fiber dialyzers at veterinary institutions, but are most adequate for hemodialysis in the dog.

Hollow fiber dialyzers are made up of about 10,000 membrane tubes, each with a wall thickness of 20 to 30 μm and an extremely small internal diameter dimension of 200 to 300 μm, arranged in a cylindrical case much like straws in a container (Figures 8A and 8B). Because each blood channel is small, there is a relatively high membrane surface area to blood volume ratio and a relatively low blood side resistance to mass transfer, thus collectively favoring uremic solute loss to dialysate. The dialyzer's internal dialysate volume is small (approximately 100 mℓ) which allows a high shear rate (breakdown and scatter) of the solution near the membranes without a high overall dialysate flow. Like the parallel plate dialyzer, then, the hollow fiber dialyzer can be run on a SPC delivery system, with a dialysate flow rate of approximately 500 mℓ/min. Hollow fiber dialyzers have been successfully employed in veterinary medicine for support of canine patients with dialysis dependent renal failure (Figures 9A, 9B, and 9C).

Hemodialyzer Surface Area

Commercially available hemodialyzers are usually rated in terms of creatinine clearance (i.e., volume of blood cleared of a particular substance — creatinine — per unit time).[26] Creatinine clearance, in turn, is influenced by blood flow, dialysate flow, membrane type, transmembrane pressure, and most importantly, surface area of the dialyzer. The larger the membrane surface area, the greater the solute clearance per minute of time of blood flow (dependent somewhat, also, on blood flow rate). However, the larger the membrane surface area, the greater the quantity of blood needed to fill the dialyzer. Membrane surface areas of dialyzers vary from 0.6 to 2.5 m^2. Put in perspective, a 50-kg Great Dane only has a body surface area of approximately 1.36 m^2. A 0.6-m^2 dialyzer requires approximately 50 mℓ of blood to prime the membranes and a 2.5-m^2 dialyzer requires about 210 mℓ. Add to this another 50 to 70 cc of blood to fill the lines; this gives a volume of extracorporeal circulated blood which may restrict canine patient size for dialysis. Recently, however, a new pediatric parallel plate hemodialyzer, by Gambro Lundia®*, has been introduced to the market and may be

* Gambro USA, 23872 Kelsey Road, Barrington, Ill.

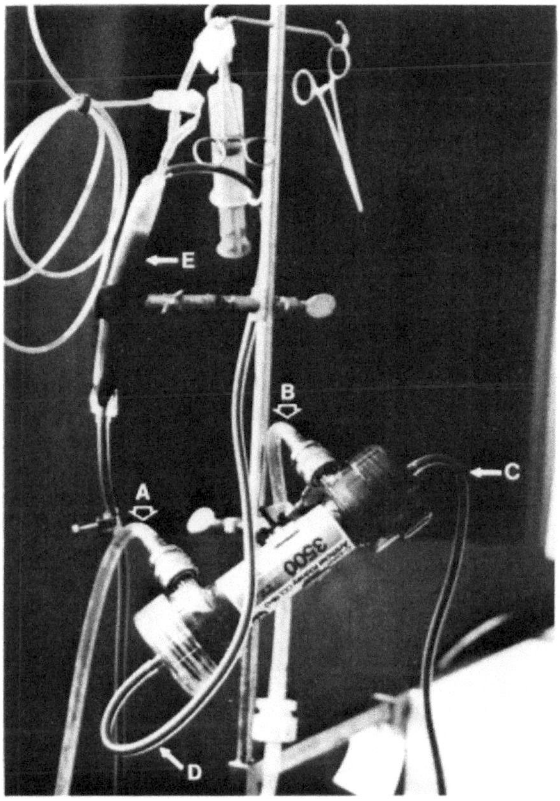

A

FIGURE 9. (A) Hollow fiber dialyzer during dialysis showing dialysate inflow (A) and outflow (B), countercurrent blood inflow (C) and outflow (D), and venous bubble trap (E). (B) Dog asleep on hemodialysis using a hollow fiber dialyzer. (C) The Controller® (Animal Restraining Devices, Inc., Farmington Hills, Mich.) restraining device for canine patients on hemodialysis.

ideal for small dogs.[27] The 0.41-m^2 dialyzer has a prime volume of only 28 m*l*, which when added to the smallest blood lines utilized, containing a volume of approximately 15 m*l*, gives a total of 43 m*l* of circulated extracorporeal blood, about 10% of the circulating blood volume in a 6-kg dog.

Clearance of creatinine vs. membrane surface area in three types of dialyzers (coil, parallel plate, and hollow fiber) is presented in Figure 10. In this graph, membrane surface areas have been restricted from 0.9 to 1.5 m^2. Creatinine clearances were obtained at blood flow rates of 200 to 300 m*l*/min and minimal ultrafiltration pressures. For each type of dialyzer, larger clearance is obtained with larger surface area. Between the various types of dialyzers, however, only slight variations are seen in clearance per unit surface area, except in the coil dialyzer with more permeable membranes (Extracorporeal*). In terms of clearance, there appears to be little basis to choose one dialyzer over another. The decision is based more on the hemodialysis system than the dialyzer.

* Extracorporeal Medical Specialties, Inc., Royal and Ross Roads, King of Prussia, Pa. 19406.

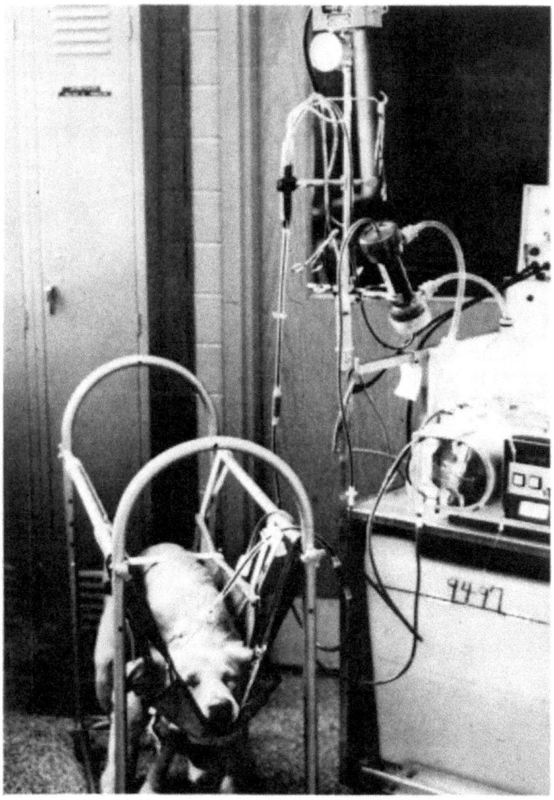

FIGURE 9B

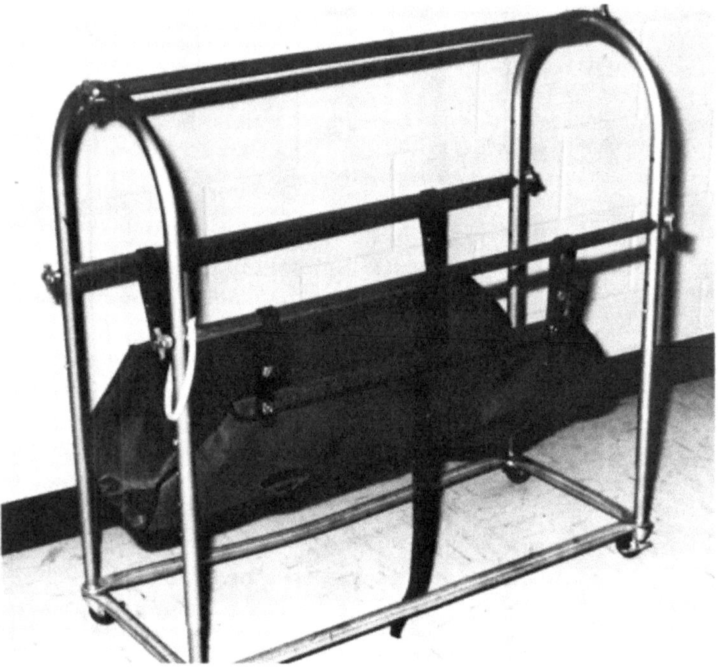

FIGURE 9C

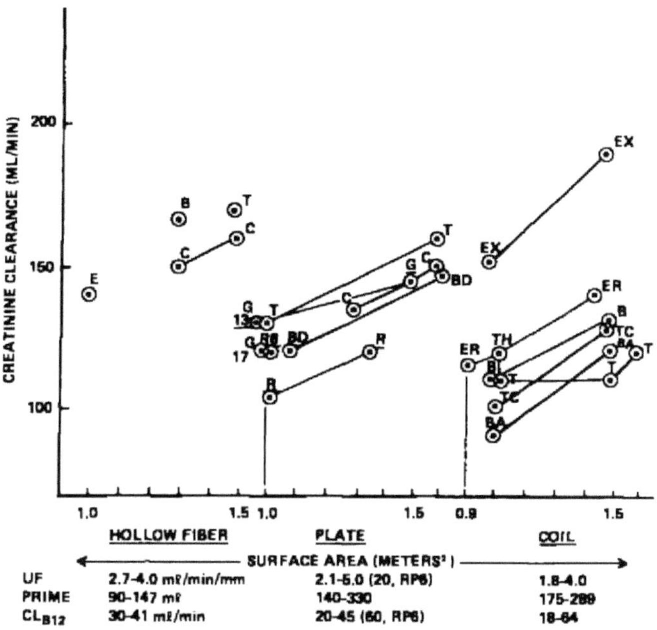

FIGURE 10. Clearance of creatinine vs. surface area in three types of dialyzers (hollow fiber, plate, and coil.) Clearances are at 200 to 300 mℓ/min blood flow and 500 mℓ/min dialysate flow (recirculated in coil dialyzer). Data is from manufacturer's literature. Superscripts indicate manufacturer: T, Travenol; B, Bentley; EX, Extracorporeal; C, Cordis Dow; G, Gambro; RG, Rhone Poulanc Model RGb; BD, Becton Dickson; ER, Erica; TH, Travenol High Filtration; TC, Travenol CD (coil dialyzer); and BA, Bentley Type A. (From Ash, S. R., in *CRC Handbook of Clinical Engineering*, Vol. 1, Feinberg, B. N., Ed., CRC Press, Boca Raton, Fla., 1980, 117. With permission.)

Hemodialysis Membranes

When two solutes separated by a semipermeable membrane possess different concentrations of solute, a concentration gradient exists. Diffusion will result in net transfer of solute from the solution with the highest concentration of solute to the solution with the lowest (Figure 1). The rate of diffusion is determined by the magnitude of the concentration gradient, temperature of the system (increased molecular movement occurring with increased temperature), and the diffusion coefficient which is the inherent ability of the solute particle, by its molecular size, shape, and ionic charge, to move through the solution.

Rate of diffusion of a solute between two compartments separated by a membrane will be further modified by the characteristics of the membrane. Most membranes in artificial kidneys (coil, parallel plate, and hollow fiber dialyzers) are constructed from cellulose which is the matrix of plant cell walls. Natural cellulose, primarily from cotton or wood, is dissolved in a copper and ammonia solution to transform it into a workable material. Cellulose is regenerated and molded from this solution into sheets or hollow fiber tubes, 10 to 20 μm thick (less than 1/1000 of an inch).[28]

Dialyzer membranes have a special kind of semipermeability. Polymers of cellulose making up the membranes are fashioned much like a plate of cooked spaghetti. When the membranes become wet, fluid is suspended between the cellulosic polymers. Small molecular substances pass around and in between the polymers and are cleared by diffusing through the fluid matrix. The average pore size in the matrix is 50 Å. Small molecules less than 300 daltons, such as urea (mol wt 60 daltons) and creatinine (mol wt 113 daltons), pass easily through the pores. Molecules between 300 to 5000 daltons, from peptide hormones up to inulin, pass slowly through the pores, and those over 10,000 daltons do not pass at all.[29] Bacteria are also too large to pass through the pores.[30]

Cellulose membranes have been used since W. J. Kolff performed the first clinical dialysis in 1944. Albeit successful (thousands of human patients with terminal renal failure have been supported by hemodialysis since 1944), the cellulosic membrane can be improved.[31] Variations in the technique of regeneration of the basic cellulose material can produce different permeabilities and clearances by altering such parameters as membrane structure, wall thickness, and pore size.[31]

Dialysate Bath

Dialysate, interfacing with blood across cellulose membranes in the hemodialyzer, must be a solution close to plasma consistency. Concentration gradients between plasma and dialysate are immediately established with the initiation of dialysis, and unrestricted removal of Na, K, Cl, and bicarbonate would spell doom for the patient. The following is a representative amount of essential ingredients per liter of dialysate that must be present:

Na^+	130 to 140 meq
Cl^-	100 to 110 meq
Mg^{2+}	2 meq
Ca^{2+}	3 to 5 meq
K^+	0 to 5 meq
Acetate	30 to 40 meq
Glucose	1000 mg

Acetate, as of 1964, has traditionally been utilized instead of bicarbonate as a buffer, since bicarbonate escapes as a volatile gas (CO_2) and precipitates Ca^{2+} and Mg^{2+} in concentrated solutions.[32] Acetate is metabolized to bicarbonate by the liver, but a spin off of this metabolism may be the synthesis of cholesterol and fats.

Dialysate is manufactured and supplied as a concentrated solution which need not be refrigerated as acetate serves as a stabilizing agent (as a vinegar). Dialysate concentrate is mixed 1 part to 34 parts water by hand in the "batch" system before dialysis or by a pump in a "proportioning" system during dialysis. In the "batch" system, a conductivity meter is used before dialysis is started to check the salt content. Omission of concentrate to the water bath would result in severe hemolysis and possible death to the patient. In the "proportioning" system, concentrated dialysate and water are separately pumped and mixed at a "Y" connector within the machine at a 1:34 ratio. A conductivity cell is utilized to control the rate of dialysate concentrate addition, through an electrical feedback loop, and another conductivity cell is used as a monitor for the final dilution.

A very serious question, arising in human medicine at this time, attacks the validity of acetate usage in dialysate concentrate as an alkalinizing agent rather than bicarbonate.[33,34] Acute and chronic renal failure cause metabolic acidosis. A major goal of hemodialysis therapy then is to correct the acidosis. However, hemodialysis itself further removes bicarbonate as well as organic acids that would be metabolized to bicarbonate from the patient. Acetate theoretically balances the patient's acidosis as well as the hemodialyzer loss by being metabolized in the liver to acetyl-CoA, which generates bicarbonate by consuming a proton from the Henderson-Hasselbalch equation.[35]

Thus, the net alkalinizing effect of hemodialysis is determined by the difference between acetate infusion and the patient's acid-base status plus removal of bicarbonate and organic acids. However, acetate infusion from dialysate may exceed metabolism to bicarbonate, leaving spin-off metabolites such as glucose, protein, fatty acids, and cholesterol.[36] This creates a predicament, in that too little acetate metabolism to bicarbonate fails to correct the metabolic acidosis, and too much acetate metabolism incorporates the genesis of unwanted metabolites, especially fats. With acetate concentrations in the bath in the order of 35 to 40mmol/ℓ, a normal hemodialysis cycle of 4 hr in man has been shown to exceed the calculated maximum utilization rate for acetate at 3.5 mmol/ℓ/kg of body weight.[37,38] Thus, although still controversial, potential unfavorable trade-offs arise with each dialysis using acetate bath, even though it generates bicarbonate better than baths containing lactate or succinate.[39]

When infused into dogs, acetate also has two major effects on the cardiovascular system that cause the blood pressure to fall: first, it causes peripheral dilation, and second, it has marked depressant effects on the myocardial contractility.[40,41] The significance of these findings has also remained controversial. Whether the amount of acetate gained by the body during a hemodialysis session causes clinical hypotension is not certain at this time. However, one report described two human patients who experienced repeated hypotensive episodes when their plasma acetate concentrations rose to greater than 15 meq/ℓ during dialysis.[42] Dogs on hemodialysis at Purdue have experienced some degree of hypotension in greater than 50% of the dialysis sessions.

Due probably more to the trade-off of lipidemia in chronically managed hemodialysis patients with the acetate bath, nephrologists are taking a hard look at reinstituting bicarbonate as the dialysate base.[33,43] Bicarbonate baths do require proportioning pumps, delicate pH monitoring, and increased nursing time, but many of the biological consequences attributed to acetate infusion have been shown in some cases to be circumvented.[33,44-47]

The effect of dialysate sodium concentration on body fluid distribution during hemodialysis is also being evaluated.[48,49] Sodium content of dialysate has been purposely kept lower (130 to 140 meq/ℓ) than that of plasma to produce a concentration gradient for sodium plus water removal from the patient. However, lowered sodium concentration in dialysate has been found to produce systemic hypotension (possibly by reducing total peripheral vascular resistance), increased renin activity, and elevated aldosterone concentration in the plasma, all of which can be reversed with high sodium dialysate (approximately 7% above plasma concentration).[48-52] The unfavorable trade-off of high sodium dialysate concentration is that of extracellular fluid volume expansion in the patient during dialysis or increased thirst resulting in greater water intake and weight gain between dialysis treatments.[53]

Increasing the sodium concentration in the dialysate may improve vascular stability by possibly minimizing changes in serum osmolality which occur as urea is removed during the dialysis procedure. In animal studies, it has been found that the addition of glycerol (350 mℓ glycerol to each gallon of dialysate concentrate) to the bath can prevent the fall in serum osmolality associated with low sodium dialysate and attendant urea removal, curbing hypotension.[54,55] In another study, it was established that low sodium dialysate could be used without the complication of hypotension and other side effects if rapid decreases in plasma osmolality and/or sodium concentrations are avoided.[56] Here, dialysate of decreasing sodium content was used, starting with 150 down to 133 meq/ℓ over a 4-hr period, rather than an abrupt and continuous bath of decreased sodium content, and fluid removal from the patients was favorable. The contribution of standard dialysate solution containing lowered sodium concentration (Diasol*) to the incidence of hypotension in dogs clinically dialyzed at Purdue needs to be determined.

* Travenol Laboratories, One Baxter Parkway, Deerfield, Ill. 60047.

Water Source and Water Requirements

Because of dilution of concentrated dialysate, hemodialysis systems require voluminous quantities of water which bring large amounts of trace substances into contact with the patient.[25] The following is a partial list of trace substances known to be in tap water and known to be relatively toxic in small concentrations of dialysate: aluminum, calcium, iron, tin, lead, copper, arsenic, mercury, nitrates, sodium, fluorine, chlorine, chloramines, hypochlorite (bleach used in sterilization of dialysis machine), formalin and soap (also used in storage and cleaning of machine), bacterial endotoxins, and viruses.

Untreated tap water has been associated with trace quantities of heavy metals or organic materials being transferred to the blood during dialysis, causing fever, hemolysis, anemia, methemoglobinemia, hypernatremia, hypertension, hypotension, metabolic bone disease, and dialysis encephalopathy.[59] Water used in the preparation of dialysate for hemodialysis, therefore, should have minimal chemical and bacterial contamination. This can be accomplished by water purification systems of which there are several types (filtration, water softening, deionization, and reverse osmosis).[57] These systems are not without cost and without proper maintenance and regeneration of ingredients involved, organic contamination can be added rather than prevented. The deionization systems replace all cations and anions in tap water with hydrogen and hydroxide ions, respectively. These ions then combine to form water.

VASCULAR ACCESS

Reliable Blood Supply

The development and preservation in a patient of a reliable blood supply is without a doubt the most important contributing factor to success in a hemodialysis program. In man, the evolution of vascular access approaches has left a plethora of arteriovenous shunt, fistula, and graft creations from which to choose.[58-64]

Early success in man was facilitated by the development of the Quinton-Scribner chronic vascular shunt which circulated blood from artery to vein through an extracorporeal circuit within connectable polytetrafluoroethylene (PTFE — a Teflon® substance) tubing.[65] Teflon® has a low thrombogenicity which permits small fibrin clots which form in the extracorporeal loop to be swept away maintaining shunt patency. However, the rigid Teflon® tubing transmitted extracorporeal motion to the tips of the shunt resulting in vessel intimal irritation, thrombosis, and eventual shunt loss. Replacing the ends of Teflon® with flexible silicone rubber (silastic tubing), which is also relatively nonthrombogenic, relieved some of this problem.

Modifications of the Quinton-Scribner shunt, which included most importantly the Thomas femoral shunt and the Buselmeier shunt, followed.[66-69] The short extracorporeal segment of the Buselmeier shunt minimizes clotting and allows auscultation of a bruit to access patency (Figure 11A). Limited experience with the Buselmeier shunt in the dog suggests it may have some application for chronic angio access (Figure 11B).

In the early 1970s, shunts were still in frequent use, but gradually during this period, the transition was made to primary use of arteriovenous (AV) fistulas.[70-73] This access involves direct anastomosis between an artery and vein to create a s.c. arterialized vein which is large and easily accessible for repeated punctures. It was found that thrombosis occurs less frequently with fistulas than with shunts due to lack of an external component. The AV fistula appears to be ideal for veterinary application, but numerous experiments with the vascular access at Purdue failed because of limited large, straight blood vessels in the dog and unsuccessful flow of blood to meet dialysis requirements (approximately 200 ml/min) after a few punctures. Also, fistulas require from 6 to 8 weeks of maturity before they can be punctured, thus precluding their use in patients with dialysis-dependent acute renal failure.

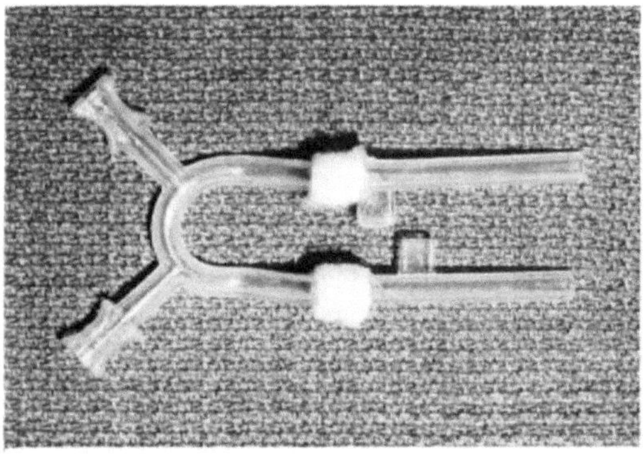

A

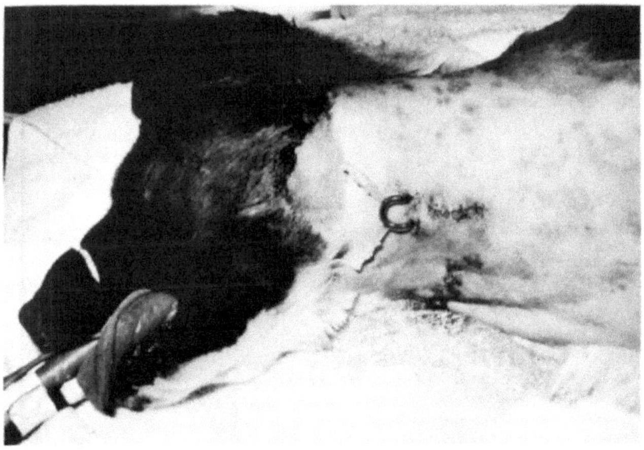

B

FIGURE 11. (A) Buselmeier shunt. (B) Buselmeier shunt between common carotid artery (dorsally) and external jugular vein (ventrally).

At the same time, in the early 1970s, grafts began to be used as a replacement for failed fistulas. The first grafts were bovine carotid artery xenografts which had poor survival as femoralsaphenous grafts (originally placed here), but good survival when placed in the upper arm or forearm.[74-77] Not long after the development of the bovine xenograft, experiments began with development of synthetic grafts. Using Dacron® (exp. Surgikos Artegraft*) or expanded, low-density PTFE (exp. Gore-Tex®**), synthetic grafts proved to offer a simpler surgical procedure, earlier usage (can be used immediately, but better after 1 to 2 weeks maturity), and longer half-life than bovine arteriografts.[78-80] Grafts show promise in veterinary medicine as reliable vascular access devices, but more investigative work in this area is needed.

There are, of course, complications which arise in any type of vascular access. Shunts, because of extracorporeal circulation, are more susceptible to thrombosis and infection.[81] Fistulas and grafts may also become clotted, but have less incidence of

* Extracorporeal Medical Specialties, Inc., Royal and Ross Roads, King of Prussia, Pa. 19406.

** W. L. Gore & Associates, Inc., 555 Paper Mill Road, Newark, Del. 19711.

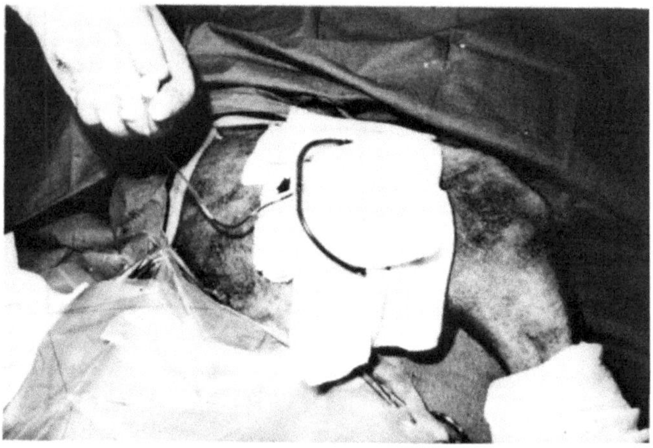

FIGURE 12. Modified Schribner shunt in dog from femoral artery to femoral vein, with shunt ends tunneled subcutaneously to exit on lateral side of thigh and connected by a "T" piece (Quinton Instrument) connector (arrow).

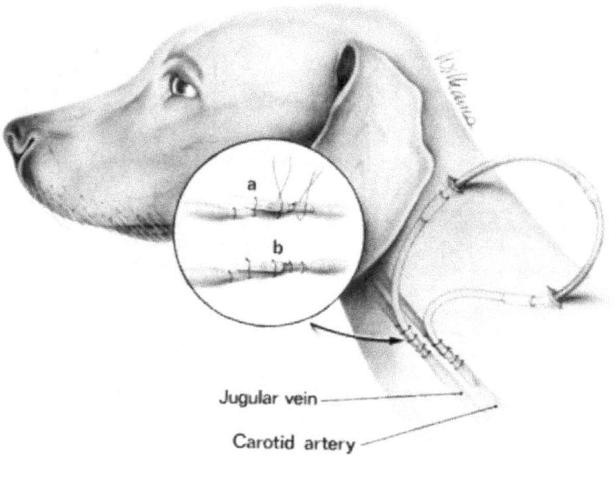

FIGURE 13. Drawing of modified Schribner shunt from common carotid artery to external jugular vein for extracorporeal circulation used at Purdue University.

infection because of s.c. burial. Fistulas and grafts, however, can bleed after puncture and do form pseudoaneurysms which generally result in loss or replacement of the access.[82-84]

For vascular access in research and clinic dogs, a modified Quinton-Scribner shunt has been successfully employed at two veterinary schools. In the hemodialysis program at the University of California, Davis, the shunt is placed between the femoral artery and vein (Figure 12).[13-15] In the program at Purdue University, the shunt is placed between the carotid artery and external jugular vein.[19] The technique for insertion is as follows (Figure 13):

1. Under general methoxyflurane gaseous anesthesia, the external jugular vein and carotid artery are identified, partially exteriorized, and tied off cranial to the vessel puncture site with 2-0 surgical silk.

2. Two 15-cm segments of sterile silastic tubing*, O.D. 1/4in., I.D. 1/8in., primed with heparinized saline (1000 units heparin to 500 m*l* of saline), are sewn into the respective vessels utilizing Teflon® vessel tips** and 2-0 silk.
3. The tubing is tunneled subcutaneously with a pair of Doyan clamps to exit through a skin stab incision on the dorsal, lateral surface of the neck (tubing is positioned as to allow the arterial end of the shunt to be caudal to the cranial venous return).
4. Shunt ends are connected with either a straight piece cannula connector** or an infusion "T" connector** (Figures 14A to 14I).

Appearance of a research dog with an AV shunt is shown in Figure 15. During storage (off dialysis), the neck is gently wrapped with Kling® gauze and elastic tape to prevent trauma and separation of shunt ends which would lead to immediate exsanguination (Figure 16).

AV Shunt Anticoagulation

Preservation of shunt patency after surgical placement becomes the next most important ingredient for successful maintenance of vascular access. In the dog, anticoagulation therapy is practiced during the lifetime of the shunt. However, different modes of therapy are applied dependent on whether the patient is on or off dialysis.

Off dialysis, shunt patency is maintained by the administration of aspirin (Ascriptin®), at a dosage of 2.5 gm/10 kg of body weight (15 mg/kg), 2 to 3 times a day.[19,20] This is an extrapolation of the accepted dosage of aspirin in the dog and low-dose aspirin administration used in man.[85-87] Aspirin suppresses platelet aggregation, preventing the initiation of clot formation at the region of intima contact with the Teflon® tips and within the silastic tube.[88,89]

On dialysis, heparin administration is employed to suppress thrombogenesis of the shunt, blood tubing, and dialyzer membranes.[90-93] As in man, dosage of heparin given can be based on monitorization of the dog's activated clotting time (ACT), which is normally between 60 and 125 sec.[94,95] Experience in hemodialysis at Purdue has revealed that an ACT of 4 min is ideal to prevent clotting of greater than 10% of the dialyzer.[18-20] The formula is as follows: 4 min − 30 sec × 1000 units of heparin per 30 kg of body weight per 1/2 hr. For each 30 sec less than 4 min, determined by an ACT, 1000 units of heparin are given for 30 kg of body weight, monitored each 1/2 hr.

Other methods for monitoring heparin therapy during dialysis are used in human medicine, including whole blood clotting time, activated partial thromboplastin time, and normograms.[96-101] However, for the dog, monitorization of the ACT seems to be efficient and is very practical. Not all patients (man or animal), due to differences in metabolism, will fit into established dose schedules for heparin during dialysis.[102] Therefore, the above discussion must serve as a guideline from which dialysis personnel must extrapolate and apply carefully.

Despite the use of heparin, platelet activation still occurs on the surface of dialyzer membranes leading to thrombus formation, microembolism, and thrombocytopenia.[103] In a study using prostacyclin (the most potent inhibitor of platelet aggregation) with heparin or prostacyclin alone, there were decreased incidences of membrane clotting and attendant thrombocytopenia.[104] Prostacyclin was given as constant infusion into the arterial blood line (to the dialyzer), during dialysis at a rate of 30 to 50 ng/kg/min with heparin and 60 to 100 ng/kg/min without heparin. The results suggested that platelet aggregation is the most important contributing factor to dialyzer

* Dow Corning Corporation, 2200 West Salzburg Road, Midland, Mich. 48640.

** Quinton Instrument Company, 2121 Terry Avenue, Seattle, Wash. 98121.

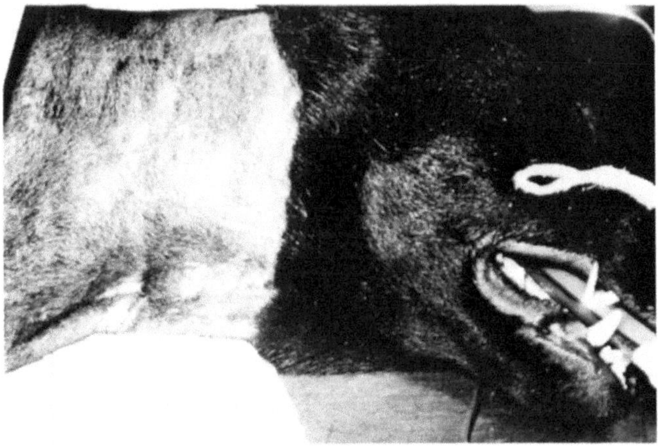

A

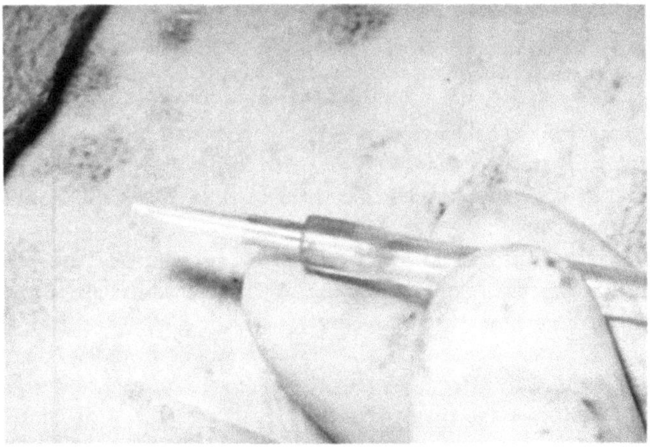

B

FIGURE 14. (A) Research dog prepared for shunt surgery. (B) Silastic tubing (Dow Corning) with Teflon® vessel tip (Quinton Instrument) which has been primed with heparinized saline. (C) Common carotid artery exteriorized with segment distal to surgical puncture site ligated and proximal segment pretied. (D) Vessel tip introduced and fixed into surgical puncture site of common carotid artery. (E) External jugular vein exteriorized and prepared for surgical puncture with cranial portion ligated and caudal portion pretied. (F) Shunt fixed into common carotid artery (above) and external jugular vein (below). (G) Shunt ends tunneled subcutaneously to exit on dorsum of neck and joined with a "T" piece connector (Quinton Instrument) (arrow). (H) Completed arteriovenous shunt surgery. (I) Dog on dialysis with shunt ends (A, artery; V, vein) disconnected to create vascular access for blood flow through a dialyzer.

membrane loss (i.e., prostacyclin does not inhibit prothrombin to thrombin formation) and that there is heparin-sparing effect with prostacyclin during extracorporeal circulation.

Anephric research dogs or uncontrolled uremic patients, first entering a hemodialysis program, will usually require less anticoagulation therapy to preserve patency of the AV shunt. This is due to the suppression of platelet aggregation attendant with

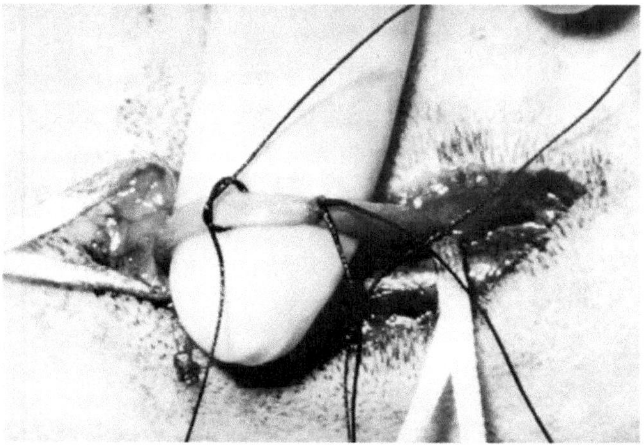

FIGURE 14C

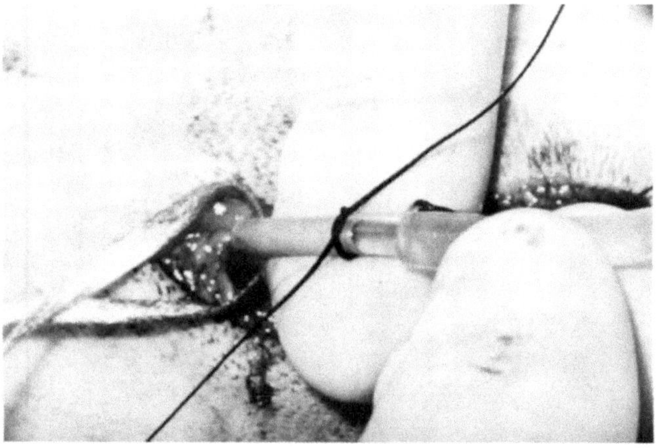

FIGURE 14D

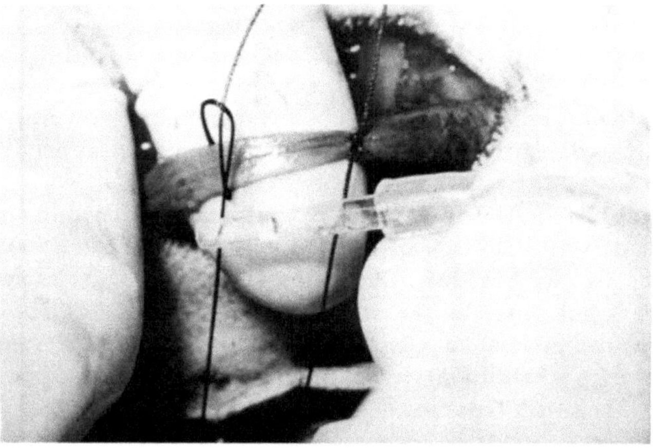

FIGURE 14E

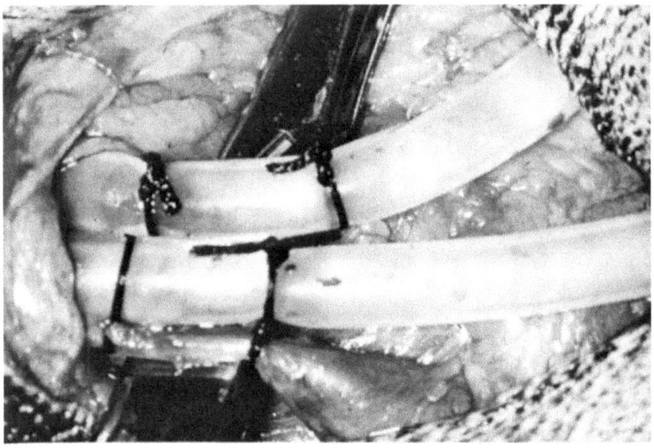

FIGURE 14F

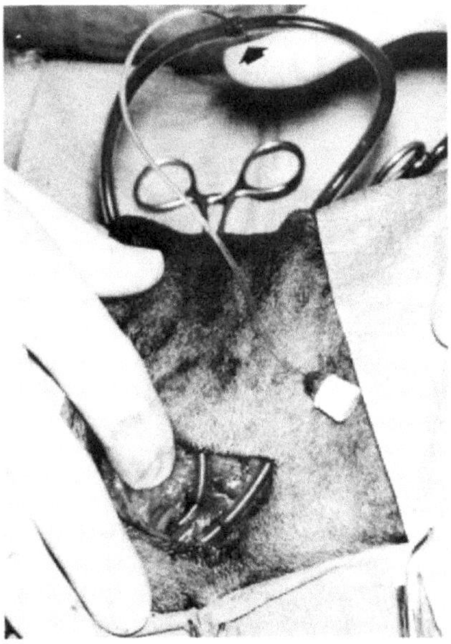

FIGURE 14G

uremia.[105] After the initiation of successful dialysis support, however, platelet function adequately returns, obligating the need for the stepped up anticoagulation schedule discussed above.

Shunts are checked daily for patency by palpation of carotid arterial pulsations through the silastic wall. Clotting of the shunt generally implies inadequate coagulation, but may be due to infection. Declotting is performed by aspiration of the clot material. If this is not possible, an embolectomy catheter* may be used to dislodge clots from either the arterial or venous arm of the shunt. The sterile embolectomy catheter consists of a small diameter flexible tube with an inflatable bulb on the end. The shunt is scrubbed with Betadine® and draped off. With surgeons gloves, the cath-

* Shiley, Incorporated, 17600 Gillette Avenue, Irvine, Calif. 92714.

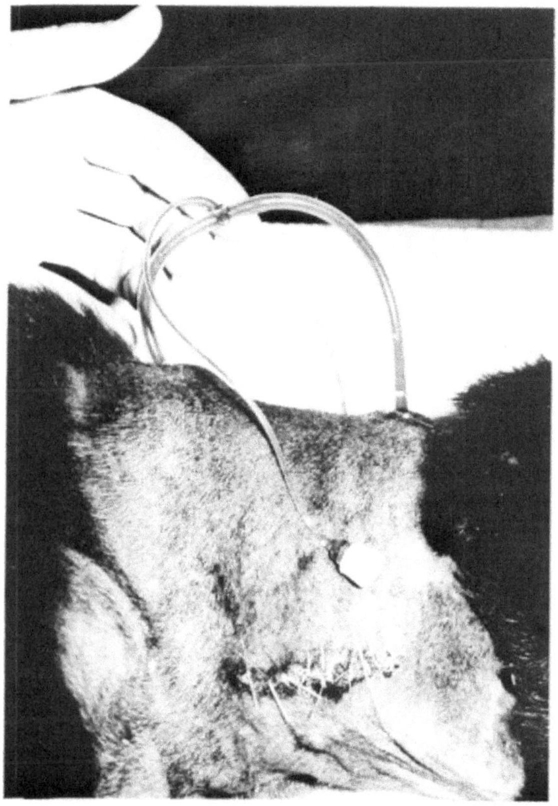

FIGURE 14H

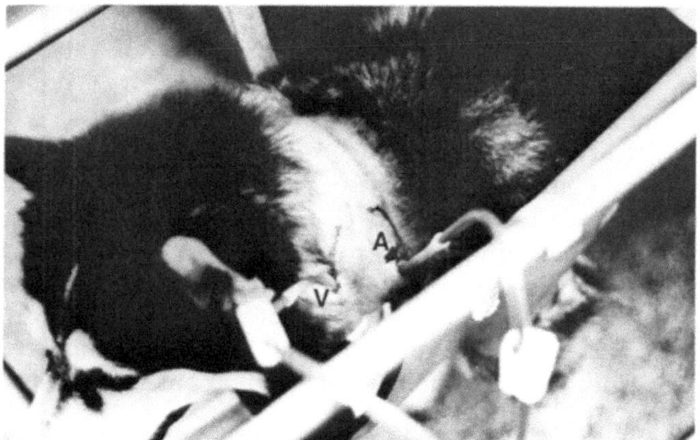

FIGURE 14I

eter is passed through the clot with the bulb deflated. The bulb is then filled with saline through its separate line and the catheter is withdrawn dislodging the clot. Once removed, whether arterial or venous line, only the venous side is flushed with heparinized saline and the shunt components are reconnected.

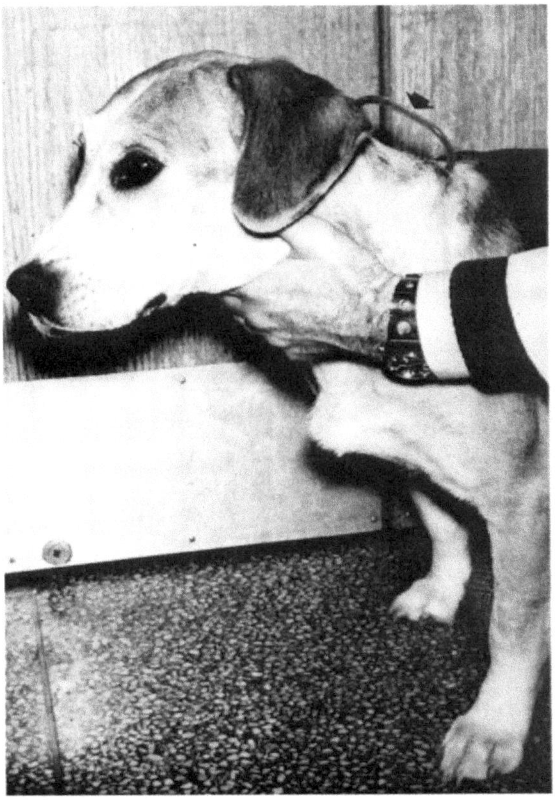

FIGURE 15. Modified Scribner shunt in a dog between dialysis sessions with shunt ends connected by a "straight" piece (Quinton Instrument) connector (arrow).

HEMODIALYSIS PROCEDURE

Technical Aspects in the Research and Clinic Dog

The procedure for hemodialysis will vary with available equipment and experience of personnel involved. Figure 5 reveals the overall scheme, and the following describes the setup and procedure, shown in Figure 9B, for a dialysis session on a research or clinical patient at Purdue University. Equipment used (Figure 2) includes a Travenol®* RSP dialysis machine, utilizing a Cordis Dow®** SPC dialysate delivery system, Cordis Dow® 3500 (1 m^2) hollow fiber dialyzer — although other hollow fiber dialyzers have been used, such as Extracorporeal*** Tri Ex-1 and Travenol® 1500 — and standard arterial and venous blood lines*, which are shortened to accommodate approximately 50 m*l* of blood.

The dialysis machine, which has been stored with a dilute (1:5) bleach solution in the dialysate bath compartment (enough to adequately cover the bottom of the tank), is rinsed with tap water and drained cleanly. The bath is then filled with 1 part dialysate concentrate to 34 parts deionized water (Culligan®† — lab system) to a total volume

* Travenol Laboratories, One Baxter Parkway, Deerfield, Ill. 60047.

** Cordis Dow Corporation, 999 Brickell Avenue, Miami, Fla. 33145.

***Extracorporeal Medical Specialties, Incorporated, Royal and Ross Roads, King of Prussia, Pa. 19406.

† Culligan International, One Culligan Parkway, Northbrook, Ill. 60062.

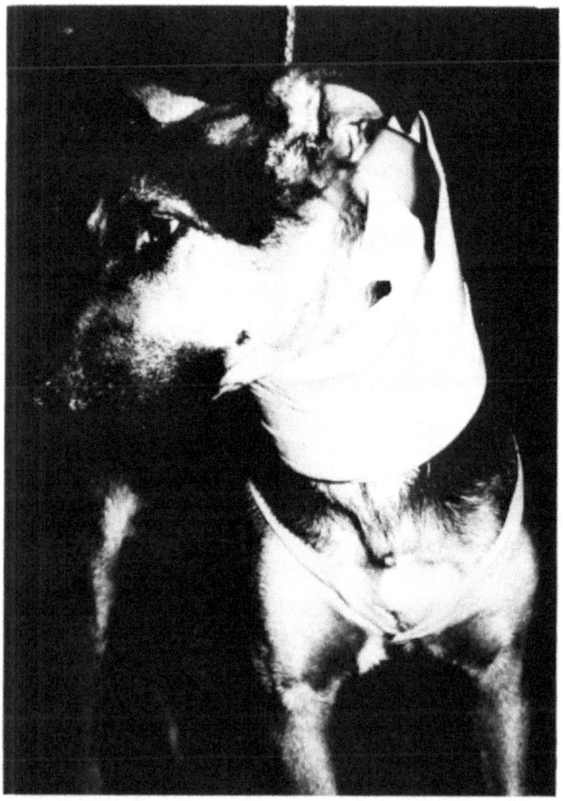

FIGURE 16. Shunt protective bandage and collar during off dialysis period.

of 120 *l*, which is marked on the bath storage tank. The solution is mixed and checked with a conductivity meter to assure that dialysate concentrate has been added in the correct proportions. Dialysate delivery lines are attached to dialysate ports on the dialyzer, and blood lines are attached to dialyzer blood ports, but the ends of the tubing which will connect to the patient's shunt are left capped for sterility. The arterial tubing is then passed through the blood pump. Ringer's solution (1 *l*), through a fluid administration set, is connected to the arterial blood tubing via a side port adapter located just in front of the portion of line entering the blood pump. The cap over the free end of the venous return line is opened for drainage and the blood pump is turned on to circulate Ringer's through the dialyzer, rinsing out the hollow fibers. Dialysate is also pumped through its dialyzer compartment during the fiber bundle rinse. However, dialysate is not discarded as an SPC, but returned to the tank via the drain tube for further use during the dialysis session.

The research dog or clinical patient is restrained in a standing sling (Figure 9C) or placed in lateral recumbency if extremely depressed. The AV shunt is scrubbed with a Betadine® solution, both arms clamped and aseptically separated. The arterial and venous arms of the shunt are connected to their respective blood lines. The arterial blood line is injected with 2000 to 3000 units of heparin, then both shunt arms are unclamped and the blood pump is started very slowly. All components of the system are visualized carefully as the blood column gradually replaces the priming solution. With no initial complications, the blood pump is increased in speed to generate approximately 200 m*l*/min of blood flow for a 35-kg dog.[14,19] The dialysate flow rate is set at 500 ml/min, giving a dialysis session of approximately 4 hr. The patient's temperature, pulse, respiratory rate (TPR), capillary refill, blood pressure, and ACT are monitored every 30 min during the procedure. Blood drawn for the ACT is taken from the arterial line (i.e., directly from the patient), and aliquots of heparin are administered, based on the formula of 1000 units given for each 30 sec less than 4 min for a 30-kg dog, discussed previously. Heparin is delivered into the arterial line, so that fresh anticoagulated blood enters the dialyzer first before returning to the patient.

Blood pressure during dialysis is measured directly from a side port in the arterial line by a pressure transducer* and a physiograph recorder.** Normal blood pressure in the dog is reported as 139 to 160/79 to 90 mmHg (systolic/diastolic).[106] After the dog is attached to the machine and all lines are in place, a resting blood pressure should be obtained before the initiation of dialysis. If the patient is hypertensive, the most probable reason is volume overload from oliguria (especially acute renal failure). However, the patient may be hypotensive from GI losses. Both conditions can be managed by manipulation of either fluid delivery or withdrawal, to or from the patient, through the dialyzer. If hypertension is present, fluid must be ultrafiltrated off during the early phase of dialysis (not to exceed 10 ml/kg/hr). Ultrafiltration proceeds until the blood pressure falls to the bottom end of normal range, reflecting a "drying out of the patient".[25] Ringer's solution is then titrated back to the patient to build the blood pressure off the floor, but maintaining it in the lower range of normal. If the patient is hypotensive, Ringer's solution can be administered during the initiation of dialysis to build the blood pressure.

Ultrafiltration can be accomplished by applying positive pressure to the venous blood side of the dialyzer or by increasing negative pressure to the dialysate side. Increasing resistance to the venous blood side (monitored through the venous pressure gauge, Figure 5) increases the ultrafiltration force from blood to dialysate in the dialyzer. Each brand and type of dialyzer (i.e., coil, parallel plate, or hollow fiber) has a recorded ultrafiltration coefficient on a package insert, which simply states how much fluid per hour can be withdrawn from a patient at 100 mmHg blood positive, or dialysate negative, pressure. Negative pressure on the dialysate side, controlled and monitored by a negative pressure gauge on the single pass converter, Figure 4, is easier to regulate and will not contribute to potential hemodialyzer hemolysis as does positive pressure on the blood side.[107]

Dogs undergoing hemodialysis at Purdue have needed assisted thermoregulation due to hypothermia.[18,19] The patient's temperature during dialysis can be significantly influenced by the temperature of the dialysate interfacing with blood circulating through the dialyzer. Hypothermic patients can be warmed by increasing the dialysate temperature, controlled by a gauge on the dialysis machine, Figure 15, from 2 to 3°C above normal (for the dog, 101°F = 38.3°C). Care must be taken to not overheat the dialyzer with dialysate, as hemolysis will ensue. At 47 to 51°C, hemolysis is apparently related to splenic trapping and destruction of altered red blood cells, and at temperatures greater than 52°C, hemolysis is immediate.[107]

Patient monitoring also includes blood sample harvesting for pre- and postdialysis serum evaluation of urea nitrogen, creatinine, sodium, potassium, calcium, and phosphorous levels. These values document the success of dialysis and assist in establishing the prescription for the quality and quantity of a dialysis session (discussed below). Blood gas determinations are generally made before and at the end of a session, but may be assayed 1 to 2 times during the course of dialysis, especially if the patient is severely acidotic (serum bicarbonate less than 10 meq/l).

At the termination of a dialysis session, the arterial shunt arm is clamped and Ring-

* Statham Laboratories, Inc., Hato Rey, Puerto Rico.

** Physiograph MK II, Narco Biosystems, Inc., Houston, Tex.

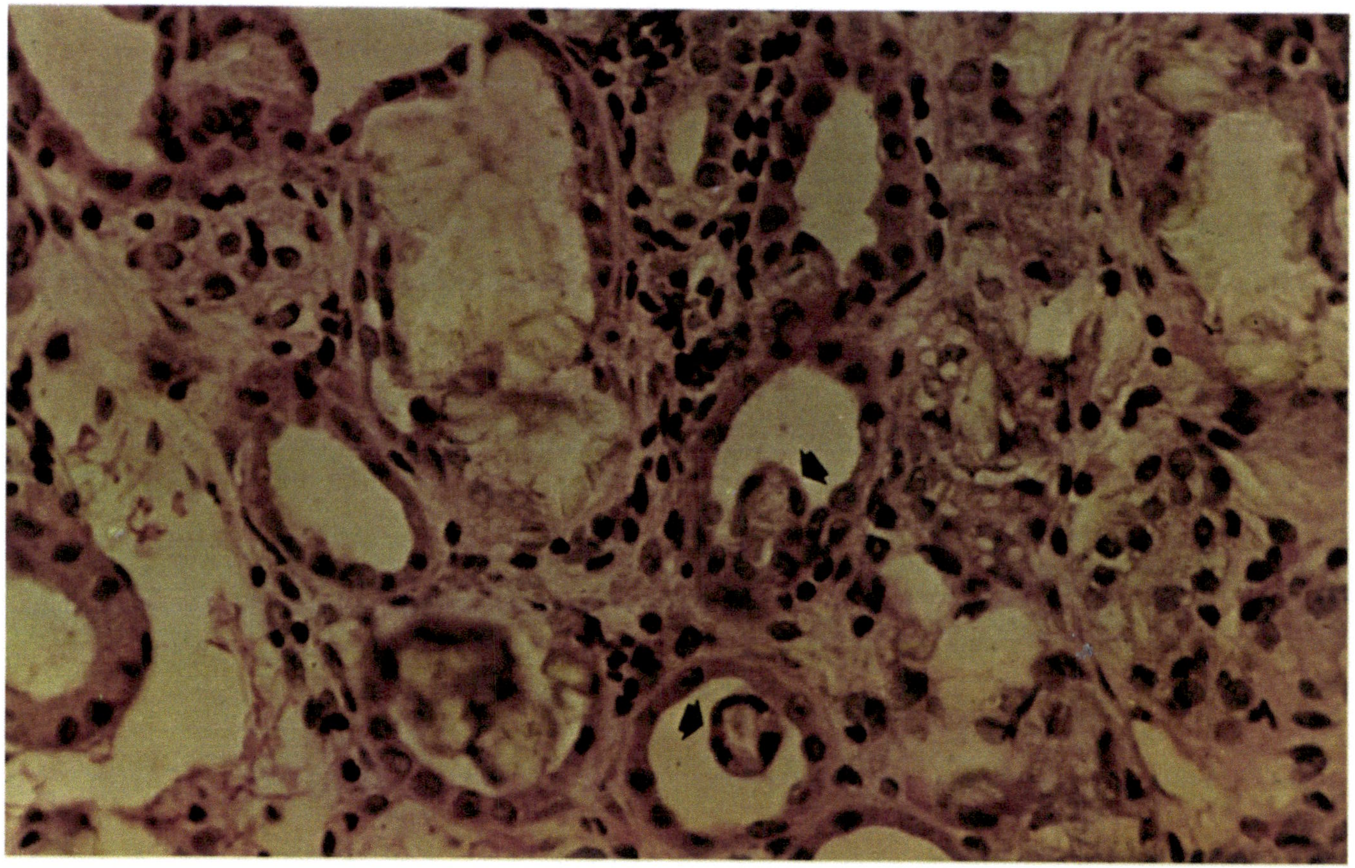

PLATE 2. Calcium oxulate crystals with patent tubular lumens and varying degrees of neoepithelization (arrows) around crystals migrating to the basal region and interstitium. (Hematoxylin - eosin; magnification × 400.) (From Thornhill, J. A., Ash, S. R., Crabtree, B. J., Riviere, J. E., and Thacker, H. L., *J. Comp. Pathol.* With permission.)

PLATE 1. Calcium oxulate crystals revealed under polarized light in a needle biopsy sample from a dog in oliguric acute renal failure following ingestion of ethylene glycol. (Magnification × 40.)

er's solution is circulated through the arterial line (from its point of entry ahead of the roller pump), thus rinsing out this portion of the tubing, the dialyzer, and the venous return line. This rinse-back technique, along with proper anticoagulation during dialysis, should keep dialyzer blood losses to less than 10 m*l* for the procedure (excluding blood drawn for monitorization).[108,109] Following the rinse, blood lines are disconnected and the AV shunt is aseptically recreated by connection of the silastic shunt arms with a Teflon®* connector.

Dialysis Prescription

The dialysis prescription is the frequency and duration of each dialysis session needed to support a dog with induced or spontaneous dialysis-dependent renal failure and is determined by parameters evaluated on an individual basis. Water balance, electrolyte balance, and level of azotemia, as well as the patient's clinical condition, all play a role in determining the dialysis schedule. Dogs have a higher minimum high biological value protein requirement (1.26 gm/kg/day) than does man (0.24 gm/kg/day) and subsequently have a higher endogenous urea generation.[110,111] The high rate of urea generation and the relatively high creatinine clearance found in the dog (3.7 ± 0.77 m*l*/min/kg),[112] result in a more frequent and more efficient dialysis requirement for this species compared to man.

The rate of urea generation can be utilized to develop a dialysis schedule for a patient by the following formula:[113]

$$G_u = (C_t - C_0)\ V/0$$

where G_u = urea generation rate (mg/min), C_t = plasma concentration of urea at time t (mg/m*l*), C_0 = plasma concentration of urea at time 0 (mg/m*l*), V = volume of total body water - urea space (m*l*), total body water (m*l*) = (0.6) (kg body weight) kg = 1000 m*l*, and 0 = time elapsed (min).

For a hypothetical 30-kg anephric research dog (zero renal function),with a rise in BUN from 20 to 160 mg/d*l* over a period of 48 hr, the following G_u can be calculated:

$$G_u = (1.6 - 0.2\ \text{mg/m}\ell) \times 18{,}000\ \text{m}\ell/2880\ \text{min}$$

$$G_u = 8.75\ \text{mg/min}$$

If the G_u is multiplied by 24 hr (1440 min) and divided by the total body water (18,000 m*l*), it is seen, assuming all other variables constant, that the BUN will continue to rise in the anephric model at a rate of 70 mg/d*l*/day.

To reduce the BUN from 160 to 20 mg/d*l* by hemodialysis in our model, a total of (1.6 − 0.2 mg/m*l*) × 18,000 m*l* = 25,200 mg urea nitrogen must be removed from the blood. Using the dialyzer urea clearance (about 140 m*l*/min at a blood pump speed of 200 m*l*/min in hollow fiber dialyzers used at Purdue)[18,19] and the patient's volume of total body water, an estimation of dialysis duration can be calculated using the following formula: urea removal = dialyzer clearance × mean urea concentration × dialysis time (t). Since the urea concentration in the serum changes continuously with time on dialysis (starting at 160 mg/d*l* to the end point of 20 mg/d*l*), an average urea concentration,

* Quinton Instrument Company, 2121 Terry Avenue, Seattle, Wash. 98121.

$$\frac{160 + 20 \text{ mg/d}\ell}{2} = 90 \text{ mg/d}\ell \ (0.9 \text{ mg/m}\ell)$$

can be used in the calculation:

$$\frac{25{,}000 \text{ mg}}{(140 \text{ m}\ell/\text{min})(0.9 \text{ mg/m}\ell)}$$

$$t = 200 \text{ min or } 3.33 \text{ hr}$$

Therefore, 3.33 hr of hemodialysis every 2nd day should theoretically keep our hypothetical model successfully supported. This method of dialysis prescription, however, should be used only as a rough guide, as urea generation during the span of dialysis has not been considered. Also, the time necessary for urea to transfer from cells to plasma (a time which lags true dialysis time) has not been entered into the calculations. However, although more exact methods of calculating dialysis time can be used, which take in the above factors, more complicated mathematical equations are required (lag mean concentrations and nomograms). For veterinary dialysis, the above formulas discussed are simpler and have easier understood clinical application.

By nature, dialysis-dependent clinic dogs have greatly reduced functional renal tissue. For that reason, endogenous kidney function in these patients contributes very little to lessening the frequency and duration of dialysis. However, in the patient with oliguric acute renal failure, where renal function might partially or completely return (if dialysis provided successful life support giving time for kidney tissue to recover from the insult), dialysis requirements would diminish commensurate with the degree of endogenous functional restoration. It is, therefore, important that each dog on dialysis be monitored with a dialysis index (prescription) to predicate the frequency and duration of each therapeutic session.

Hemodialysis Complications and Consequences

Potentially there are immediate and long-term complications and consequences attendant with hemodialysis therapy that may be encountered. Of clinical importance are cardiovascular compromise (especially hypotension), dialyzer blood loss, infection or thrombus formation leading to loss of the shunt, blood tubing, or dialyzer, metabolic waste (such as water, sodium, potassium, calcium, bicarbonate, or B-complex vitamin loss), ascites, disequilibrium syndrome, and staff or technician error in the mechanical operation of dialysis equipment, all of which have been extensively reviewed.[114-116] Drs. Daugirdas and Ing cover dialysis-induced hypotension in Chapter 2 of this text and, of importance in setting up a dialysis research program with the dog, the complication of infection and thrombus formation will be discussed here.

Infection is a common complication in patients undergoing hemodialysis. In a study at the Peter Bent Brigham Hospital in Boston, from 1964 to 1975, 9.5 % of the patients with chronic renal failure and 10.9% with acute renal failure, supported by hemodialysis, had bacteremic infection.[117] In fact, infection (of especially vascular access and venipuncture sites) is the main reason chronic renal failure patients are hospitalized and is the leading cause of death in patients with acute renal failure.[118-120] Gram positive (primarily *Staphlococcus*) is the most common organism type isolated in patients with chronic renal failure, and Gram negative (primarily *Escherichia coli* and *Pseudomonas*) is the most common in patients with acute renal failure.[117,121-123]

Of interest is the fact that there has been little change in the pattern of bacteremic infection in human patients with chronic or acute renal failure in the 20 to 25 years active hemodialysis has been available.[124] While the technical aspects of hemodialysis have been proficiently cultivated, the problem of sepsis in the dialysis patient remains one that is unsolved.

Utilizing an AV shunt in man and the dog, and recognizing the almost inevitable complication of infection attendant with this particular type of vascular access, there is little hesitation to practice prophylactic antibiotic therapy (48 hr before and 72 to 96 hr after surgical placement). For, once infection is established, primarily at skin sites of the shunt (usually *Staphlococcus* sp.), the tubing loosens, tissue around the vessel tips becomes septic and inflamed, and eventually the access may be lost.[61,64] Remember, antibiotic therapy in uremic patients usually requires dosage adjustment and regulation to the individual's degree of renal function.[125]

Thrombosis is the most common complication of AV extracorporeal circulation and the most common cause of shunt failure and hemodialysis shutdown.[58,61,64] A thrombus may form acutely when flow rate through the shunt drops sufficiently (average flow rate is approximately 200 to 300 mℓ/min) or gradually as clot material grows into the shunt suggesting intimal hyperplasia. Thrombosis usually begins on the venous side of the shunt at the region of the vessel tip where exposure to high pressure initiates intimal hyperplasia and fibrosis. Thrombosis is usually preceded by slowness of the flow rate and increasing venous resistance registered during dialysis. Early recognition of the problem can lead to successful bougie of the tubing and immediate vessel region (as discussed above — AV Shunt Anticoagulation).

Thrombosis of the dialyzer is also a contributing factor to loss of hemodialysis efficiency. Even with proper anticoagulation, off and on dialysis, platelet activation can occur on the surface of dialyzer membranes leading to thrombus formation.[103,126] However, with the dialysis protocol used on dogs at Purdue, less than 10% of the dialyzer fiber bundle has clotted off during dialysis.

CLINICAL EXPERIENCE IN THE DOG

Dialysis-Dependent Uremia

The rationale of dialysis treatment for uremia is based on the concept that there are retained solutes in the blood, either physiological metabolites normally excreted by the kidney or ingested substances, which exhibit concentration dependent toxicity resulting in the multisystem abnormalities of the uremic syndrome.[127-130] In veterinary and human medicine, nondialytic management for uremia has been reviewed.[131-133] Dialysis has its place for the management of uremia when conservative medical management fails to ameliorate the signs and symptoms of the disorder.[134,135] In acute renal failure, attendant with oliguria and rapidly progressive azotemia, as will be discussed below in the dog, dialysis therapy must be practiced early and often to be successful.

Acute renal failure (ARF) represents a pathophysiologic insult to the kidney, characterized by rapid deterioration of renal function, progressive uremia, and usually oliguria.[136-139] As biochemically devastating as the disorder of ARF is, however, in the dog, hemodialysis is probably better suited for life support, than in chronic renal failure (CRF). The obvious reason is the prognosis accompanying each disorder. Patients with oliguric ARF have a poor short-term but favorable long-term prognosis, if life can be supported by dialysis and if endogenous kidney function partially or fully returns (usually 10 to 21 days in man), releasing the obligation for dialysis.[140,141] On the other hand, CRF has a favorable short-term and unfavorable long-term prognosis, obligating long-standing maintenance dialysis or transplantation due to the relentless development of the "end-stage" kidney disease state. In the canine dialysis model for human disorders, then, a more concentrated effort can be expended dealing with dialysis related problems with an oliguric ARF dog.

The fact that kidney function potentially returns in ARF, if patients can be successfully life supported, is not to say that the success rate is going to be high. In man, up to 50% of oliguric ARF patients who are undergoing hemodialysis still die from sepsis, tissue catabolism, hemorrhage, acidosis, electrolyte disturbance, and other complications attendant with the disorder.[142] For this reason, once prerenal causes have been ruled out and oliguric ARF has been established in patients, a more vigorous approach of daily or alternate-day dialysis, rather than customary thrice weekly, has been initiated by many nephrologists.[143,144]

A dog with established oliguric ARF (i.e., less than 0.6 m*l*/kg/day of urine production, urine specific gravity from 1.007 to 1.022, azotemia, a urinary creatinine to plasma creatinine ratio of less than 5:1, and a urinary sodium greater than 25 meq/*l* — all done on a spot check urine analysis) becomes a candidate for hemodialysis. A modified Scribner shunt from the common carotid artery to external jugular vein (Figure 12) is surgically placed and dialysis proceeds as previously discussed (see Dialysis Procedure).

Maintaining caloric requirements to prevent accelerated tissue catabolism and resultant accumulation of metabolic wastes in dogs with dialysis dependent ARF proves to be a challenge. Since these patients are usually anorectic and vomiting the first day, until dialysis begins to control the degree of uremia, fluid balances are given priority attention. However, if after 2 to 3 days on dialysis, anorexia continues, gastric intubation via pharyngotomy or oral routes allows administration of high caloric density fats, carbohydrates, and small amounts of high biological value protein. However, the mixture must be in low water content to prevent fluid overload and must be given in small feedings spaced throughout the day to prevent the recurrence of vomiting. The administration of cimetidine (50 mg/10 kg i.v., b.i.d, in uremia, 50 mg/10 kg every 24 hr orally in the evening as maintenance, withdrawn after 3 to 4 weeks), to block H_2 receptors on parietal cells, has shown promise in curbing GI upsets in dialyzed dogs.[145] The dosage must be given after dialysis due to dialyzer clearance.

The use of insulin and total parenteral nutrition (TPN) can provide balanced nutrition and inhibit protein catabolism.[146,147] Insulin and glucose together have been shown to produce striking inhibition of protein breakdown and endogenous urea synthesis.[146] Mixed properly, TPN solutions (Travenol®*, Aminosyn**) usually provide 1 kcal/m*l* and should be administered through a central vein.[148] A 10-kg dog, requiring roughly 75 kcal/kg body weight, would need from 750 to 1000 m*l* of TPN solution a day for maintenance. To this 1000 m*l* of TPN fluid is added 10 units of regular insulin.

Outside of the potential added risk of infection predisposed by TPN solution, the obvious problem in an oliguric ARF patient is fluid overload and pulmonary edema. Dogs on TPN therapy must have cardiopulmonary evaluations daily, along with weight monitorization. Remember, if fluid overload occurs, diuretics will be ineffective, and the only recourse is increased ultrafiltration during dialysis to remove the excess water. Therefore, blood pressure, including a resting level at the beginning of dialysis, must be closely monitored.

There is little doubt nutritional therapy plays an important role in the successful management of acutely uremic patients on dialysis.[149,150] In dogs that will eat, a low-protein diet is not practiced. Instead, large quantities of high biological protein (primarily eggs) and calories (creams, gravies, jelly, candy, cookies, etc.) are given. Clinically, patients appear to do better (more alert and active) fed this way, letting dialysis

* Travenol Laboratories, One Baxter Parkway, Deerfield, Ill. 60047.

** Abbott Laboratories, Abbott Park, North Chicago, Ill. 60064.

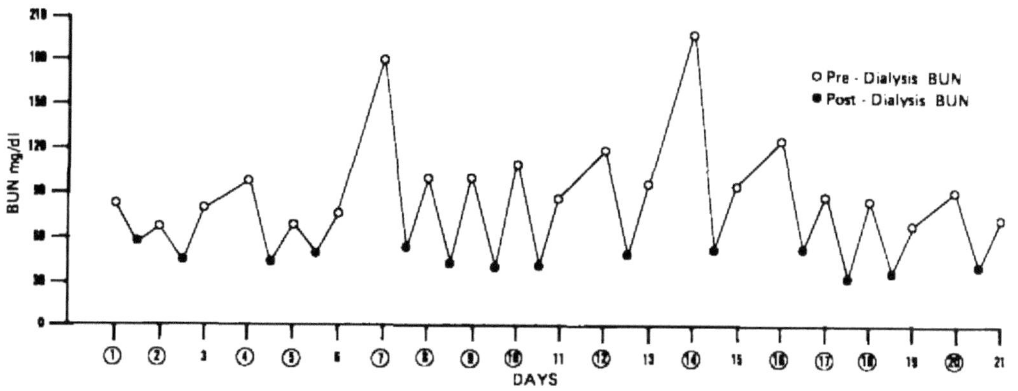

FIGURE 17. Predialysis and postdialysis BUN values on a 4-year-old, male Golden Retriever with oliguric acute renal failure induced by ethylene glycol intoxication. Days dialyzed are circled.

take care of the excess protein, supporting similar results of dietary management of uremic children on hemodialysis.[151]

Dogs with established ARF are placed on antibiotic therapy immediately. A standard dose of cephalothin is given intravenously to establish a minimal therapeutic concentration of the drug in the plasma. Further doses are then given based on the "constant dose-interval extension" method, which extends the dosage interval dependent on the degree of residual renal function.[125] Due to the acute nature of the disorder and the priority of dialysis it creates, a creatinine clearance to measure this degree of renal function is usually performed on the third or fourth hospitalized day. Cephalothin doesn't require dosage adjustments until renal function is less than 20% (and then, the interval is usually doubled) and is dialyzable.[152-155] Therefore, in ARF, the standard dose is given twice daily, with the morning dose given immediately after dialysis on days of the procedure, until a creatinine clearance of the patient predicates otherwise.[125]

Hemodialysis Support of a Dog with Ethylene Glycol-Induced ARF

In Figure 17, life support and control of uremia by hemodialysis in a 4-year-old, male golden retriever with oliguric ARF, induced by ethylene glycol intoxication, is illustrated. The patient was dialyzed 14 out of the 21 hospitalized days, with no recovery from oliguria that is traditionally seen in man with similar nephrotoxicity.[156,157] A needle biopsy on the 15th day (Plates 1 and 2)* confirmed the presence of calcium oxalate crystals in the kidney sample, involving tubular lumens and interstitium with intact basement membranes and healthy tubular structure.[175] Because of the favorable histopathological tissue review, intermittent dialysis was practiced until the 21st day, when because of the continued presence of scanty urine production and attendant guarded prognosis, the owner elected for euthanasia.

Hemodialysis Support of a Dog with Heat Stroke-Induced ARF

In Figure 18, successful life support by hemodialysis in a 2 1/2-year-old, male golden retriever with oliguric ARF, predisposed by heat stroke-induced rhabdomyolysis and myoglobinuria, is illustrated. The patient was nearly over dialyzed the 1st day, dropping the BUN from 320 to 86 mg/dℓ, however, no disequilibrium syndrome developed. Dialysis continued daily for the next 3 days when the patient's urine production began to increase. Remaining dialysis dependent over the next 12 days, the procedure was

* Color plates appear after page 24.

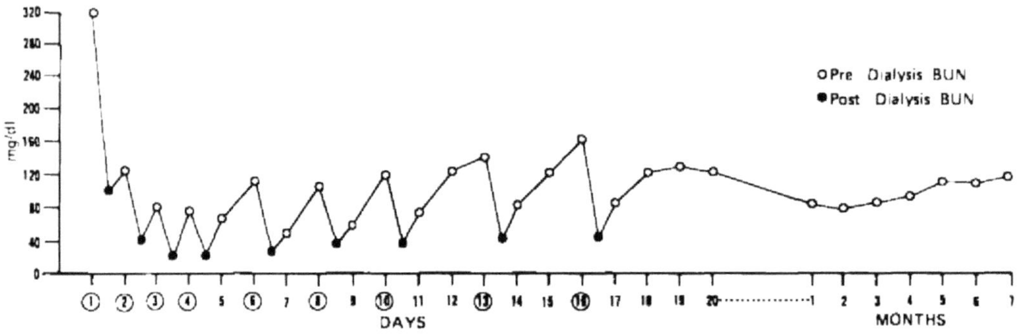

FIGURE 18. Predialysis and postdialysis BUN values while in hospital and BUN values while an outpatient, on a 2 1/2-year-old, male Golden Retriever with oliguric acute renal failure induced by heat stroke. Days dialyzed are circled.

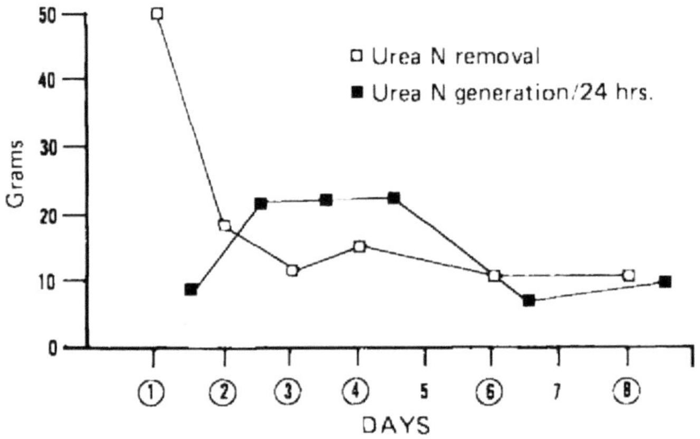

FIGURE 19. Urea nitrogen removal and generation during the first 8 days in the hospital for the patient presented in Figure 18. Days dialyzed are circled.

able to be practiced on a more spread out schedule as renal function steadily improved. The patient was discharged on the 20th hospitalized day, but at that time the endogenous creatinine clearance was only approximately 8% of normal.[112] Renal function increased to about 14% of normal 1 month after the insult, then deteriorated over the next few months to an "end-stage" state, at which time the patient was euthanatized, rather than be allowed to go back on dialysis. Although discouraging, hemodialysis was initially successful and provided life support during the fatal oliguric stage of the disorder, until endogenous renal function returned.

In Figure 19, urea removal and generation in the above myoglobinuric ARF dog, during the oliguric and early diuretic recovery period, is displayed. Calculations are based on the patient's plasma level of urea (milligrams per deciliter) and deciliter space of total body water (urea space) at the beginning and end of dialysis for urea removal (does not include urine loss) and end of dialysis to beginning of next dialysis for urea generation (does not include generation during the 3-hr dialysis session). The 1st day, 50 gm of urea was removed from the patient and 9 gm was generated before the next dialysis. At that time (day 2), 18.5 gm of urea was removed and 22.5 gm was generated before the next dialysis, etc. As noted, urea generation over a 48-hr period dropped considerably from the end and beginning of dialysis between the 6th and 8th day, representing a return of endogenous renal function, although total GFR was still reduced as witnessed by the increase in BUN during this period (Figure 18), obligating continued dialysis support.

NEW TRENDS IN DIALYSIS

Although giant strides have been made in hemodialysis, especially over the last 10 to 15 years, still some uremic patients that are dialyzed are not well.[158] The inadequacy of dialysis seen in these patients, and the reason they are ill, may be due to decreased clearance of middle molecules.[159-163]

New generations of dialyzers more than likely will incorporate sorbents in the dialysate side to more efficiently remove middle molecular weight substances (500 to 3000 daltons) from the plasma. Hemofiltration (a technique where plasma is filtered through membranes rather than dialyzed) studies, using sorbents as the interfacing solution, have shown favorable clearance data for small and middle molecular weight substance removal.[164,165] Sorbents studied consisted of zirconium phosphate (PO_4) as a cation exchanger (for ammonium, K, Ca, and Mg removal), zirconium oxide as an anion exchanger (to remove PO_4), activated charcoal to remove nonionic organic compounds of molecular weight up to 5000 daltons, and urease to hydrolyze urea.

Using the dog as a dialysis model, investigators at Purdue have been developing a self-contained, portable kidney for man which utilizes much of the same sorbent combination discussed above.[166-168] The dialyzer operates off a single vein access, which gives it great potential as a support device for acute or chronic renal failure in the dog.

Another area of great importance to man with application for the dog is removal of excess fluid from maintenance hemodialysis patients. With the ever present threat of hypotension during ultrafiltration accompanying hemodialysis, a new technique of fluid removal shows promise in curbing the cardiovascular compromise. It involves the use of ultrafiltration without dialysis, i.e., blood is coursed through the dialyzer, but dialysate is not pumped; the driving force for ultrafiltration being provided by a hydrostatic pressure gradient applied across the dialyzer membrane.[169-174] These studies have revealed that with this method of ultrafiltration, practiced on days between hemodialysis sessions or for 1 to 2 hr before the initiation of hemodialysis, hypotension can be minimized because urea is not being removed and osmol shifts are less likely to occur.

SUMMARY

Hemodialysis is technically a demanding procedure to perform. Dogs on dialysis must have vital signs monitored at all times and dialysis personnel must be trained to handle emergencies that might arise. Because of this technical demand, especially in ARF, dogs make an excellent model for complications encountered with dialysis dependent-like disorders in man, and, although dogs have been successfully supported on short-term dialysis programs, the procedure to date is relatively untested as a life sustaining therapeutic modality for dogs with dialysis dependent CRF. Much more could be learned about the long-term effects of maintenance dialysis in man, as well as the clinical benefit gained for the dog with "end-stage" kidney disease, if chronic dialysis programs were established for this species. With the development of newer portable artificial kidneys, such as sorbent based models, which function off of a single vein access (no shunt, fistula, or graft required), and greatly reduce the technical aspects and cost of the procedure, maintenance dialysis for the dog may possibly be more readily feasible.

REFERENCES

1. Abel, J. J., Roundtree, L. G., and Turner, B. B., On the removal of diffusable substances from the circulating blood by means of dialysis, *Trans. Assoc. Am. Physicians*, 28, 41, 1913.
2. Kolff, W. J., Berk, H. T. J., ter-Welle, M., van der Leg, J. W., van Dijk, E. C., and van Nordwijk, J., The artificial kidney: a dialyzer with a great area, *Acta Med. Scand.*, 117, 121, 1944.
3. Schlesinger, K., The history of nephrology and dialysis: the 1940's, *Contemp. Dial.*, 1(3), 17, 1980.
4. Alwall, N., On the organization of treatment with the artificial kidney and clinical nephrology in the 1940's and following decades. A contribution to the history of medicine. I, *Dial. Transplant.*, 9, 307, 1980.
5. Alwall, N., On the organization of treatment with the artificial kidney and clinical nephrology in the 1940's and following decades. A contribution to the history of medicine. II. *Dial. Transplant.*, 9, 475, 1980.
6. Schlesinger, K., The history of nephrology and dialysis: the 1950's, *Contemp. Dial.*, 1(3), 21, 1980.
7. Swanbrow, D., The history of nephrology and dialysis: the 1960's, *Contemp. Dial.*, 1(3), 26, 1980.
8. Lettieri, C., The history of nephrology and dialysis: the 1970's, *Contemp. Dial.*, 1(3), 35, 1980.
9. U.S. Department of Health and Human Services, Kidney and Urinary Tract Diseases and Treatment, National Institute of Arthritis, Metabolism and Digestive Diseases, National Institutes of Health, Public Health Service, Bethesda, Md., 1980.
10. Houck, C. R., Problems in maintenance of chronic bilaterally nephrectomized dog, *Am. J. Physiol.*, 176, 175, 1954.
11. Butler, H. C., Advanced therapy for renal failure, in *Proc. 35th Annu. Meeting*, American Animal Hospital Association, Mishawaka, Ind., 1968, 174.
12. Butler, H. C., Renal support systems and transplantation, in *Proc. 38th Annual Meeting*, American Animal Hospital Association, Mishawaka, Ind., 1971, 162.
13. Parker, H. R., Gourley, I. M., and Bell, R. L., Current developments in peritoneal and hemodialysis, Gaines 22nd Vet. Symp., Stillwater, Okla., September 1972, 3.
14. Gourley, I. M., Parker, H. R., Bell, R. L., and Ishizaki, G., Responses of nephrectomized dogs during hemodialysis, *Am. J. Vet. Res.*, 34, 1421, 1973.
15. Cowgill, L. D. and Bovee, K. C., The feasibility and efficiency of hemodialysis in acutely uremic dogs, in *Proc. 42nd Annu. Meet.*, Vol 2, American Animal Hospital Association, Mishawaka, Ind., 1975, 165.
16. Cowgill, L. D. and Bovee, K. C., Current status of hemodialysis and renal transplantation, in *Current Veterinary Therapy*, Vol. 6, Kirk, R. W., Ed., W. B. Saunders, Philadelphia, 1977, 1149.
17. Cowgill, L. D., Current status of veterinary hemodialysis, in *Current Veterinary Therapy*, Vol. 7, Kirk, R. W., Ed., W. B. Saunders, Philadelphia, 1980, 1111.
18. Dhein, C. R., Hemodialysis in the dog: a review, *Compend. Contin. Educ. Pract. Vet.*, 3, 1031, 1981.
19. Ash, S. R., Dhein, C. R., Rebar, A., and Thornhill, J. A., Dialytic support of dogs with clinically occurring renal failure: a realistic model of acute renal failure in man, *Clin. Exp. Dial. Apheresis*, 6, 25, 1982.
20. Bower, J. D., Holbert, R. D., Pearsons, J. Q., Bengis, R., and Berndt, W. D., Chronic hemodialysis in the anephric goat, *Nephron*, 25, 34, 1980.
21. Friedman, E. A. and Lundin, A. P., III, Dialysate delivery: historical, theoretical, and practical aspect, *Kidney Int.*, 18, S-33, 1980.
22. Gordon, A., Lewin, A., Marantz, L. B., and Maxwell, M. H., Sorbent regeneration of dialysate, *Kidney Int.*, 10, S277, 1976.
23. Henderson, L. W., Redy® or not, *Am. Soc. Artif. Intern. Organs J.*, 2, 49, 1979.
24. Ackerman, R. A., Crosby, S. C., Morefield, E. F., and Stevenson, R. S., Dialysate delivery and regenerative system, *Trans. Am. Soc. Artif. Intern. Organs*, 25, 398, 1979.
25. Ash, S. R., Principles and practice of hemodialysis therapy, in *CRC Handbook of Clinical Engineering*, Vol. 1, Feinberg, B. N., Ed., CRC Press, Boca Raton, Fla., 1980, 177.
26. Farrell, P. C., Assessing dialyzer performance, *Dial. Transplant.*, 9, 644, 1980.
27. Ebben, J., Mauer, M., Davin, T., Otieno, I., and Kjellstrand, C., A new pediatric hemodialyzer, *Dial. Transplant.*, 9, 735, 1980.
28. Klein, E., Autian, J., Bower, J. D., Buffaloe, G., Centella, L. J., Colton, C. K., Darby, T. D., Farrell, P. C., Holland, F. F., Kennedy, R. S., Lipps, B., Jr., Mason, R., Nolph, K. D., Villarroel, F., and Wathen, R. L., Evaluation of hemodialyzers and dialysis membranes, clinical evaluation, *Artif. Organs*, 3, 47, 1979.
29. Klein, E., Holland, F. F., and Eberle, K., Advances in dialysis: hemodialysis membranes, *Kidney Int.*, 18, S-19, 1980.
30. Bermick, J. J. and Port, F. K., Absence of bacteremia and endotoxemia despite contaminated dialysate, *Clin. Nephrol.*, 14, 13, 1980.

31. Klinkmann, H., Holtz, M., Falkenhagen, D., Ahrenholz, P., Bartsch, D., Bossin, F., Gensrich, H. J., Grobe, V., and Schoner, H. W., Development and assessment of new cellulose membranes for blood purification, *Kidney Int.*, 18, S-26, 1980.
32. Mion, C. M., Hegstrom, R. M., Goen, S. T., and Scribner, B. H., Substitution acetate for sodium bicarbonate in the bath fluid for hemodialysis, *Trans. Am. Soc. Artif. Intern. Organs*, 10, 110, 1964.
33. Van Stone, J. C., A critical comparison of bicarbonate and acetate dialysates, *Contemp. Dial.*, 1(1), 25, 1980.
34. Friedman, E. A. and Lewis, E. J., Discussion of dialysate composition: acetate versus bicarbonate, *Kidney Int.*, 18, S-75, 1980.
35. Vreman, H. J., Assomull, V. M., Kaiser, B. A., Blaschke, T. F., and Weiner, M. W., Acetate metabolism and acid-base homeostasis during hemodialysis: influence of dialyzer efficiency and rate of acetate metabolism, *Kidney Int.*, 18, S-62, 1980.
36. Davidson, W. D., Morin, R. J., Srikantaiah, M., and Bassist, L., The role of acetate in dialysate for hemodialysis, *Proc. Am. Contract. Conf. Artif. Kidney Program*, National Institute of Arthritis, Metabolism and Digestive Diseases, Bethesda, Md., Vol. 11, 75, 1978.
37. Wingert, K. J. and Weiner, M. W., Acetate transfer across membranes of artificial kidneys in vitro, *Kidney Int.*, 15, 593, 1979.
38. Lewis, E. J., Tolchin, N., and Roberts, J. L., Estimation of the metabolic conversion of acetate to bicarbonate during hemodialysis, *Kidney Int.*, 18, S-51, 1980.
39. Kirkendol, P. L., Starrs, J., and Gonzalez, F. M., The effects of acetate, lactate, succinate and gluconate or plasma pH and electrolyte in dogs, *Trans. Am. Soc. Artif. Intern. Organs*, 26, 323, 1980.
40. Kirkendol, P. L., Robie, F. M., Gonzalez, F. M., and Devia, C. J., Cardiac and vascular effects of infused sodium acetate in dogs, *Trans. Am. Soc. Artif. Intern. Organs*, 24, 714, 1978.
41. Kirkendol, P. L., Devia, C. J., Bower, J. D., and Holbert, R. D., A comparison of the cardiovascular effects of sodium acetate, sodium bicarbonate and other potential sources of fixed base in hemodialysis solutions, *Trans. Am. Soc. Artif, Intern. Organs*, 23, 399, 1977.
42. Novello, A., Kelsch, R. C., and Easterling, R. E., Acetate intolerance during hemodialysis, *Clin. Nephrol.*, 5, 29, 1976.
43. Ibels, L. S., Simons, L. A., King, J. O., Williams, P. F., Neale, F. C., and Stewart, J. H., Studies on the nature and causes of hyperlipidemia in uremia, maintenance dialysis and renal transplantation, *Q. J. Med.*, 44, 601, 1975.
44. Graefe, U., Milutimovich, J., Follette, W. C., Vizzo, J. E., Babb, A. L., and Scribner, B. H., Less dialysis-induced morbidity and vascular instability with bicarbonate in dialysate, *Ann. Intern. Med.*, 88, 332, 1978.
45. Sargent, J. A. and Gotch, F. A., Bicarbonate and carbon dioxide transport during dialysis, *Am. Soc. Artif. Intern. Organs J.*, 2, 61, 1979.
46. Cambi, V., Arisi, L., Biasini, A., Bono, F., David, S., Savazzi, G., and Migone, L., Improvement of intradialytic tolerance with low-volume dialysis and bicarbonate buffering of dialysate, *Kidney Int.*, 18, S-56, 1980.
47. Kirkendol, P. L., Pearson, J. E., and Gonzalez, F. M., A potential source of fixed base in dialysate solutions: a comparison of the cardiovascular effects of sodium succinate versus bicarbonate and acetate, *Am. Soc. Artif. Intern. Organs J.*, 3, 57, 1980.
48. Van Stone, J. C., Bauer, J., and Carey, J., The effect of dialysate sodium concentration on body fluid distribution during hemodialysis, *Trans. Am. Soc. Artif. Intern. Organs*, 26, 383, 1980.
49. Van Stone, J. C., Baner, J., and Carey, J., The effect of dialysate sodium concentrations on body fluid compartment volume, plasma renin activity, and plasma aldosterone concentration in chronic hemodialysis patients, *Am. J. Kidney Dis.*, 2, 58, 1982.
50. Port, F. K., Johnson, W. J., and Klass, D. W., Prevention of dialysis disequilibrium syndrome by use of high sodium concentration in the dialysate, *Kidney Int.*, 3, 327, 1973.
51. Van Stone, J. C. and Cook, J., Decreased postdialysis fatigue with increased dialysate sodium concentration, in *Proc. Clin. Transpl. Forum*, Vol. 8, 1978, 152.
52. Ogden, D. A., A double blind crossover comparison of high and low sodium dialysis, in *Proc. Clin. Transpl. Forum*, Vol. 8, 1978, 157.
53. Wilkinson, R., Barber, S. G., and Robson, V., Cramps, thirst and hypertension in hemodialysis patients — the influence of dialysate sodium concentrations, *Clin. Nephrol.*, 7, 101, 1977.
54. Van Stone, J. C., Meyer, R., Murrin, C., and Cook, J., Hemodialysis with glycerol dialysate, *Trans. Am. Soc. Artif. Intern. Organs*, 25, 354, 1979.
55. Van Stone, J. C., Carey, J., Meyer, R., and Murrin, C., Hemodialysis with glycerol containing dialysate, *Am. Soc. Artif. Intern. Organs J.*, 2, 119, 1979.

56. Chen, W-T., Ing, T. S., Daugirdas, J. T., Humayun, H. M., Brescia, D. J., Gandhi, V. C., Hano, J. E., and Kheirbek, A. O., Hydrostatic ultrafiltration during hemodialysis using decreasing sodium dialysate, *Artif. Organs*, 4, 190, 1980.
57. Bok, D. J. and Grondin, G. D., Water treatment in dialysis — current problems and potentials, *Contemp. Dial.*, 1(2), 28, 1980.
58. Butt, K. M. H., Blood access, *Clin. Nephrol.*, 9, 138, 1978.
59. Butt, K. M. H., Pascua, L. J., Hardy, M. A., Lawton, R. L., and Uldall, P. R., Vascular access update, *Trans. Am. Soc. Artif. Intern. Organs*, 25, 526, 1979.
60. Lawton, R. L., Vascular access, always one more arrow in the quiver, *Dial. Transplant.*, 9, 10, 1980.
61. Ishihara, A. M., The current state-of-the-art for vascular access in hemodialysis, *Contemp. Dial.*, 1(1), 29, 1980.
62. Friedman, E. A. and Lundin, A. P., III, Dialysate delivery: historical, theoretical, and practical aspect, *Kidney Int.*, 18, S-33, 1980.
63. Latimer, R. G., Gebhart, W. F., Freidell, H. V., Murray, J. J., and Fisher, M. B., Comparison of chronic hemodialysis angioaccess procedures, *Dial. Transplant.*, 9, 499, 1980.
64. Bukhari, H. I., Wheeler, C. G., Hempel, G. K., Hunt, D. G., and Anderson, J. F., Angioaccess for hemodialysis techniques, problems and future trends, *Contemp. Dial.*, 1(3), 66, 1980.
65. Quinton, W., Dillard, D., and Scribner, B. H., Cannulation of blood vessels for prolonged hemodialysis, *Trans. Am. Soc. Artif. Intern. Organs*, 6, 104, 1960.
66. Thomas, G. I., A large vessel applique A-V shunt for hemodialysis, *Trans. Am. Soc. Artif. Intern. Organs*, 15, 288, 1969.
67. Thomas, G. I., Improvisations with the Thomas femoral shunt, *Dial. Transplant.*, 9, 1152, 1980.
68. Buselmeier, T. J., Kjellstrand, C. M., Santiago, E. A., Simmons, R. L., and Najarian, J. S., A new subcutaneous arteriovenous shunt: applicable in cases where the standard Quinton-Scribner shunt and arteriovenous fistula have failed, *Surgery*, 73, 512, 1973.
69. Buselmeier, T. J., Simmons, R. L., Najarian, J. S., Duncan, D. A., von Hartitzsch, B., and Kjellstrand, C. M., The clinical application of a new prosthetic arteriovenous shunt, *Nephron*, 12, 22, 1973.
70. Brescia, M. J., Cimino, J. E., Appel, K., and Hurwich, B. J., Chronic hemodialysis using venipuncture and a surgically-created arteriovenous fistula, *N. Engl. J. Med.*, 275, 1089, 1966.
71. Byrne, J. P., Stevens, L. E., Weaver, D. H., Maxwell, J. G., and Reemtsma, K., Advantages of surgical arteriovenous fistulas for hemodialysis, *Arch. Surg.*, 102, 359, 1971.
72. Kuruvila, K. C. and Beven, E. G., Arteriovenous shunts and fistulas for hemodialysis, *Surg. Clin. North Am.*, 51, 1219, 1971.
73. Thompson, B. W., Internal arteriovenous fistula for hemodialysis, *Am. J. Surg.*, 124, 785, 1972.
74. Foran, R. F., Saphenous vein and bovine arterial grafts: a comparative study, *Dial. Transplant.*, 4, 34, 1975.
75. Rosenberg, N., The bovine arterial graft and its several applications, *Surg. Gynecol. Obstet.*, 142, 104, 1976.
76. Rolley, R. T., Sterioff, S., and Williams, G. M., Arteriovenous fistulas for dialysis using modified bovine arteries, *Surg. Gynecol. Obstet.*, 142, 700, 1976.
77. Sannella, N. A., Mehigan, J. T., Pennell, J. P., Halpren, B., and Fogarty, T. J., Initial approach to blood access for the chronic hemodialysis patient — the bovine heterograft, *Trans. Am. Soc. Artif. Intern. Organs*, 22, 394, 1978.
78. Kaplan, M. S., Mirahmadi, K. S., Winer, R. L., Gorman, J. T., Dabirvaziri, N., and Rosen, S. M., Comparison of PTFE and bovine grafts for blood access in dialysis patients, *Trans. Am. Soc. Artif. Intern. Organs*, 22, 388, 1976.
79. Baker, L. D., Johnson, J. M., and Goldfarb, D., Expanded polytetrafluoroethylene (PTFE) subcutaneous arteriovenous conduit: an improved vascular access for chronic hemodialysis, *Trans. Am. Soc. Artif. Intern. Organs*, 22, 382, 1976.
80. Flores, L., Dunn, I., Frumkin, E., Forte, R., Requena, R., Ryan, J., Knopf, M., Kirschner, J., and Levowitz, B. S., Dacron arteriovenous shunts for vascular access in hemodialysis, *Trans. Am. Soc. Artif. Intern. Organs*, 19, 33, 1973.
81. Reckard, C. R., Buselmeier shunt ports: a unique source for recurrent bacteremia, *Dial. Transplant.*, 9, 1156, 1980.
82. Abhondzadeh, L., Wilson, S. E., Williams, R., and Owens, M. J., Infection of materials used in vascular access surgery: an evaluation of dacron, bovine heterograft, Teflon, and human umbilical vein grafts, *Dial. Transplant.*, 9, 697, 1980.
83. Burbridge, G. E., Biggers, J. A., Remmers, A. R., Jr., Lindley, J. D., Sarles, H. E., and Fish, J. C., Late complications and results of bovine xenografts, *Trans. Am. Soc. Artif. Intern. Organs*, 22, 377, 1976.
84. Haimov, M., Baez, A., Neff, M., and Slifkin, R., Complications of arteriovenous fistulas for hemodialysis, *Arch. Surg.*, 110, 708, 1975.

85. Yeary, R. A. and Brant, R. J., Aspirin dosages for the dog, *J. Am. Vet. Med. Assoc.*, 167, 63, 1975.
86. Davis, L. E., Clinical pharmacology of salicylates, *J. Am. Vet. Med. Assoc.*, 176, 65, 1980.
87. Harter, H. R., Burch, J. W., Majerus, P. W., Stanford, N., Delmez, J. A., Anderson, C. B., and Weerts, C. A., Prevention of thrombosis in patients on hemodialysis by low-dose aspirin, *N. Engl. J. Med.*, 301, 577, 1979.
88. O'Grady, J., Aspirin, a paradoxical effect on bleeding time, *Lancet*, 780, 1978.
89. Woods, H. F. and Weston, M. J., The use of antiplatelet agents during hemodialysis, *Dial. Transplant.*, 8, 958, 1979.
90. Rosenberg, R. D., Actions and interactions of antithrombin and heparin, *N. Engl. J. Med.*, 292, 146, 1975.
91. Levin, R. D., Kwaan, H. C., and Ivanovich, P., Changes in platelet function during hemodialysis, *J. Lab. Clin. Med.*, 92, 779, 1978.
92. Grabowski, E. F., Herther, K. K., and Didisheim, P., Human vs. dog platelet adhesion to cuprophane under controlled conditions of whole blood flow, *J. Lab. Clin. Med.*, 88, 368, 1976.
93. Lee, E. S. and Kim, S. W., Absorbed glycoproteins in platelet adhesion into polymer surfaces: significance of terminal galactose units, *Trans. Am. Soc. Artif. Intern. Organs*, 25, 124, 1976.
94. Schanklin, N., Evaluation of an activated clotting time technique for use in hemodialysis, *Dial. Transplant.*, 9, 955, 1980.
95. Byars, D., Ling, G. V., Ferris, N. A., and Keeton, K. S., Activated coagulation time (ACT) of whole blood in normal dogs, *Am. J. Vet. Res.*, 37, 1359, 1976.
96. Kennedy, C. C. and Rocks, M. J., Bedside control of heparin therapy by a simple whole blood clotting method, *J. Clin. Pathol.*, 26, 893, 1972.
97. Shapiro, W. B., Faubert, P. F., Porush, J. G., and Shyan-Yih, C. M., Low-dose heparin in routine hemodialysis monitored by activated partial thromboplastin time, *Artif. Organs*, 3, 73, 1979.
98. Shapiro, W. B., Faubert, P. F., Shyan-Yih, C. M., and Porush, J. G., Low-dose heparin in the high-risk bleeding hemodialysis patient monitored by activated partial thromboplastin time, *Dial. Transplant.*, 9, 322, 1980.
99. Ward. R. A. and Farrell, P. C., Precise anticoagulation for routine hemodialysis using nomograms, *Trans. Am. Soc. Artif. Intern. Organs*. 24, 439, 1978.
100. Buccianti, G., Pogliani, E., Valenti, G., Cofrancesco, E., Miradoli, R., and Vigo, A., Clinical monitoring of heparin during hemodialysis: a new method, *Dial. Transplant.*, 8, 1118, 1979.
101. Ward, R. A. and Wathen, R. L., Precise heparization for hemodialysis of nonuremic subjects, *Am. Soc. Artif. Intern. Organs J.*, 3, 147, 1980.
102. Lindsay, R. M., Variable heparin requirements during hemodialysis — why?, *Am. Soc. Artif. Intern. Organs J.*, 3, 81, 1980.
103. Addonizio, V. P., Jr., Colman, R. W., and Edmunds L. H., Jr., The effect of blood flow rate and circuit surface area on platelet loss during extracorporeal circulation, *Trans. Am. Soc. Artif. Intern. Organs*, 24, 650, 1978.
104. Woods, H. F., Ash, G., Weston, M. J., Bunting, S., Moncada, S., and Vane, J. R., Prostacyclin can replace heparin in hemodialysis in dogs, *Lancet*, 2, 1075, 1978.
105. Naets, J. P., Hematologic disorders in renal failure, *Nephron*, 14, 181, 1975.
106. Spangler, W. L., Gribble, D. H., and Weiser, M. G., Canine hypertension: a review, *J. Am. Vet. Med. Assoc.*, 1970, 995, 1977.
107. Schnett, H. and Port, R. K., Hemolysis in hemodialysis patients, *Dial. Transplant.*, 9, 345, 1980.
108. Lindsay, R. M., Burton, J. A., King, P., Davidson, J. F., Boddy, K., and Kennedy, A. C., The measurement of dialyzer blood loss, *Clin. Nephrol.*, 1, 24, 1973.
109. Lindsay, R. M., Burton, J. A., Edward, N., Dargie, H. J., Prentice, C. R. M., and Kennedy, A. C., Dialyzer blood loss, *Clin. Nephrol.*, 1, 29, 1973.
110. Bovee, K. C., What constitutes a low protein diet for dogs with chronic renal failure?, *J. Am. Anim. Hosp. Assoc.*, 8, 246, 1972.
111. Osborne, C. A. and Polzin, D. J., Dietary management of renal failure: why? when? how?, *D. V. M. Magazine*, 10(5), 26, 1979.
112. Bovee, K. C. and Joyce, T., Clinical evaluation of glomerular function: 24-hour creatinine clearance in dogs, *J. Anim. Vet. Med. Assoc.*, 174, 488, 1979.
113. Sargent, J. A., Kinetic modeling in the guidance of dialysis therapy, *Dial. Transplant.*, 8, 1101, 1979.
114. Lazarus, J. M., Complications in hemodialysis: an overview, *Kidney Int.*, 18, 78, 1980.
115. Kjellstrand, C. W., Current problems in long-term hemodialylsis, *Dial. Transplant.*, 9, 295, 1980.
116. Wathen, R., Keshaviah, P., and Shapiro, F., Unsolved technical problems of maintenance dialysis, *Clin. Nephrol.*, 9, 174, 1978.
117. Nsouli, K. A., Lazarus, J. M., Schoenbaum, S. C., Gottlieb, M. N., Lowrie, E. G., and Shocair, M., Bacteremic infection in hemodialysis, *Arch. Intern. Med.*, 139, 1255, 1979.

118. Swan, R. C. and Merrill, J. P., The clinical course of acute renal failure, *Medicine*, 32, 215, 1953.
119. Montgomerie, J. Z., Kalmanson, G. M., and Goze, L. B., Renal failure and infection, *Medicine*, 47, 1, 1968.
120. Kaslow, R. A. and Zellner, S. R., Infection in patients on maintenance hemodialysis, *Lancet*, 2, 117, 1972.
121. Kirmani, N., Tauzon, C. U., Murry, H. W., Parrish, A. E., and Sheagren, J. W., *Staphylococcus aureus* carriage rate of patients receiving long-term hemodialysis, *Arch. Intern. Med.*, 138, 1657, 1978.
122. Keane, W. F., Shapiro, F. L., and Raij, L., Incidence and type of infection occurring in 445 chronic hemodialysis patients, *Trans. Am. Soc. Artif. Intern. Organs*, 23, 41, 1977.
123. MacGregor, R. R. and Beaty, H. N., Evaluation of positive blood cultures: guidelines for early differentiation of contaminated from valid positive cultures, *Arch. Intern. Med.*, 130, 84, 1972.
124. Gelfand, M. C., Bacteremia in the hemodialyzed patient: a persistent problem in spite of technological progress, *Arch. Intern. Med.*, 139, 1223, 1979.
125. Riviere, J. E. and Coppoc, G. L., Dosage of antimicrobial drugs in patients with renal insufficiency, *J. Am. Vet. Med. Assoc.*, 178, 70, 1981.
126. Mockros, L. F., Doerr, D. E., and Zukerman, L., Coagulopathy induced by extracorporeal shear, *Trans. Am. Soc. Artif. Intern. Organs*, 25, 139, 1979.
127. Gotch, F. A., Progress in hemodialysis, *Clin. Nephrol.*, 9, 144, 1978.
128. Stein, I. M., Cohen, B. D., and Kornhauser, R. S., Guanidinosuccinic acid in renal failure, experimental azotemia and inborn errors of the urea cycle, *N. Engl. J. Med.*, 280, 926, 1969.
129. Cohen, B. D., Uremia: pillar of salt of column or nitrogen, *Dial. Transplant.*, 9, 535, 1980.
130. Lowrie, E. G., Steinberg, S. M., Galen, M. A., Gagneux, S. A., Lazarus, J. M., Gottlieb, N., and Merrill, J. P., Factors in the dialysis regimen which contribute to alterations in the abnormalities of uremia, *Kidney Int.*, 10, 409, 1976.
131. Osborne, C. A., Low, D. G., and Finco, D. R., Reversible versus irreversible renal disease in the dog, *J. Am. Vet. Med. Assoc.*, 115, 2062, 1969.
132. Bovee, K. C., The uremic syndrome: patient evaluation and treatment, *Compend. Contin. Educ. Small Anim. Pract.*, 1, 279, 1979.
133. Gulyassy, P. F., Nondialytic management of chronic renal failure, *Kidney*, 5(4), 1, 1972.
134. Manis, T. and Friedman, E. A., Dialytic therapy for irreversible uremia (first of two parts), *N. Engl. J. Med.*, 301, 1260, 1979.
135. Manis, T. and Friedman, E. A., Dialytic therapy for irreversible uremia (second of two parts), *N. Engl. J. Med.*, 301, 1321, 1979.
136. Flamenbaum, W. and Kaufman, J. S., Acute renal failure, *Kidney*, 9, 21, 1976.
137. Levinsky, N. G., Pathophysiology of acute renal failure, *N. Engl. J. Med.*, 296, 1453, 1977.
138. Schrier, R. W., Acute renal failure, *Kidney Int.*, 15, 205, 1979.
139. DiBartola, S. P., Acute renal failure: pathophysiology and management, *Compend. Contin. Educ. Small Anim. Pract.*, 2, 952, 1980.
140. Hall, J. W., Johnson, W. J., Maher, F. T., and Hunt, J. C., Immediate and long-term prognosis in acute renal failure, *Ann. Intern. Med.*, 73, 515, 1970.
141. Lynn, K. L., Neale, T. J., Bailey, R. R., and Little, P. J., Oliguria and its sequelae, *N. Z. Med. J.*, 86, 563, 1977.
142. Swartz, R. D., Volk, T. W., Brain, A. J. S., and Hsu, C. H., Complications of hemodialysis and peritoneal dialysis in acute renal failure, *Am. Soc. Artif. Intern. Organs J.*, 3, 98, 1980.
143. Conger, J. D., The use of prophylactic dialysis in acute renal failure, *Contemp. Dial.*, 1(2), 36, 1980.
144. Kleinknecht, D., Jungers, P., Chanard, J., Barbanel, C., and Caneval, D., Uremic and non-uremic complications in acute renal failure: evaluation of early and frequent dialysis in prognosis, *Kidney Int.*, 1, 190, 1972.
145. Thornhill, J. A., Control of vomiting in the uremic patient, in *Current Veterinary Therapy*, Vol. 8, Kirk, R. W., Ed., W. B. Saunders, Philadelphia, 1983, 1022.
146. Anthony, M. J., Woolfson, M. D., Heatley, M. D., and Allison, S. P., Insulin to inhibit protein catabolism after injury, *N. Engl. J. Med.*, 300, 14, 1979.
147. Abel, R. M., Beck, C. H., and Abbott, W. M., Improved survival from acute renal failure after treatment with intravenous essential amino acids and glucose, *N. Engl. J. Med.*, 288, 695, 1973.
148. Carter, J. M. and Freedman, A. B., Total intravenous feeding in the dog, *J. Am. Vet. Med. Assoc.*, 171, 71, 1977.
149. Asbach, H. W., Stoeckel, H., and Schuler, H. W., The treatment of hypercatabolic acute renal failure by adequate nutrition and hemodialysis, *Acta Anaesth. Scand.*, 18, 255, 1974.
150. Baek, S. M., Makabali, G. G., and Bryan-Brown, C. W., The influences of parenteral nutrition on the course of acute renal failure, *Surg. Gynecol. Obstet.*, 141, 405, 1975.

151. Conley, S. B., Rose, G. M., Robson, A. M., and Bier, D. M., Effects of dietary intake and hemodialysis on protein turnover in uremic children, *Kidney Int.*, 17, 837, 1980.
152. Yamasaku, F., Tsuichida, R., and Usuda, Y., A study of the kinetics of cephalosporins in renal impairment, *Postgrad. Med. J.*, 46, 57, 1970.
153. Powers, T. E. and Gary, R. C., Pharmacotherapeutics of new penicillins and cephalosporins, *J. Am. Vet. Med. Assoc.*, 176, 1054, 1980.
154. Anderson, R. J., Bennett, W. M., Gambertoglio, J. G., and Schrier, R. W., Fate of drugs in renal failure, in *The Kidney*, 2nd ed., Brenner, B. M. and Rector, F. C., Eds., W. B. Saunders, Philadelphia, 1981, 2659.
155. Richard, C. J., Dialyzability of drugs, *Dial. Transplant.*, 10, 474, 1981.
156. Schreiner, G. E., Maher, J. F., Marc-Aurele, J., Knowlan, D., and Alvo, M., Ethylene glycol. Two indications for hemodialysis, *Trans. Am. Soc. Artif. Intern. Organs*, 5, 81, 1959.
157. Collins, D. M., Hennes, D. M., Holzgang, C. R., Gourley, R. T., and Porter, G. A., Recovery after prolonged oliguria due to ethylene glycol intoxication, *Arch. Intern. Med.*, 125, 1059, 1970.
158. Kjellstrand, C. M., Arieff, A. I., Friedman, E. A., Furst, P., Henderson, L. W., and Massry, S. G., Inadequacy of dialysis: why patients are not well, *Trans. Am. Soc. Artif. Intern. Organ*, 25, 518, 1979.
159. Bergstrom, J. and Furst, P., Uremic molecules, *Clin. Nephrol.*, 5, 143, 1976.
160. Nolph, K. D., Short dialysis, middle molecules and uremia, *Ann. Intern. Med.*, 86, 93, 1977.
161. Milutinovic, J., Babb, A. L., Eshbach, J. W., Follette, W. C., Graefe, U., Straud, M. J., and Scribner, B. H., Uremic neuropathy: evidence of middle molecule toxicity, *Artif. Organs*, 2, 45, 1978.
162. Kjellstrand, C. M., The clinical significance of middle molecules, *Dial. Transplant.*, 8, 860, 1979.
163. Chapman, G. V., Ward, R. A., and Farrell, P. C., Separation and quantification of the "middle molecules" in uremia, *Kidney Int.*, 17, 82, 1980.
164. Klein, E., Holland, F. F., Eberle, K., Morton, F. C., and Cabasso, I., Sorbent-filled hollow fibers for hemopurification, *Trans. Am. Soc. Artif. Intern. Organs*, 24, 127, 1978.
165. Shapiro, W. B., Faubert, P. F., Fein, P. A., Tzeng, T., Salsman, K., and Porush, J. G., Hemofiltration with sorbent recycling of ultrafiltrate in uremic dogs, *Trans. Am. Soc. Artif. Intern. Organs*, 24, 185, 1978.
166. Ash, S. R., Barile, R. G., Thornhill, J. A., Sherman, J. D., and Want, N-H. L., In vivo evaluation of calcium-loaded zeolites and urease for urea removal in hemodialysis, *Trans. Am. Soc. Artif. Intern. Organs*, 26, 111, 1980.
167. Ash, S. R., Barile, R. G., Wilcox, P. G., Wright, D. L., Thornhill, J. A., Dhein, C. R., Kessler, D. P., and Wang, N-H. L., The sorbent suspension reciprocating dialyzer: a device with minimal sorbent saturation, *Am. Soc. Artif. Intern. Organs J.*, 1, 28, 1981.
168. Barile, R. G., Wang, N-H. L., Blake, D. E., Belcastro, P. F., Gupta, S., Regnier, F. E., Thornhill, J. A., Kessler, D. P., and Ash, S. R., A reciprocating single-needle hemodialyzer with bidirectional flow sorbent suspension, *Artif. Organs*, 6, 267, 1982.
169. Bergstrom, J., Asaba, H., Furst, P., and Oules, R., Dialysis, ultrafiltration, and blood pressure, *Proc. Eur. Dial. Transplant. Assoc.*, 13, 293, 1976.
170. Ing, T. S., Vilbar, R. M., Shin, K. D., Viol, G. W., Bunsal, V. K., Geis, W. P., and Hano, J. E., Predialytic isolated ultrafiltration, *Dial. Transplant.*, 7, 557, 1978.
171. Wehle, B., Asaba, H., Castenfors, J., Furst, P., Gunnersson, B., Shaldon, S., and Bergstrom, J., Hemodynamic changes during sequential ultrafiltration and dialysis, *Kidney Int.*, 15, 411, 1979.
172. Hampl, H., Praeper, H., Unger, U., and Kessel, M. G., Hemodynamics during hemodialysis, sequential ultrafiltration, and hemofiltration, *J. Dialysis*, 3, 51, 1979.
173. Keshaviah, P., Ustruep, K., and Constantini, E., The influence of ultrafiltration and diffusion on cardiovascular parameters, *Trans. Am. Soc. Artif. Intern. Organs*, 26, 328, 1980.
174. Ing, T. S., Ghen, W. T., Daugirdas, J. T., Kwaan, H. C., and Hano, J. E., Isolated ultrafiltration and new techniques of ultrafiltration during dialysis, *Kidney Int.*, 18, S-77, 1980.
175. Thornhill, J. A., Ash, S. R., Crabtree, B. J., Riviere, J. E., and Thacker, H. L., Proposed mechanism for re-establishment of tubular luminal patency in obstructive calcium oxalate nephrosis due to ethylene glycol intoxication in the dog, *J. Comp. Pathol.*, in review.

HEMODYNAMIC STUDIES IN THE CONSCIOUS, HEMODIALYZED DOG

John T. Daugirdas and Todd S. Ing

INTRODUCTION

Although development of the human artificial kidney was based on animal experiments,[1] chronic hemodialysis of animals for either investigational or therapeutic purposes is, at present, uncommon. Pioneering studies of the hemodialyzed, uremic dog by Grollman et al.,[2] Skeggs et al.,[3,4] and Houck[5] have not been much enlarged upon by modern investigators. Apart from factors of cost and labor, hemodialyzed dogs are prone to anorexia and vomiting.[4-9] Vascular access is not easy to maintain. During hemodialysis, hemolysis, hypotension, and clotting of the dialyzer have been frequently encountered.[6-8] If these problems could be solved, hemodialysis-maintained dogs might be used to investigate many fascinating aspects of uremia, including cardiovascular hemodynamics, erythropoiesis, nitrogen metabolism, endocrine response, and osteodystrophy. The hemodialyzed animal is also ideally suited for study of the acute effects of electrolyte changes on cardiovascular and neurologic function. Recently, nephrectomized, hemodialyzed goats or sheep[10-12] have been proposed as experimental uremic animals. Ruminants apparently tolerate uremia better than dogs and may be easier to support with hemodialysis. However, the fact that the dog tolerates uremia poorly makes it an ideal species in which to evaluate the complications of uremia and in which to compare the benefits of various renal replacement therapies.

The purpose of the present communication is to describe our methods of preparing uremic dogs and of maintaining them on hemodialysis and to describe preliminary results of hemodynamic studies in these animals. Our experience is derived from animals dialyzed for 2 to 3 weeks. Substantial problems were encountered even over this brief time interval. Additional difficulties may well arise when attempting to maintain uremic animals for more extended periods.

LABORATORY METHODS

Nonuremic Dogs

Initially, nonuremic dogs were studied in an attempt to perfect technical aspects of performing hemodialysis and to obtain control data for comparison with the uremic condition.

Vascular Access

Femoral Arteriovenous Shunt

Vascular access for dialysis of these animals was by way of a silicone elastomer shunt connecting the femoral artery with the ipsilateral femoral vein. To minimize the risk of dislodgement by the animal biting or scratching, the shunt is placed so as to exit the skin on the dog's back. The method of implanting the shunt is as follows. Pentobarbital anesthesia is used during surgery. The dog is ventilated using room air administered by a Harvard® ventilator. The arteriovenous (AV) shunt tubing is of silicone elastomer (0.104-in. I.D., Dow Corning Corp., Midland, Mich.). Prior to use, a cuff of USCI® Teflon® felt (#H8015-007837, C. R. Bard, Inc., Billerica, Mass.) is glued (medical adhesive, silicone type A, catalog #891, Dow Corning Corp.) to the shunt tubing in the style of a Tenckhoff peritoneal catheter. The exit sites of the shunt are

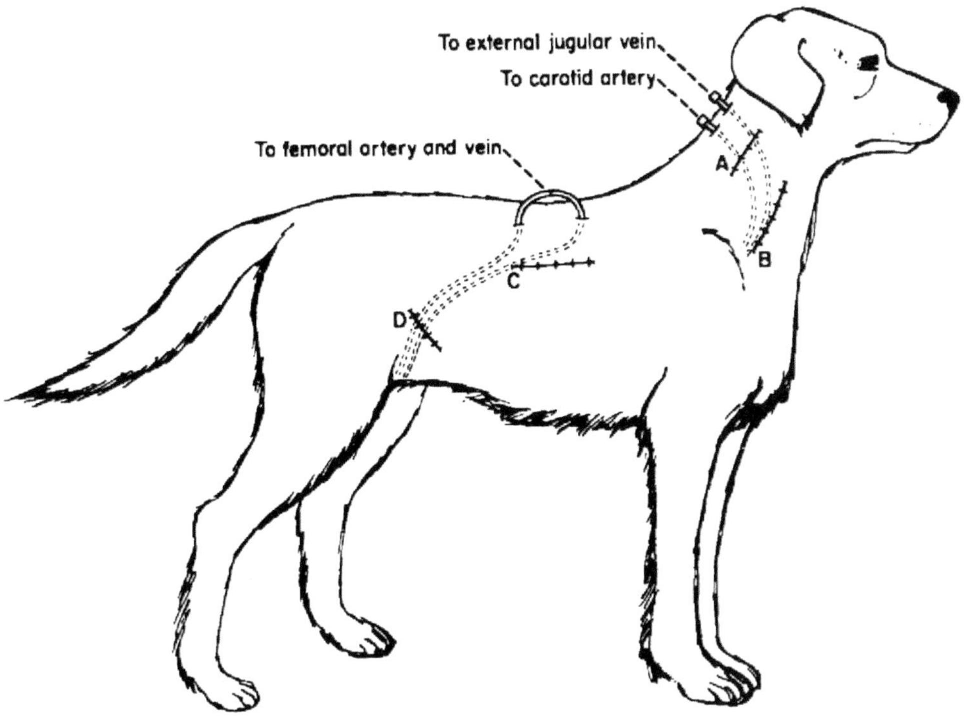

FIGURE 1. Placement of femoral artery to femoral vein shunt and neck vascular access tubes leading to the carotid artery and jugular vein.

placed first, using a small access incision (marked C in Figure 1). Working through the access incision, each felt cuff is secured to the underside of the dermis at the skin exit site with a suture of 3-0 chromic gut. Fixation here prevents motion at the skin exit site, limiting the spread of possible exit-site infection. Meticulous technique must be used during this part of the surgery. Contamination of the felt cuffs, e.g., by making contact with the surface of the dog's skin, must be avoided. The cephalad skin exit site of the shunt should be 5 cm behind the posterior scapular margin and 2 cm to the right of the dorsal thoracic spines. The caudad exit site is separated from the cephalad site by 5 cm. When properly placed, the cephalad and caudad shunt tubings, linked by means of a small cannula connector, form a smooth arc, without kinking. The arc should lie nearly flat against the skin. Proper lie is facilitated by leaving a large loop of the cephalad tubing under the skin before routing this tubing caudally.

After placement at the skin exit sites, the shunt tubings are passed through an intermediate incision opposite the iliac spine (marked D in Figure 1) to an incision in the groin (Figure 2). The tubings are connected to the femoral vessels by means of Teflon® cannula connectors (3.4-mm I.D., 4.0-mm O.D., Quinton Instruments Co., Seattle, Wash.), trimmed to two thirds of their original length. It is important that these tips be of adequate diameter, to minimize the incidence of late obstruction. The femoral vessels are entered 4 to 5 cm below the inguinal ligament, at the point of origin of the first major medial branch vessels. The relatively distal entry point into the femoral vessels prevents interference of the inguinal ligament with the shunt vessel tips when the dog sits or squats. A common reason for shunt clotting is kinking of the shunt tubing loop in the groin when the animal squats, a problem that can be prevented by making the loop large (Figure 2) before routing the tubing to the femoral vessels.

Immediately prior to surgery, 80 mg of gentamicin and 300,000 units of procaine/

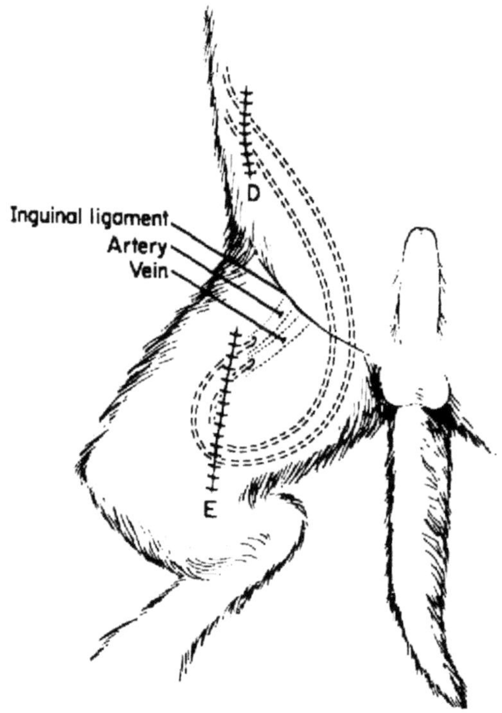

FIGURE 2. Detail showing the insertion of the shunt tubings into the femoral artery and vein.

benzathine penicillin are administered intramuscularly. The same dose of antibiotics is then given daily for at least 1 week postoperatively.

Neck Vascular Access Tubes

In our model, the common carotid artery and external jugular vein are also accessed, for the purpose of hemodynamic monitoring during dialysis. The technique of placing the neck vascular access tubes is as follows. Relatively thick, 1/8-in. I.D., 1/4-in. O.D. silicone elastomer tubing (Dow Corning Corp.) is used. Teflon® felt cuffs are glued to the silicone elastomer tubes prior to implantation. Placement is through a separate access incision (Figure 1), and the cuffs are sutured to the underside of the dermis, as described above for the femoral shunt catheters. Connection of the tubes to the vessels is made by means of plastic connectors. The type of tip used is important. Standard Teflon® cannula connectors were found to damage the intima of the vessels because of their relatively sharp edges. Much better results were obtained with use of polyethylene connectors supplied with Gambro® shuntlines (#S-194, Gambro Inc., Barrington, Ill). After placement, the lines are filled with heparinized saline. A plastic tip is also inserted into the free end of the silicone elastomer tube, allowing the access tube to be capped with a red rubber cap (Cobe Laboratories, Lakewood, Colo.) of the type used to close off a Tenckhoff peritoneal dialysis catheter.

Catheter Care

Povidone iodine ointment is applied to all tubing skin exit sites every 24 to 48 hr. Protection of the extracorporeal part of the AV shunt against manipulation is necessary only in certain types of dogs. Dogs of the "German shepherd" phenotype tend to be more curious about the shunt and may even bite at the shunt unless it is protected by a jacket. We prefer dogs of the "hound" phenotype because of their placid disposition. Hounds seem to ignore the shunt, despite lack of a jacket, and none of our hounds has ever bitten or separated the extracorporeal shunt tubing. If the AV shunt clots, declotting can usually be achieved by passing a 4F balloon-tipped catheter down the arterial and venous limbs. In the nonuremic animal, the tendency towards clotting can be greatly reduced by assuring proper hydration. Dogs' drinking habits are often inhibited when automatic cage watering systems are in place, and it is best to offer the animals fresh water in a pan daily. We normally do not administer aspirin to improve shunt patency, as recommended by Thornhill (see Chapter 1), because its antiprostaglandin effects might conceivably interfere with certain hemodynamic responses. Nonuremic dogs are routinely phlebotomized down to a hematocrit of 30 to 35%. This degree of anemia, by decreasing blood viscosity, not only facilitates ultrafiltration during "dialysis," but may also decrease the incidence of clotting of the AV shunt.

Since dogs will normally scratch at the neck incisions, the neck access tubes are protected by a gauze dressing. In particularly rambunctious animals, the gauze dressing can be reinforced with one or two rolls of plaster-impregnated gauze. Every 24 to 48 hr the neck dressing is removed and, using sterile technique, the neck access tubes are irrigated with heparinized saline.

Alternate Techniques

Despite the relative simplicity of the methods outlined above, when longer periods of study are desired, our techniques of vascular access and vascular monitoring may well have to be modified. The longevity of our femoral AV shunt past 3 weeks has not been tested. Using aspirin anticoagulation (see Chapter 1), Thornhill et al. have been able to achieve prolonged patency (up to many months) of carotid-jugular silicone elastomer AV shunts. As an alternative to our carotid artery access tube, chronic arterial lines can be inserted into any of a number of arteries for blood pressure monitoring. Cardiac output can be followed using flow probes inserted at the aortic root, and cardiac filling pressures and contractility can be monitored by catheters implanted in the appropriate chambers of the heart during prior thoracotomy.

Dialysis Technique

Dialyzers

Many types of dialyzers can be used. We began by using a parallel-plate dialyzer, the RP607 (Hospal, Inc., Littleton, Colo.), which has a polyacrylonitrile membrane with a surface area of $0.7m^2$. At the present time, we prefer the more efficient RP1210 (Hospal, Inc.), with a similar membrane, and a surface area of $1m^2$.

Parallel-plate dialyzers are not ideally suited for hemodynamic research, since their filling volume changes with the transmembrane pressure applied. However, change in filling volume with ultrafiltration is not a major problem with the highly permeable polyacrylonitrile dialyzers. We avoid use of hollow fiber dialyzers when possible because, in a substantial percentage of animals, clotting of many dialyzer fibers occurs despite seemingly adequate heparinization.

Extracorporeal Blood Circuit

Pediatric blood lines (TS-111, Extracorporeal, Inc., King of Prussia, Pa.) are normally used to make the connection between dog and dialyzer. The filling volume of the blood lines is approximately 70 mℓ. The blood lines and the blood compartment of the dialyzer are primed with 0.9% saline solution, and 5000 units per liter of heparin are added to the solution when priming hollow fiber dialyzers. When performing hemodynamic studies (see below), the dialysate compartment of the dialyzer is also primed with sterile saline, and then the dialysate inlet and outlet tubings are clamped.

Before connecting the animal to the dialyzer, the femoral arterial and venous shunt lines are clamped, the connection between them is opened, and 4000 to 8000 units of heparin (for a 25-kg dog) are injected into the venous shunt line. The arterial and venous shunt lines are then reconnected to each other, and shunt flow is reestablished to assure mixing of the heparin. After several minutes, the shunt limbs are again clamped, and the loop is separated. Saline is injected into each limb to remove static blood from the clamped shunt lines (this blood could clot despite heparinization if hookup to the dialyzer is delayed). Connections to the dialyzer are then made and blood flow is initiated. Blood flow rate is increased to the desired level over a period of 5 min. Dogs larger than 25 kg tolerate a blood flow rate of 200 mℓ/min without difficulty. We like to precede most hemodynamic studies by 1 hr of extracorporeal circulation without weight removal, to allow for stabilization of hemodynamic indexes and hematocrit. During this period, the inlet and outlet tubings to the dialysate compartment, which has been primed with sterile saline, are kept clamped to prevent ultrafiltration. When using a parallel-plate dialyzer which has been primed with sterile saline as described above, it is important to relieve pressure in the dialysate compartment at the time that blood flow is initiated (by temporarily unclamping the dialysate outlet port). Otherwise pressure can build up in the blood compartment soon after initiation of blood flow, causing leakage at the blood inlet port.

Clotting time is monitored every 30 to 60 min. We use clotting tubes containing siliceous earth as a clotting activator (ACTest®, TRIMED, Huntington Beach, Calif.) Sufficient heparin must be given to extend the normal clotting time of 1 to 2 min in these tubes to 4 to 7 min. Proper deworming of the animals is important since intestinal infestation with parasites predisposes to GI bleeding during heparinization.

When dialysate is not being circulated, the dialyzer will function as a radiator, cooling the extracorporeal blood so that returning blood temperature is approximately 35°C. Conscious dogs soon begin to shiver under these circumstances. To prevent heat loss when dialysate is not being circulated, the dialyzer can be immersed in a water bath kept at 39 to 42°C. This will result in a returning blood temperature of 37 to 39°C.

Dialysate

In normal dogs, the object of experiments is usually to perform dialysis with either no change in plasma solute composition (control studies) or to selectively change one or more plasma solute levels. In the control situation a dialysate is required that closely resembles normal dog plasma. The latter normally contains approximately 147 m*M/ℓ* sodium, 4.4 m*M/ℓ* potassium, 1.3 m*M/ℓ* ionized calcium, 0.4 m*M/ℓ* ionized magnesium, 26 m*M/ℓ* bicarbonate, 118 m*M/ℓ* chloride, 88 mg/dℓ dextrose, and 18 mg/dℓ urea nitrogen.[13-15] A dialysate having these concentrations can be mixed using a "batch" system: 100 kg of deionized water are weighed into a mobile plastic tank using a platform balance (Scaletronix, Inc., Carol Streams, Ill.). The required amounts of urea, dextrose, sodium bicarbonate, and chlorides of sodium, potassium, magnesium, and calcium are then added. The chloride concentration in the dialysate exceeds the normal range in plasma since the dialysate contains no proteins. Calcium and magnesium salts are not normally soluble in alkaline bicarbonate solutions. For this reason, when mixing bicarbonate-buffered dialysate, solutes are added to the water in a specific order. Sodium chloride and sodium bicarbonate are dissolved first. Then 3 to 4 m*M/ℓ* hydrochloric acid is added. The hydrogen ion in the added acid will consume an equivalent amount of bicarbonate, generating carbon dioxide. Typically the partial pressure of CO_2 in the dialysate will increase to 70 to 100 mmHg, and the pH will decrease to 7.1 to 7.2. Calcium and magnesium are then added. At this pH, these divalent cations will be soluble and stay in solution. One disadvantage of a batch-mixed dialysate is that the chemically generated CO_2 is slowly lost to the atmosphere with time, increasing the dialysate pH. Ultimately calcium and magnesium can precipitate if dialysate pH increases to more than 7.8.[16] To prevent this complication, bicarbonate-buffered dialysate should be prepared immediately prior to use. When dialysis machines are used that agitate the dialysate as it is being heated (e.g., Travenol RSP® Artificial Kidney, Travenol Laboratories, Deerfield, Ill.), degassing of CO_2 is greatly accelerated and must be compensated for by bubbling 10% CO_2 gas (the remaining gas being nitrogen or nitrogen plus 20% oxygen) through the dialysate during dialysis. Since the blood temperature of nonuremic dogs averages 38.0 to 38.5°C,[17] optimum dialysate temperature should be 39.0 to 40.0°C.

Control of Ultrafiltration Rate

When blood flow is 200 m*l*/min and dialysate flow is 400 m*l*/min, pressure at the dialyzer blood outlet port will ordinarily range from 20 to 100 mmHg. When using a "single-pass" converter, or a dialysate delivery system without added inflow resistance or added "negative pressure", pressure at the dialysate outlet port will usually be slightly negative. The pressure differential between the blood and dialysate compartments of the dialyzer will result in a hydrostatic pressure gradient across the dialyzer membrane, which in turn will cause convective transfer of fluid from the blood to the dialysate compartment. The ultrafiltration rate will also depend upon the ultrafiltration coefficient of the dialyzer used. If more rapid ultrafiltration is desired, a screw clamp can be applied to the blood outlet tubing, distal to the venous filter and air trap. to raise the hydrostatic pressure in the blood compartment. Alternatively, the negative pressure in the dialysate compartment can be increased by increasing dialysate inflow resistance with an appropriately designed dialysate pump. When ultrafiltration is not desired, a screw clamp is applied to the dialysate outlet tubing distal to the pressure monitor and is tightened to raise the hydrostatic pressure at the dialysate outlet tubing to a level similar to that at the blood outlet tubing.

We perform dialysis with the dog supported in a sling (Alice King Chatham Medical Arts, Los Angeles, Calif.), with the sling on a platform balance. Weight change is used to follow the progress of ultrafiltration, adjusting pressures at the dialysate and blood outlet tubings as needed.

Hemodynamic Monitoring

In our first studies, blood pressure during dialysis was monitored by inserting a stopcock in the arterial line between the shunt and dialyzer. When a blood pressure measurement was desired, the blood pump was stopped, and the arterial blood was rerouted via the stopcock to a pressure transducer. The dialyzer blood inlet and outlet tubings were clamped at this time to prevent partial emptying of the dialyzer blood compartment back into the dog through the venous limb of the femoral shunt. The method was tedious, since it required stopping of blood flow. Also, only mean arterial blood pressure could be reliably obtained since the long, compliant length of tubing between the pressure transducer and femoral artery distorted actual values for systolic and diastolic pressures.

We now routinely insert an access tube in the neck leading to the right carotid artery as described above. Just prior to dialysis we uncap the access tube and thread a catheter down through the carotid artery into the aorta for monitoring of blood pressure. If desired, a transducer catheter, (Millar Instruments, Inc., Houston, Tex.) can often be manipulated into the left ventricle through the carotid artery access tube for monitoring left ventricular pressure and its first derivative with respect to time, dP/dt. In all studies a thermodilution Swan-Ganz® catheter is inserted through the access tube leading to the external jugular vein for monitoring right heart pressures and thermodilution cardiac output. In larger dogs (greater than 25 kg), formaldehyde-soaked, (7F) Swan-Ganz® catheters intended for human monitoring can be used and reused many times. However, it must be kept in mind that the distance between the proximal and distal ports of these catheters designed for the adult human is greater than the optimal distance for use in dogs. When the 7F Swan-Ganz® catheter tip is in the proper position in the dog's pulmonary artery, the proximal port may still be in the jugular or innominate vein. In such an instance, the respiratory variation of the central venous pressure tracing recording from the proximal port will be dampened or absent, and the recorded value for the central venous pressure may be falsely elevated. Aberrant thermodilution cardiac output measurements can then result due to incomplete or delayed delivery of injected thermodiluent fluid to the heart. In such instances, and especially, in smaller dogs, pediatric (5F) Swan-Ganz® catheters, or catheters designed for dogs, should be used.

After each study, the monitoring catheters are removed from the neck access tubes and the latter are filled with heparinized saline. A limiting feature of this method is that repeated catheter insertion ultimately causes scarring and closure of the jugular vein and carotid artery. When closure becomes imminent, patency of the vessels can be extended by daily dilatation with a balloon-tipped catheter.

Determination of Plasma Volume

When plasma volume measurement is desired, Evans® blue dye or ^{125}I-tagged albumin can be injected into the venous limb of the femoral shunt with dilution samples drawn from the arterial limb of the shunt at appropriate time intervals. We splenectomize our dogs at time of shunt placement since splenectomy has been reported to result in a constant ratio of large-vessel to total-body hematocrit under a number of different hemodynamic conditions (anemia, volume depletion, etc.).[18] To determine changes in blood volume during dialysis, one can use the following method, which avoids the problem of trapping of radioisotope in the dialyzer. After an initial hour of extracorporeal circulation without weight removal, the dialyzer is temporarily disconnected without returning its blood, and 5 to 10 μCi of ^{125}I-tagged albumin is injected intravenously and plasma volume is determined by dilution. We have found a mixing period of 15 min to be adequate. During the mixing period, the dialyzer blood inlet and outlet tubings are connected to each other and the extracorporeal blood is recirculated to prevent clotting. After the mixing period, the dialyzer is reconnected to the dog's circulation, and the desired ultrafiltration procedure is performed. At the end of ultrafiltration, the dialyzer is again disconnected from the circulation. Postultrafiltration plasma volume is determined by injection of a second, larger dose of ^{125}I-tagged albumin. Since refilling occurs between the end-point of ultrafiltration and the time of tracer injection, and also between the time of tracer injection and the time of sampling (15 min), simultaneous hematocrits are determined with each sample to permit appropriate correction in background radioactivity counts and a correction for refilling. The change in plasma volume as a result of ultrafiltration is then calculated by subtracting the postultrafiltration from the preultrafiltration plasma volume.

Uremic Dogs

Method of Inducing Uremia

Uremia can be induced either by bilateral ureteral ligation or by bilateral nephrectomy. We perform either surgery at the same time as splenectomy, AV shunt placement, and insertion of the neck access tubes.

Maintenance Hemodialysis

Dialysis is begun on the second postoperative day. Heparinization is performed as described above for normal dogs, although the required initial dose is less (3000 to 5000 units). We use the RP1210 dialyzer for maintenance dialysis of uremic dogs. Dialysate should contain 140 m*M*/*ℓ* sodium, even though the normal dog plasma sodium level averages 145 to 148 m*M*/*ℓ*. Higher dialysate sodium levels cause thirst and exacerbate hypertension. Similar to Langston and colleagues,[19] we found that uremic dogs, when allowed to drink freely, will dilute their plasma sodium level to 136 to 142 m*M*/*ℓ*. Acetate-buffered dialysate (38 m*M* acetate per liter) is reasonably well tolerated by most animals. Otherwise, dialysate contains 2.0 m*M*/*ℓ* potassium, 0.4 m*M*/*ℓ* magnesium, 1.3 m*M*/*ℓ* calcium, and 125 mg/d*ℓ* dextrose, with the remaining anion being chloride. Dogs weighing 25 kg or more tolerate a blood flow rate of 200 m*ℓ*/min and a dialysate flow rate of 400 m*ℓ*/min without obvious adverse effects, provided that predialysis plasma urea nitrogen level is less than 100 mg/d*ℓ*. At higher predialysis plasma urea nitrogen values, shorter dialysis at lower blood and dialysate flows should be performed in order to avoid disequilibrium syndrome. We normally perform maintenance dialysis for 3 hr and remove fluid at a rate of 600 to 800 m*ℓ*/hr, monitoring weight loss using a platform balance. Fluid removal is continued until the mean arterial pressure decreases to 90 mmHg. The remainder of dialysis is performed without further ultrafiltration. At the end of dialysis, the blood in the dialyzer circuit is returned using approximately 300 m*ℓ* of 0.9% saline. The dialyzer is removed and both compartments are thoroughly rinsed with tap water. The dialyzer is then filled with 2% formaldehyde and stored. Polyacrylonitrile dialyzers can also be sterilized with sodium hypochlorite, commercial bleach. Prior to the next dialysis, the dialysate compartment is rinsed with at least 5 *ℓ* of dialysate and the blood compartment is rinsed with 2 *ℓ* of saline to remove residual formaldehyde or bleach. Using either new or reused polyacrylonitrile dialyzers, we have not observed hemolysis nor early hypotension during dialysis as described by Gourley and co-workers.[7] When acetate-buffered dialysate is used, blood pressure does decrease initially by approximately 10 to 20 mmHg, as noted by Keshaviah et al.[20] in anesthetized dogs, but blood pressure usually stabilizes within the first half-hour.

The amount of hemodialysis required by anephric or ureter-ligated dogs is unknown. Initially, we performed dialysis 3 to 4 times per week, 3 hr per session. Despite the fact that plasma urea nitrogen rarely exceeded 120 mg/d*ℓ*, all of the animals became anorectic within several days. Vomiting was common, and the animals were in markedly negative nitrogen balance as evidenced by an average weight loss of 0.5 kg/day. In an effort to obtain clinical improvement, we increased dialysis frequency to 7 days/week, 3 hr per session. With this schedule, predialysis plasma urea nitrogen levels range from 60 to 90 mg/d*ℓ*. However, the more intensive dialysis regimen has failed to prevent anorexia, and vomiting continues to occur in a substantial percentage of the animals treated.

Feeding

Some type of gavage feeding is mandatory for maintaining dogs with chronic renal failure. We normally feed the dogs once daily after hemodialysis. A large, 34 Fr. Edlich gastric lavage tube (Monoject, St. Louis, Mo.) is inserted through a mouth speculum while the animal is still in the support sling. Over 5 to 10 min approximately 400 m*ℓ* of a blenderized meal is introduced into the stomach. For a 25-kg dog, the contents of the meal are as follows: dog chow, 250 g; Dextrose, 200 g; and Water, 400 m*ℓ*. The dog chow used is "Canine Lab Chow #5006" from Ralston Purina Co., St. Louis, Mo. The above meal (chow plus dextrose) contains approximately 18 m*M* sodium, 64 g of protein, 22 g of fat, 366 g of carbohydrate, and 1900 kcal. An attempt to increase caloric value by adding several tablespoonfuls of corn oil resulted in an increased incidence of vomiting. Two multivitamin tablets are also added. Dogs that have been vomiting often do not tolerate such a large feeding at one sitting. In such animals, one third to one half of the feeding is given prior to dialysis. A number of animals do achieve a stable "dry weight" when fed in this fashion. However, other animals still lose weight at an alarming rate, frustrating attempts at gavage feeding by frequent vomiting. Thornhill and colleagues (Chapter 1) have suggested that vomiting in uremic dogs is due largely to a low gastric pH. In their experience, when uremic dogs are treated with cimetidine, vomiting is greatly reduced and the dogs continue to eat. We have chosen to avoid cimetidine because of possible hemodynamic effects, but use of this agent or some other method of reducing gastric acidity may ultimately be necessary to achieve stable nutritional status in dialyzed dogs. Recently, Pils and colleagues[21,22] attempted to solve the nutritional problem of hemodialyzed dogs rather heroically by administering parenteral nutrition. In their report it is unclear whether or not animals maintained in such a fashion achieved stable weight and nonnegative nitrogen balance. The complexity of performing parenteral nutrition in addition to daily dialysis prompts one to look for a simpler solution.

Hypertension

In our hemodialyzed, uremic dog model we were unable to achieve a normal blood pressure. Hypertension was found even in the nephrectomized animals, in whom free access to water was denied and in whom a relatively modest amount of fluid was given to return blood postdialysis and to solubilize a daily gavage feeding. Early investigators who attempted to maintain uremic dogs on hemodialysis also noted that the animals often became hypertensive.[5,19,23] The controversy at the time was whether or not hypertension in the hemodialyzed dog model was due entirely to overhydration.[5,23] The question remains unanswered to this day. Langston and colleagues[19] claimed that uremic dogs did not develop hypertension if salt intake was restricted, but animals in their experiments retained substantial residual renal function and did not require dialysis. In a second experiment from the same laboratory by Norman and co-workers,[23] it was shown that volume-loading alone, even in the face of salt restriction, could produce hypertension in uremic, hemodialyzed sheep. In the comparison group, which was salt-loaded, but in which volume was restricted, hypertension did not develop. However, the comparison group retained some renal function and did not require dialysis. Hence, it was not convincingly demonstrated that hypertension could be prevented in hemodialyzed animals by appropriate fluid restriction. In most human patients with end-stage renal disease, hypertension seems to be due to overhydration, since dehydration by means of extracorporeal ultrafiltration often normalizes blood pressure. In a minority, blood pressure remains elevated despite fluid removal. In this subgroup, some, but not all, investigators have found plasma renin activity to be elevated.[24] However, patients in end-stage renal failure have other reasons to have elevated blood pressures. Plasma aldosterone levels may be inappropriately high in nonnephrectomized chronic dialysis patients,[25] with possible resultant hypertension. Levels of "natriuretic factor", a substance with vasoconstrictive properties,[26] are elevated in uremic patients.[27,28] Finally, elevated baseline norepinephrine values may reflect sympathetic nervous system overactivity in uremia.[29] A sympathetic component to the hypertension of dialysis patients was suggested in studies by McGrath and colleagues,[30] who found that administration of sympathetic blocking agents reduced elevated peripheral vascular resistance.

HEMODIALYSIS-INDUCED HYPOTENSION

One of the principal reasons why we attempted to study uremic, hemodialyzed dogs was to obtain more information about hemodialysis-induced hypotension. Further understanding of the exact hemodynamic mechanisms underlying the occurrence of hypotension during hemodialysis may lead to the development of new, better-tolerated dialysis techniques and might also shed light on previously unrecognized factors affecting response of the cardiovascular system to hypovolemia.

Hemodynamic Mechanisms

In patients, if fluid is removed at an excessive rate during dialysis, blood volume will fall, leading to decreased cardiac filling, decreased cardiac output, and hypotension. However, refractory, or so-called "dialysis-induced" hypotension is not always related to the amount of fluid removed or to the decrease in blood volume.[32-34] As a result, many such patients experience falls in blood pressure during dialysis before adequate amounts of fluid have been removed. From hemodynamic studies in humans, the mechanism of hypotension during hemodialysis appears to be decreased total peripheral resistance (TPR).[32-34] In addition to impairment of constriction of the resistance vessels, constriction of the venous system may also be impaired during dialysis causing a premature decrease in cardiac filling and cardiac output.[35] Cardiac contractility, on the other hand, usually increases during dialysis, probably due to an increase in plasma ionized calcium levels and to a decrease in plasma potassium values.[36,37]

Isolated Ultrafiltration

Interestingly, fluid can be removed from hypotension-prone uremic patients with much less hypotension using a procedure called isolated ultrafiltration.[38-40] During isolated ultrafiltration blood is circulated through the dialyzer in usual fashion. Dialysate, however, is not circulated. Instead a hydrostatic pressure is applied across the dialyzer membrane, causing an ultrafiltrate of plasma to enter the dialysate compartment from where it is allowed to drain. During isolated ultrafiltration, diffusion is not allowed to occur and waste products, other than those that are dissolved in the small volume of ultrafiltrate, are not removed from the blood. When detailed hemodynamic measurements were performed in patients undergoing isolated ultrafiltration, total peripheral resistance was found to increase, suggesting that the resistance vessels were constricting appropriately during fluid removal.[32-34] During isolated ultrafiltration, there was also indirect evidence for venoconstriction during fluid removal[41] resulting in better cardiac filling and better maintenance of the cardiac output. The improved tolerance to fluid removal by isolated ultrafiltration led to the notion that the process of dialysis itself was in some fashion impairing constriction of arteries, arterioles, and possibly also the capacitance vessels during fluid removal.

Theories as to Causation of Hemodialysis-Induced Hypotension

A number of ideas were promulgated to explain why hemodialysis might result in hypotension: the acetate theory, the catecholamine hypothesis, the osmolality hypothesis, the potassium hypothesis, and, most recently, the temperature theory.

Acetate

Acetate is commonly used as a dialysate buffer base since dialysis solutions containing bicarbonate, calcium, and magnesium are difficult to prepare and store. Normally, acetate is promptly metabolized by the body to form bicarbonate. However, during dialysis, detectable plasma acetate levels occur, and acetate is a known cardiodepressant and vasodilator.[42-44] Plasma acetate levels are highest in a subgroup of dialysis patients who seem to have decreased metabolic capacity for acetate.[45] High plasma acetate levels also are more common when high-efficiency, large-surface-area dialyzers are used.[45,46] Additionally, when acetate dialysate is used, a moderate degree of hypoxemia may occur, which of itself might impair cardiovascular response;[47] Hypoxemia during acetate dialysis has been thought to be related to decreased minute ventilation due to loss of CO_2 drive as CO_2 is removed by dialysis.[47] Recently, metabolism of acetate during dialysis has been shown to cause hypoxemia by changing the respiratory quotient.[48] Whatever the mechanisms for acetate-induced hypotension, several investigators have shown that when bicarbonate is used in place of acetate in the dialysate, response of the resistance vessels to fluid removal is improved, and blood pressure is better maintained.[49,50] The matter remains controversial, however, and patient selection may be important, since other investigators could not detect impaired blood pressure response with use of acetate-buffered dialysate.[51-54]

Vasoactive Substances

The finding that TPR response during bicarbonate dialysis is still relatively depressed led to other explanations for hemodialysis-induced hypotension. One hypothesis focused on plasma catecholamine levels. The observation was made that plasma catecholamine concentrations failed to increase during fluid removal by hemodialysis, whereas they did rise when fluid was removed by isolated ultrafiltration.[55] Failure of catecholamine values to increase was thought to reflect a dialysis-induced suppression of the sympathetic nervous system.

Acute Changes in Plasma Solute Levels

Urea

Impaired vascular response during hemodialysis has also been ascribed to acute decreases in plasma osmolality, which occur principally because of decreases in the plasma urea level. Kopp and Grossman[56] noted a reduced incidence of side effects when they added urea to the dialysate and thus minimized abrupt changes in plasma urea level. Chan and co-workers[57] recently showed that high levels of urea sensitized blood vessels to the effects of angiotensin II. On the other hand, a number of investigators have found no relationship between changes in plasma urea levels and blood pressure during dialysis.

Sodium

Impaired vascular response during dialysis might also be due to decreases in plasma sodium levels, which occur when dialysate that is hyponatric to plasma is used. It has long been known that by increasing dialysate sodium concentration, one can reduce the incidence of disequilibrium syndrome.[58-60] Fluid removal by ultrafiltration is also less likely to cause hypotension.[61] However, the protective effects of high-sodium dialysis could be due merely to better osmotic maintenance of the plasma volume, with accelerated entry of intracellular and interstitial fluid into the circulation.[62] Another hypothesis is that plasma sodium concentration changes during dialysis may affect vascular reactivity. Any of three mechanisms may be invoked. Logically, low-sodium dialysis might suppress, and high-sodium dialysis might stimulate, the secretion of vasopressin. Salt-loading has been shown to increase the sensitivity of blood vessels to catecholamines.[63] Lastly, salt-loading may promote secretion of a yet-to-be-identified "natriuretic factor", which may also have vasoconstrictive properties.[26] One hemodynamic study, in which vascular resistance during low-sodium dialysis was measured, suggested that hypotension during hyponatric dialysis was due principally to decreased total peripheral resistance.[64] On the other hand, a number of physiologic studies evaluating the effects of plasma osmolality and plasma sodium concentration on the resistance of isolated vascular beds suggest that acute decreases in plasma sodium level raise, rather than lower, vascular resistance.[65] In fact, acute increases in plasma sodium concentration cause vasodilatation.[65,66]

Potassium, Calcium, Magnesium

Changes in potassium, calcium, and magnesium occurring during dialysis might also conceivably affect vascular tone, but in this instance the effects may be to enhance rather than impair blood pressure response. The salutory effects of increases in plasma ionized calcium levels and decreases in plasma potassium levels on cardiac contractility have already been mentioned.[36,37] Similarly, decreases in plasma potassium and magnesium concentrations and increases in plasma ionized calcium values (all of which commonly occur during hemodialysis) are known to constrict, rather than dilate vascular beds.[67] One possible adverse hemodynamic effect attributable to decreases in plasma potassium levels during dialysis relates to the interaction between plasma potassium levels and aldosterone. Normally, fluid removal during dialysis engenders hypovolemia- or hypotension-mediated aldosterone release. Henrich and colleagues[68] have shown, however, that decreases in plasma potassium levels during dialysis can prevent the "normal" increase in plasma aldosterone level, a phenomenon which can be reversed by dialyzing against a potassium-supplemented dialysate.

Extracorporeal Blood Temperature

Recently Maggiore and co-workers[69,70] suggested that a major cause of dialysis-induced hypotension might be the relatively high temperature of dialysate. They found that core body temperature increased by an average of 1°C during dialysis, possibly due to warming of blood in the extracorporeal circuit by contact with warmed dialysate. When dialysate temperature was set to 35 instead of the usual 38°C, the increase in body temperature did not occur and blood pressure was better maintained. Furthermore, it was found that during the newer treatment methods of isolated ultrafiltration or hemofiltration, cooling of blood took place in the extracorporeal circuit, possibly accounting for better maintenance of the blood pressure during these procedures. When the temperature of blood returning to the body was maintained at the same level, the incidence of hypotension during isolated ultrafiltration or during hemofiltration was found to be similar to that during conventional dialysis.[69,70] Maggiore's temperature hypothesis does find support in the physiology literature. When conscious volunteers are exposed to heat such that core body temperature rises, cardiac output and rate increase markedly and total peripheral resistance falls.[71,72] Others have emphasized the abolition of the normal venoconstrictor response during exercise when a heat stress is simultaneously applied.[73]

ULTRAFILTRATION HEMODYNAMICS IN CONSCIOUS DOGS

In an attempt to sort out the importance of the above factors in causing hemodialysis-induced hypotension, we began examining their effects on the response to volume depletion in conscious dogs.

Extracorporeal Circulation Without Weight Removal

Before study of volume depletion during dialysis can be meaningfully interpreted, one must first measure the hemodynamic effects of simply circulating blood through the dialyzer and extracorporeal circuit. We performed such an experiment in eight nonuremic animals. During the first hour of extracorporeal circulation, using a circuit with a total volume of approximately 190 m*l*, mean arterial pressure, cardiac index, central venous pressure, and hematocrit all showed slight, but significant decreases, the result of exteriorization of a certain volume of blood into the extracorporeal circuit. Subsequently, however, hematocrit and all hemodynamic values were all stable.[74] Similarly, after the first hour, extracorporeal circulation had no identifiable effect on blood gases or acid-base values and no effect on plasma osmolality or plasma electrolyte levels. In uremic animals, extracorporeal circulation without weight removal using a circuit of slightly larger volume (210 m*l*) yielded similar results.

Isolated Ultrafiltration

Nonuremic Animals

Having demonstrated that the mere act of circulating blood through a dialyzer was of little hemodynamic importance, we next evaluated response to fluid removal using this system.[74] Isolated ultrafiltration was performed in 10 alert, nonuremic, splenectomized dogs at a rate of 3.3% body weight per hour, until mean arterial pressure decreased to less than 80 mmHg. This hypotensive end-point was reached after 119 ± 42 (SD) min. As a result of isolated ultrafiltration, with the weight of the animals decreasing from 25.4 ± 3.8 to 23.8 ± 3.3 kg, mean arterial pressure decreased from 103 ± 10.6 to 77.2 ± 1.6 mmHg. Cardiac index fell from 4.32 ± 0.65 to 1.75 ± 0.38 $l \cdot min^{-1} \cdot m^{-2}$. The proportionate fall in cardiac index was much greater than the fall in blood pressure, attesting to a marked increase in total peripheral resistance, from 1906 ± 278 to 3472 ± 507 $dyn \cdot sec \cdot cm^{-5} \cdot m^{-2}$. Cardiac contractility was not measured in these experiments, but it was assumed that the fall in cardiac output was due principally to a decrease in plasma volume and cardiac filling pressures; i.e., as a result of isolated ultrafiltration, hematocrit increased by 7.3 ± 3.4%, central venous pressure fell from 0.6 ± 0.9 to −2.5 ± 1.0 mmHg, and pulmonary capillary wedge pressure fell from 3.7 ± 1.3 to 1.8 ± 1.1 mmHg.[74] Pulse rate, measured from the femoral arterial pressure tracing, increased from 106 ± 11 to 125 ± 27 beats per minute, but overall, the increase in heart rate was not statistically significant. In fact, in a number of animals, pulse rate actually decreased slightly during isolated ultrafiltration. In several such animals, the pulse rate was compared with a simultaneously determined electrocardiogram, and no evidence of pulsus alternans or heart block was seen. Marked increases in concentrations of several vasoactive hormones were noted as a result of isolated ultrafiltration. Plasma renin activity increased from 0.46 ± 0.32 to 19.5 ± 14.2 $ng \cdot AI \cdot ml^{-1} \cdot hr^{-1}$. Plasma norepinephrine values increased from 190 ± 76 to 1832 ± 884 pg/m*l*, and plasma epinephrine levels increased from 104 ± 70 to 4775 ± 3255 pg/m*l*. Marked increases were also noted in plasma aldosterone and vasopressin concentrations.

Isolated ultrafiltration resulted in few changes in plasma solute composition, as expected. Total levels of protein-bound solutes such as calcium and magnesium increased as a result of concentration of plasma proteins. Isolated ultrafiltration and hypotension induced a metabolic acidosis, which was almost completely compensated for by increased ventilation, as evidenced by hypocapnia and increased blood arterial oxygen tension.[74]

Uremic Animals

Our model allowed us to perform a very interesting experiment. We removed fluid by isolated ultrafiltration from each of seven nonuremic dogs and measured hemodynamic indexes and plasma levels of various vasoactive hormones. We then ligated both ureters, maintained each animal on dialysis for 1 to 2 weeks, and repeated the isolated ultrafiltration study. We also performed isolated ultrafiltration in 15 additional nonuremic and 15 uremic dogs, permitting unpaired comparison of hemodynamic response between these larger groups.[75]

The purpose of these experiments was to assess to what extent acute uremia affects the cardiovascular system, especially, the hemodynamic response to volume depletion. After bilateral ureteral ligation, we suspected that renin response to volume depletion might be impaired. The renin-angiotensin system has been shown to be important in maintaining blood pressure in a number of animal models.[76,77] On the other hand, high levels of plasma urea in the ureter-ligated dogs might increase blood vessel sensitivity to angiotensin II, as has been suggested from work in cats.[57] Parathyroid hormone, a peptide with strong vasodilator properties,[78] accumulates readily in acute renal failure[79] and might conceivably destabilize the normal vasoconstrictor response to hypovolemia. In patients with end-stage renal failure, defects in regulation of heart rate in response to Valsalva maneuver,[80-84] vasodilator-induced hypotension,[80,82,85,86] vasoconstrictor-induced increases in blood pressure,[80,84,86,87] and to immersion of the face in ice water (diving reflex)[83,88] have been described. It is not known how soon after onset of uremia these changes begin to occur.

Rather surprisingly, we found that hemodynamic responses to volume depletion by isolated ultrafiltration in ureter-ligated, hemodialyzed dogs were similar to those in nonuremic controls.[75] Initial mean arterial pressures of the uremic dogs (132 ± 9 mmHg) were higher than those of controls (106 ± 12), making precise comparison of the two groups difficult. However, as blood pressure fell during ultrafiltration, uremic animals achieved the same high levels of peripheral resistance as their nonuremic counterparts, despite marked attenuation of plasma renin activity. In other, nephrectomized animals, with presumably no activity of the renin-angiotensin system, hemodynamic response to ultrafiltration was also unimpaired, further supporting the notion that the renin-angiotensin system is of minor importance in maintaining blood pressure during volume depletion in the acutely uremic dog model. Interestingly, the pulse rate was not at all fixed in the acutely uremic, hemodialyzed animals. In fact, the pulse rate tended to accelerate more uniformly in the ureter-ligated or nephrectomized dogs, such that, by the time the end-point mean arterial pressure (averaging 76 to 78 mmHg) had been reached, pulse rate in the uremic dogs averaged 150 ± 38 beats per minute.

In the uremic animals, baseline plasma norepinephrine levels were higher than in nonuremic controls, similar to findings reported by some, but not all, investigators in chronically uremic humans.[29,89,90] During volume depletion, plasma norepinephrine concentrations increased to a similar extent in both ureter-ligated and nonuremic animals. However, levels of plasma epinephrine were increased to a lesser degree in the uremic animals, and the ratio of plasma epinephrine to norepinephrine at end-point was lower. We believe that lower plasma epinephrine values in uremic animals at end-point are consistent with loss of angiotensin-stimulated release of epinephrine from the adrenal medulla during shock, as previously reported by Harrison and co-workers.[91]

Isolated Ultrafiltration vs. Hemorrhage

In a small number of nephrectomized animals (n = 5), we compared the hemodynamic response to isolated ultrafiltration with the response to hemorrhage.[75] In the hemorrhage experiments, blood was removed from the arterial limb of the femoral AV shunt at 20 m*l*/min. Again, the end-point was when mean arterial pressure had decreased to less than 80 mmHg. In these animals, initial levels of total peripheral resistance were quite high prior to both studies. As a result of hemorrhage, total peripheral resistance remained high, but did not increase from baseline values, whereas further increases in TPR were noted during isolated ultrafiltration.[75] These results emphasize that great caution must be used before making inferences about "vascular tone" based on calculations of "total peripheral resistance". Vascular resistance is only one component of total peripheral resistance, blood viscosity and capillary resistance being other important factors.[92] Blood viscosity depends upon hematocrit, plasma fibrinogen concentration, and blood flow rate (the last two affecting shear forces).[93] During isolated ultrafiltration, both hematocrit and plasma protein concentration increase, whereas during hemorrhage, erythrocytes and plasma proteins are diluted by refilling of the vascular compartment from other fluid spaces. Thus, isolated ultrafiltration and hemorrhage tend to have opposite effects on blood viscosity.

Circulation of Plasma-Resembling, Bicarbonate-Buffered Dialysate

Before assessing the effects of changing various plasma solutes, we measured the consequences of simply dialyzing nonuremic animals against a dialysate whose composition closely resembled normal dog plasma (see the section entitled "Dialysate"). Circulation of dialysate began after stabilizing baseline hemodynamic values by 1 hr of extracorporeal circulation without weight removal, and 2 hr of dialysis of nonuremic dogs against the above dialysate caused no detectable changes in cardiac index, blood pressure, hematocrit, cardiac filling pressures, or heart rate. Similarly, dialysis without ultrafiltration had no effect on blood gases and acidity, plasma osmolality, or on plasma levels of sodium, potassium, chloride, or bicarbonate. Plasma renin, noradrenaline, and adrenaline concentrations were largely unchanged.[74]

In a second study, isolated ultrafiltration (ultrafiltration without dialysis) was compared to dialytic ultrafiltration[94] (ultrafiltration during dialysis) in ten nonuremic, splenectomized animals.[74] Again, a plasma-resembling, bicarbonate-buffered dialysate was used. Addition of dialysis to ultrafiltration resulted in no major differences in hemodynamic response to fluid removal. Similarly, addition of dialysis to ultrafiltration failed to blunt the rise otherwise seen in plasma levels of norepinephrine, epinephrine, renin activity, and aldosterone. Dialyzer clearance of norepinephrine in these experiments was found to average 46 ± 8.0 mℓ of plasma per minute. Thus, loss of vasoactive substances during dialysis was shown to be of little hemodynamic importance in the normal animal.

The study was repeated in uremic animals[95] for two reasons. First, a larger dialyzer and higher blood flow rates were used to achieve a norepinephrine clearance by the dialyzer of approximately 81 ± 7.8 mℓ of plasma per minute. Second, whereas normal animals might have sufficient reserve to overcome removal of catecholamines by the dialyzer, in uremic animals with potential suppression of sympathetic activity, dialyzer loss of catecholamines could be clinically important. When the experiment was performed, however, nearly identical hemodynamic responses to fluid removal were found whether or not a plasma-resembling dialysate was circulated, and increases in plasma norepinephrine concentrations during both methods of volume depletion were comparable.[95]

Effect of a Decreasing Plasma Urea Level

In a number of ureter-ligated dogs, dialysate composition was again adjusted to approximate that of each animal's predialysis plasma, but urea was omitted from the dialysate. This allowed acute decreases in plasma urea level (and osmolality) to occur during fluid removal. We found that the treatments in which plasma urea levels were allowed to decrease resulted in a slight tendency toward decreased levels of total peripheral resistance and plasma norepinephrine at end-point. However, none of these differences was statistically significant.[96]

Effect of Hyponatremia

Circulation of Hyponatric Dialysate

In nonuremic animals, we performed dialysis at constant weight against a dialysate that resembled plasma, with the exception that the dialysate sodium concentration was 120 mM/ℓ. The control group underwent dialysis without ultrafiltration against a dialysate isonatric with respect to plasma. The isonatric dialysis treatment had no hemodynamic effect. In contrast, use of hyponatric dialysate caused marked decreases in blood pressure. The principal reason for the decreased blood pressure appeared to be decreased cardiac output and cardiac filling pressures, which in turn were related to marked osmotic contraction of the plasma volume. Clear-cut impairment of total peripheral resistance response was not found.[97]

Preultrafiltration Hyponatremia and Hypernatremia

To more precisely evaluate the effects of hyponatremia on hemodynamic response to volume depletion, the preultrafiltration plasma sodium level of ureter-ligated or nephrectomized dogs was manipulated by appropriately changing the sodium level in the dialysate, feeding a high- or low-sodium diet, and by controlling water intake. Using such techniques, we were able to perform isolated ultrafiltration in dogs in which hyponatremia (128 ± 2.9 m*M/ℓ*) or slight hypernatremia (150 ± 3.2 m*M/ℓ*) had been maintained for 48 hr prior to volume depletion. Each dog served as its own control. Despite control of water intake, the animals with the higher plasma sodium levels had slightly higher blood pressures before isolated ultrafiltration. The higher blood pressures seemed to correlate well with higher initial blood volumes and cardiac outputs in the hypernatremic condition and were not due to elevated levels of total peripheral resistance. At end-point, levels of total peripheral resistance after isolated ultrafiltration were greater in the high-sodium condition than in the low-sodium condition, suggesting either increased vascular reactivity, a cardiac effect, or increased resistance in the microvasculature.

Effect of Extracorporeal Blood Temperature

In a follow-up to experiments of Maggiore and colleagues,[99] we examined the effects of extracorporeal blood temperature on the hemodynamic response to isolated ultrafiltration in nephrectomized, hemodialyzed dogs. Each of ten dogs was subjected to isolated ultrafiltration twice, serving as its own control. On one occasion, using a water bath system, the temperature of blood returning from the dialyzer to the dog was adjusted to be 39.5°C ("warm" isolated ultrafiltration). On another occasion, extracorporeal blood temperature was kept at 36.5°C ("cold" isolated ultrafiltration). During "warm" isolated ultrafiltration, total peripheral resistance did not increase as fluid was removed and as mean arterial pressure decreased to below 80 mmHg. During "cold" isolated ultrafiltration, total peripheral resistance did increase, a highly significant difference when compared with the "warm" studies. However, the overall importance of the difference in total peripheral resistance found was questioned, since the amount of fluid that could be removed by "cold" isolated ultrafiltration before blood pressure fell was not significantly greater than that removed by "warm" isolated ultrafiltration. Changes in plasma levels of norepinephrine were comparable during the "cold" and the "warm" procedures.

Effect of Acetate-Buffered Dialysate

Most of our dogs seemed to tolerate acetate-buffered maintenance dialysis fairly well. Acetate metabolism and uptake during dialysis in dogs has been studied by others: plasma acetate levels attained are similar to those found in humans.[100] Keshaviah and colleagues,[20] in a preliminary report, suggested that acetate-buffered dialysis causes early hypotension in pentothal/chloralose anesthetized, acutely uremic dogs, but control studies using bicarbonate-buffered dialysis were not described. We recently studied the effects of acetate in conscious, nonuremic dogs.[101] Acetate-buffered (40 m*M*), isonatric dialysate was used to perform diffusion dialysis (zero ultrafiltration rate). The same animals were also dialyzed using bicarbonate-buffered dialysate. Dialysis was begun after 1 hr of extracorporeal circulation without weight removal. We found it important to control both ambient temperature and the temperature of the blood returning from the extracorporeal blood circuit to avoid artefactual increases in the total peripheral resistance due to animal cooling. No hemodynamic changes were found when bicarbonate-buffered dialysate was used. However, with acetate, blood pressure fell approximately 8%. Cardiac output tended to be increased, and the fall in blood pressure was thus associated with a moderate (15 to 20%) fall in total peripheral resistance. Pulse rate and pulmonary artery pressure were also increased. Additionally, with acetate-buffered dialysis, we found dramatic, acute (1 to 2 min) increases in mean pulmonary artery pressure (Δ PAP: 10 ± 3.1 mmHg, $p < 0.001$) and cardiac output (Δ CO: 2.5 ± 1.3 l/min, $p < 0.001$), without associated changes in mean arterial pressure. These phenomena were not observed during bicarbonate-buffered dialysis (Δ PAP: 0.5 ± 1.0, NS; Δ CO: 0.3 ± 0.5, NS).

Effect of Chemical Sympathectomy

It has been suggested that patients with autonomic dysfunction are especially prone to dialysis hypotension.[80] In an attempt to induce susceptibility to dialysis hypotension in dogs, we treated animals with 6-hydroxydopamine, a drug that destroys sympathetic nerve terminals. Efficacy of sympathectomy was confirmed by lack of blood pressure response to injection of 500 μg/kg tyramine. In such dogs, acetate-buffered dialysis was associated with severe, early hypotension, followed by transient, rebound hypertension,[102] phenomena not seen in nonsympathectomized, conscious dogs dialyzed with acetate.[101]

Alternate Buffer Bases

With very few exceptions,[103] thinking about dialysate buffer bases has been bipolar, focusing on acetate vs. bicarbonate. In fact, a number of other bicarbonate-generating bases such as *DL*-lactate, pyruvate, and succinate are available. We have recently developed a rat tail artery strip model for quantitating the vasodilator effect of acetate and of alternative buffer bases.[104,105] Using this model, we have found that lactate and pyruvate have less intrinsic vasodilator activity than acetate.[105,106] The conscious dog model seems to be an excellent method to evaluate both the bicarbonate-generating ability and hemodynamic effects of alternate buffer bases in a dialysis setting.

Summary of Results from the Dog Model

In our model, extracorporeal circulation per se is associated with only minor hemodynamic effects, including a dilution of the hematocrit and an initial, slight fall in the blood pressure.[74,75] We have not seen acute increases in the pulmonary artery pressure after initiating extracorporeal circulation as described by Walker and colleagues,[107] perhaps, because we have always used dialyzers made of either polyacrylonitrile or of cellulose acetate. Circulation of a plasma-resembling dialysate is not harmful,[74,95] and the effect of loss of vasoactive substances to the dialysate is apparently negligible. The fall in plasma urea associated with dialysis does not alter the rate of plasma volume contraction and has no detectable effect on peripheral vascular resistance.[96] In contrast, the fall in plasma sodium level that occurs when hyponatric dialysate is used causes hypotension due to rapid osmotic contraction of the blood volume. Total peripheral resistance does not seem to be impaired.[97] Heating the blood in the extracorporeal circuit is associated with a fall in total peripheral resistance.[99] Preexisting hyponatremia may impair hemodynamic response slightly due to a reduced total peripheral resistance.[98] Finally, use of acetate as a dialysate buffer base is associated with a fall in total peripheral resistance, and increases in pulmonary artery pressure and pulse rate.[101,102]

The causative factors of dialysis hypotension identified in the animal model, i.e., temperature, use of hyponatric dialysate, and use of acetate-buffered dialysate, are all avoidable. One can easily cool the dialysate slightly and keep the dialysate sodium level at the plasma concentration. Avoiding the use of acetate presents more of a challenge, because of problems and expense associated with preparing and storing bicarbonate-buffered dialysate. Whether alternative dialysate bases such as succinate[103] or DL-lactate[105,108,109] will prove useful remains to determined.

Applicability of the Dog Model to Human Hemodialysis-Induced Hypotension

In humans, the relative importance of the many potential causes of hemodialysis-induced hypotension has not been precisely defined. Accurate hemodynamic studies in patients are difficult to perform and usually entail insertion of a Swan-Ganz catheter, an invasive procedure with a number of potential complications.[110] Also, severe, untoward reactions have been reported in dialysis patients injected with indocyanine green dye,[111,112] the substance utilized to determine cardiac output in the most popular alternative method. Possible etiologic factors of dialysis hypotension should ideally be varied one at a time, a procedure which often departs considerably from conventional dialysis and which may be difficult for patients to accept. It is highly desirable in most hemodynamic studies to include measurements of body fluid spaces using radioisotopes, substances which patients may be reluctant to accept in an experimental setting. Finally, there appears to be considerable intersubject susceptibility to dialysis-associated hypotension. An animal model is needed wherein susceptibility to dialysis-associated hypotension can be studied, and, ultimately, artificially induced.[102] For all of these reasons, we have pursued studies in the conscious dog model.

The present set of studies was carried out in nonuremic, and in acutely uremic, healthy dogs, and it may be argued that results do not generalize to chronically uremic humans. However, findings in our animal model parallel those emerging from careful studies in humans. We believe that the described animal model has been useful in pinpointing both the relative importance and the mechanisms of proposed etiologic factors for dialysis hypotension. The model may continue to be of assistance in assessing the benefits of potential modifications of the standard dialysis procedure.

REFERENCES

1. McBride, P. T., Genesis of the Artificial Kidney, Travenol Laboratories, Inc., Deerfield, Ill., 1979.
2. Grollman, A., Muirhead, E. E., and Vanatta, J., Role of the kidney in pathogenesis of hypertension as determined by a study of the effects of bilateral nephrectomy and other experimental procedures on the blood pressure of the dog, *Am. J. Physiol.*, 157, 21, 1949.
3. Skeggs, L. T., Jr., Leonards, J. R., and Heisler, C. R., Artificial kidney. II. Construction and operation of an improved continuous dialyzer, *Proc. Soc. Exp. Biol. Med.*, 72, 539, 1949.
4. Leonards, J. R. and Heisler, C. R., Artificial kidney. IV. Maintenance of life in bilaterally nephrectomized dogs and its relation to malignant hypertension, *Am. J. Physiol.*, 167, 553, 1951.
5. Houck, C. R., Effect of hydration and dehydration on hypertension in the chronic bilaterally nephrectomized dog, *Am. J. Physiol.*, 176, 183, 1954.
6. Butler, H. C., Hemodialysis in veterinary practice, *Arch. Am. Coll. Vet. Surg.*, 1, 5, 1971.
7. Gourley, I. M., Parker, H. R., Bell, R. L., and Ishizaki, G., Responses of nephrectomized dogs during hemodialysis, *Am. J. Vet. Res.*, 34, 1421, 1973.
8. Cowgill, L. D. and Bovee, K., The feasibility and efficiency of hemodialysis in acutely uremic dogs, Proc. 42nd Annu. Meet. Am. Anim. Hosp. Assoc., 1975, 165.
9. Thornhill, J. A., Peritoneal dialysis in the dog and cat: an update, *J. Comp. Cont. Educ. Pract. Vet.*, 3, 20, 1981.
10. Hamilton, R. W., Epstein, P. E., Henderson, L. W., Edelman, N. H., and Fishman, A. P., Control of breathing in uremia: ventilatory response to CO_2 after hemodialysis, *J. Appl. Physiol.*, 41, 216, 1976.
11. Bower, J. D., Holbert, R. D., Pearsons, J. E., Jones, Q., Bengis, R., and Berndt, W. O., Chronic hemodialysis inthe anephric goat, *Nephron*, 25, 34, 1980.

12. Eschbach, J. W., Adamson, J. W., and Dennis, M. B., Physiologic studies in normal and uremic sheep. I. The experimental model, *Kidney Int.*, 18, 725, 1980.
13. Baer, J. E., Peck, H. M., and McKinney, S. E., Some normal laboratory values in the dog, *Proc. Soc. Exp. Biol. Med.*, 95, 80, 1957.
14. Snyder, J. W., Liu, S. K., and Tashjian, R. J., Blood chemical and cellular changes in canine dirofilariasis, *Am. J. Vet. Res.*, 28, 1705, 1967.
15. Killen, D. A., Grogan, E. L., II, Gower, R. E., and Collins, H. A., Response of canine plasma-ionized calcium and magnesium to the rapid infusion of acid-citrate-dextrose (ACD) solution, *Surgery*, 70, 736, 1971.
16. Merrill, J. P., Thorn, G. W., Walter, C. W., Callahan, E. J., III, and Smith, L. H., Jr., The use of an artificial kidney. I. Technique, *J. Clin. Invest.*, 29, 412, 1950.
17. Short, C. E. and Paddleford, R. R., Malignant hyperthermia in the dog, *Anesthesiology*, 39, 462, 1973.
18. Reeve, E. B., Gregersen, M. I., Allen, T. H., Sear, H., and Walcott, W. W., Effects of alteration in blood volume and venous hematocrit in splenectomized dogs on estimates of total blood volume with P^{32} and T-1824, *Am. J. Physiol.*, 175, 204, 1953.
19. Langston, J. B., Guyton, A. C., Douglas, B. H., and Dorsett, P. E., Effect of changes in salt intake on arterial pressue and renal function in partially nephrectomized dogs, *Circ. Res.*, 12, 508, 1963.
20. Keshaviah, P., Himes, M., Ilstrup, K., and Shapiro, F., Hypotension during iso-osmotic dialysis, *Kidney Int.*, 21 (Abstr.), 170, 1982.
21. Pils, P., Draibe, S., Liard, J. F., and Man, N. K., Parenteral nutrition and repeated hemodialysis of bilaterally nephrectomized conscious dogs. I. Experimental model, *Res. Exp. Med.*, 177, 119, 1980.
22. Pils, P., Draibe, S., Jungers, P., and Man, N. K., Parenteral nutrition and repeated hemodialysis of bilaterally nephrectomized conscious dogs. II. Metabolic studies, *Res. Exp. Med.*, 178, 71, 1980.
23. Norman, R. A., Jr., Coleman, T. G., Wiley, T. L., Jr., Manning, R. D., Jr., and Guyton, A. C., Separate roles of sodium ion concentration and fluid volumes in salt-loading hypertension in sheep, *Am. J. Physiol.*, 229, 1068, 1975.
24. Acosta, J. H., Hypertension in chronic renal disease, *Kidney Int.*, 22, 702, 1982.
25. Farinelli, A., Squerzanti, R., De Paoli Vitali, E., Bagni, B., Cavallini, A. R., Gilli, P., and Bedani, P. L., Response of plasma aldosterone to sequential ultrafiltration, dialysis and conventional hemodialysis, *Nephron*, 26, 274, 1980.
26. Plunkett, W. C., Hutchins, P. M., Gruber, K. A., and Buckalew, V. M., Jr., Evidence for a vascular sensitizing factor in plasma of saline-loaded dogs, *Hypertension*, 4, 581, 1982.
27. Bourgoignie, J. J., Hwang, K. H., Ipakchi, E., and Bricker, N. S., The presence of a natriuretic factor in urine of patients with chronic uremia. The absence of the factor in nephrotic uremic patients, *J. Clin. Invest.*, 53, 1559, 1974.
28. Kaplan, M. A., Bourgoignie, J. J., Rosecan, J., and Bricker, N. S., The effects of the natriuretic factor from uremic urine on sodium transport, water, and electrolyte content, and pyruvate oxidation by the isolated toad bladder, *J. Clin. Invest.*, 53, 1568, 1974.
29. Izzo, J. L., Jr., Izzo, M. S., Sterns, R. H., and Freeman, R. B., Sympathetic nervous system hyperactivity in maintenance hemodialysis patients, *Trans. Am. Soc. Artif. Intern. Organs*, 28, 604, 1982.
30. McGrath, B. P., Tiller, D. J., Bune, A., Chalmers, J. P., Korner, P. I., and Uther, J. B., Autonomic blockade and the Valsalva maneuver in patients on maintenance hemodialysis: a hemodynamic study, *Kidney Int.*, 12, 294, 1977.
31. Jones, E. O., Ward, M. K., Hoenich, N. A., and Kerr, D. N. S., Separation of dialysis and ultrafiltration — does it really help?, *Proc. Eur. Dial. Transplant Assoc.*, 14, 160, 1977.
32. Wehle, B., Asaba, H., Castenfors, J., Fürst, P., Gunnarsson, B., Shaldon, S., and Bergström, J., Hemodynamic changes during sequential ultrafiltration and dialysis, *Kidney Int.*, 15, 411, 1979.
33. Hampl, H., Praeprer, H., Unger, V., Fischer, C., Resa, I., and Kessel, M., Hemodynamic changes during hemodialysis, sequential ultrafiltration, and hemofiltration, *Kidney Int.*, 18(Suppl.10), S-83, 1980.
34. Keshaviah, P., Ilstrup, K., Constantini, E., Berkseth, R., and Shapiro, F., The influence of ultrafiltration and diffusion on cardiovascular parameters, *Trans. Am. Soc. Artif. Intern. Organs*, 26, 328, 1980.
35. Chaignon, M., Chen, W. T., Tarazi, R. C., Nakamoto, S., and Bravo, E. L., Blood pressure response to hemodialysis, *Hypertension*, 3, 333, 1981.
36. Chaignon, M., Chen, W. T., Tarazi, R. C., Nakamoto, S., and Salcedo, E., Acute effects of hemodialysis on echocardiograph-determined cardiac performance: improved contractility resulting from serum increased calcium with reduced potassium despite hypovolemic-reduced cardiac output, *Am. Heart J.*, 103, 374, 1982.
37. Henrich, W. L. and Nixon, J. V., Importance of changes in ionized calcium and bicarbonate on left ventricular contractility during hemodialysis, *Kidney Int.*, 23 (Abstr.), 150, 1983.

38. Kobayashi, K., Shibata, M., Kato, K., Kato, S., Nakamura, S., Kurachi, K., Yasuda, B., Ota, K., Maeda, K., Imai, T., Kawaguchi, S., Tsutsui, S., Shimizu, K., Yamazaki, C., Manji, T., and Nomura, T., Studies on the development of a new method of controlling the amount and contents of body fluids (extracorporeal ultrafiltration method: ECUM) and the application of this method for patients receiving long-term hemodialysis, *Jpn. J. Nephrol.*, 14, 1, 1972.
39. Ing, T. S., Ashbach, D. L., Kanter, A., Oyama, J. H., Armbruster, K. F. W., and Merkel, F. K., Fluid removal with negative-pressure hydrostatic ultrafiltration using a partial vacuum, *Nephron*, 14, 451, 1975.
40. Bergström, J., Asaba, H., Fürst, P., and Oulès, R., Dialysis, ultrafiltration and blood pressure, *Proc. Eur. Dial, Transplant Assoc.*, 13, 293, 1976.
41. Chen, W. T., Chaignon, M., Omvik, P., Tarazi, R. C., Bravo, E. L., and Nakamoto, S., Hemodynamic studies on chronic hemodialysis patients with hemofiltration/ultrafiltration, *Trans. Am. Soc. Artif. Intern. Organs*, 24, 682, 1978.
42. Bauer, W. and Richards, D. W., Jr., A vaso-dilator action of acetates, *J. Physiol.*, 66, 371, 1928.
43. Kirkendol, P. L., Devia, C. J., Bower, J. D., and Holbert R. D., A comparison of the cardiovascular effects of sodium acetate, sodium bicarbonate, and other potential sources of fixed base in hemodialysis solutions, *Trans. Am. Soc. Artif. Intern. Organs*, 23, 399, 1977.
44. Frohlich, E. D., Vascular effects of the Krebs intermediate metabolites, *Am. J. Physiol.*, 208, 149, 1965.
45. Vreman, H. J., Assomull, V. M., Kaiser, B. A., Blaschke, T. F., and Weiner, M. W., Acetate metabolism and acid-base homeostasis during hemodialysis: influence of dialyzer efficiency and rate of acetate metabolism, *Kidney Int.*, 18(Suppl.10), S62, 1980.
46. Graefe, U., Milutinovich, J., Follette, W. C., Vizzo, J. E., Babb, A. L., and Scribner, B. H., Less dialysis-induced morbidity and vascular instability with bicarbonate in dialysate, *Ann. Intern. Med.*, 88, 332, 1978.
47. Sherlock, J., Ledwith, J., and Letteri, J., Hypoventilation and hypoxemia during hemodialysis: reflex response to removal of CO_2 across the dialyzer, *Trans. Am. Soc. Artif. Intern. Organs*, 23, 406, 1977.
48. Ikeda, T., Hirosawa, Y., Aizawa, Y., Shibata, A., Gejyo, F., and Ei, K., Effect of acetate upon arterial gases, *J. Dial.*, 3, 134, 1979.
49. Barcenas, C. G., Olivero, J., and Ayus, J., Bicarbonate dialysis is hemodynamically superior to acetate dialysis, *Kidney Int.*, 16(Abstr.), 881, 1979.
50. Iseki, K., Onoyama, K., Maeda, T., Shimamatsu, K., Harada, A., Fujimi, S., and Omae, T., Comparison of hemodynamics induced by conventional acetate hemodialysis, bicarbonate hemodialysis, and ultrafiltration, *Clin. Nephrol.*, 14, 294, 1980.
51. Schick, E. C., Idelson, B. A., Liang, C. S., Redline, R. C., and Bernard, D. B., Comparison of the hemodynamic response to hemodialysis with acetate or bicarbonate, *Trans. Am. Soc. Artif. Intern. Organs*, 29, 25, 1983.
52. Aizawa, Y., Ohmori, T., Imai, K., Nara, Y., Matsuoka, M., and Hirasawa, Y., Depressant action of acetate upon the human cardiovascular system, *Clin. Nephrol.*, 8, 477, 1977.
53. Weitzman, R. E., Gorbaty, I., and Davidson, W. D., The effect of bath composition on blood pressure and vasoactive hormone levels during hemodialysis, *Kidney Int.*, 14(Abstr.), 690, 1978.
54. Borges, H. F., Fryd, D. S., Rosa, A. A., and Kjellstrand, C. M., Hypotension during acetate and bicarbonate dialysis in patients with acute renal failure, *Am. J. Nephrol.*, 1, 24, 1981.
55. Baldamus, C. A., Ernst, W., Fassbinder, W., and Koch, K. M., Differing haemodynamic stability due to differing sympathetic response: comparison of ultrafiltration, haemodialysis and haemofiltration, *Proc. Eur. Dial. Transplant Assoc.*, 17, 205, 1980.
56. Kopp, K. F. and Grossman, D. R., Method and apparatus for a quantitative control of urea extraction for hemodialylsis without disequilibrium syndrome, *Proc. Eur. Dial. Transplant Assoc.*, 2, 7, 1965.
57. Chan, J. C. M., Corley, K. C., and Mauck, H. P., The effect of urea on angiotensin II and blood pressure, *Int. J. Pediatr. Nephrol.*, 1, 173, 1980.
58. Stewart, W. K., Fleming, L. W., and Manuel, M. A., Benefits obtained by the use of high sodium dialysate during maintenance haemodialysis, *Proc. Eur. Dial. Transplant Assoc.*, 9, 111, 1972.
59. Port, F. K., Johnson, W. J. and Klass, D. W., Prevention of dialysis disequilibrium syndrome by use of high sodium concentration in the dialysate, *Kidney Int.*, 3, 327, 1973.
60. Wilkinson, R., Barber, S. G., and Robson, V., Cramps, thirst and hypertension in hemodialysis patients — the influence of dialyzate sodium concentration, *Clin. Nephrol.*, 7, 101, 1977.
61. Locatelli, F., Costanzo, R., Di Filippo, S., Pedrini, L., Marai, P., Pozzi, C., Ponti, R., Sforzini, S., and Redaelli, B., Ultrafiltration and high sodium concentration dialysis: pathophysiological correlation, *Proc. Eur. Dial. Transplant Assoc.*, 15, 253, 1978.

62. Van Stone, J. C., Bauer, J., and Carey, J., The effect of dialysate sodium concentration on body fluid compartment volume, plasma renin activity, and plasma aldosterone concentration in chronic hemodialysis patients, *Am. J. Kidney Dis.*, 2, 58, 1982.
63. Rankin, L. I., Luft, F. C., Henry, D. P., Gibbs, P. S., and Weinberger, M. H., Sodium intake alters the effects of norepinephrine on blood pressure, *Hypertension*, 3, 650, 1981.
64. Wehle, B., Asaba, H., Castenfors, J., Gunnarsson, B., and Bergström, J., Influence of dialysate composition on cardiovascular function in isovolaemic dialysis, *Proc. Eur. Dial. Transplant Assoc.*, 18, 153, 1981.
65. Stainsby, W. N. and Fregly, M. J., Effect of plasma osmolality on resistance to blood flow through skeletal muscle, *Proc. Soc. Exp. Biol. Med.*, 128, 284, 1968.
66. Stainsby, W. N. and Barclay, J. K., Effect of infusions of osmotically active substances on muscle blood flow and systemic blood pressure, *Circ. Res.*, 18-19(Suppl.I.), I-33, 1971.
67. Haddy, F. J., Scott, J. B., Florio, M. A., Daugherty, R. M., Jr., and Huizenga, J. N., Local vascular effects of hypokalemia, alkalosis, hypercalcemia, and hypomagnesemia, *Am. J. Physiol.*, 204, 202, 1963.
68. Henrich, W. L., Katz, F. H., Molinoff, P. B., and Schrier, R. W., Competitive effects of hypokalemia and volume depletion on plasma renin activity, aldosterone, and catecholamine concentrations in hemodialysis patients, *Kidney Int.*, 12, 279, 1977.
69. Maggiore, Q., Pizzarelli, F., Zoccali, C., Sisca, S., Nicolò, F., and Parlongo, S., Effect of extracorporeal blood cooling on dialytic arterial hypotension, *Proc. Eur. Dial. Transplant Assoc.*, 18, 597, 1981.
70. Maggiore, Q., Pizzarelli, F., Sisca, S., Zoccali, C., Parlongo, S., Nicolò, F., and Creazzo, G., Blood temperature and vascular stability during hemodialysis and hemofiltration, *Trans. Am. Soc. Artif. Intern. Organs*, 28, 523, 1982.
71. Rowell, L. B., Brengelmann, G. L., and Murray, J. A., Cardiovascular responses to sustained high skin temperature in resting man, *J. Appl. Physiol.*, 27, 673, 1969.
72. Rowell, L. B., Brengelmann, G. L., Blackmon, J. R., and Murray, J. A., Redistribution of blood flow during sustained high skin temperature in resting man, *J. Appl. Physiol.*, 28, 415, 1970.
73. Rowell, L. B., Murray, J. A., Brengelmann, G. L., and Kraning, K. K., II, Human cardiovascular adjustments to rapid changes in skin temperature during exercise, *Circ. Res.*, 24, 711, 1969.
74. Daugirdas, J. T., Chen, W. T., Ing, T. S., Vestal, R. E., Popli, S., Hano, J. E., and Jacobs, H. K., Haemodynamics of isolated versus dialytic ultrafiltration in the unanaesthetized, non-uraemic dog, *Proc. Eur. Dial. Transplant Assoc.*, 18, 139, 1981.
75. Daugirdas, J. T., Ing, T. S., Chen, W. T., Vestal, R. E., Izzo, J. L., Jr., Hano, J. E., and Norusis, M. J., Hemodynamic responses to volume depletion in acutely uremic dogs, *Am. J. Physiol.*, 247, H229, 1984.
76. Borucki, L. J., Levenson, D., and Hollenberg, N. K., Cardiovascular responses to blockade of angiotensin alpha-adrenergic receptors, *Am. J. Physiol.*, 235, F199, 1978.
77. Badder, E. M., Seaton, J., and Harrison, T. S., Cardiac output, vascular resistance and catecholamine release in hemorrhaged, anephric dogs, *Trans. Am. Soc. Artif. Intern. Organs*, 27, 330, 1981.
78. Pang, P. K. T., Tenner, T. E., Jr., Yee, J. A., Yang, M., and Janssen, H. F., Hypotensive action of parathyroid hormone preparations on rats and dogs, *Proc. Natl. Acad. Sci. U.S.A.*, 77, 675, 1980.
79. Kovithavongs, T., Becker, F. O., and Ing, T. S., Parathyroid hyperfunction in acute renal failure — serial studies in man, *Nephron*, 9, 349, 1972.
80. Kersh, E. S., Kronfield, S. J., Unger, A., Popper, R. W., Cantor, S., and Cohn, K., Autonomic insufficiency in uremia as a cuase of hemodialysis-induced hypotension, *N. Engl. J. Med.*, 290, 650, 1974.
81. Ewing, D. J. and Winney, R., Autonomic function in patients with chronic renal failure on intermittent haemodialysis, *Nephron*, 15, 424, 1975.
82. Lilley, J. J., Golden, J., and Stone, R. A., Adrenergic regulation of blood pressure in chronic renal failure, *J. Clin. Invest.*, 57, 1190, 1976.
83. Zuccalà, A., Degli Esposti, E., Sturani, A., Chiarini, C., Santoro, A., Catizone, L., and Zucchelli, P., Autonomic function in hemodialyzed patients, *Int. J. Artif. Organs*, 1, 76, 1978.
84. Nies, A. S., Robertson, D., and Stone, W. J., Hemodialysis hypotension is not the result of uremic peripheral autonomic neuropathy, *J. Lab. Clin. Med.*, 94, 395, 1979.
85. Lazarus, J. M., Hampers, C. L., Lowrie, E. G., and Merrill, J. P., Baroreceptor activity in normotensive and hypertensive uremic patients, *Circulation*, 47, 1015, 1973.
86. Tomiyama, O., Shiigai, T., Ideura, T., Tomita, K., Mito, Y., Shinohara, S., and Takeuchi, J., Baroreflex sensitivity in renal failure, *Clin. Sci.*, 58, 21, 1980.
87. Pickering, T. G., Gribbin, B., and Oliver, D. O., Baroreflex sensitivity in patients on long-term haemodialysis, *Clin. Sci.*, 43, 645, 1972.

88. Zucchelli, P., Sturani, A., Chiarini, C., Degli Esposti, E., Santoro, A., and Zuccalà, A., Hemodialysis, in contrast to CAPD, improves parasympathetic function in ESRD patients, *Trans. Am. Soc. Artif. Intern. Organs*, 29, 617, 1983.
89. Brecht, H. M., Ernst, W., and Koch, K. M., Plasma noradrenaline levels in regular haemodialysis patients, *Proc. Eur. Dial. Transplant Assoc.*, 12, 281, 1975.
90. Atuk, N. O., Bailey, C. J., Turner, S., Peach, M. J., and Westervelt, F. B., Jr., Red blood cell catechol-O-methyl transferase, plasma catecholamines and renin in renal failure, *Trans. Am. Soc. Artif. Intern. Organs*, 22, 195, 1976.
91. Harrison, T. S., Birbari, A., and Seaton, J. F., Reinforcement of reflex epinephrine release by angiotensin II, *Am. J. Physiol.*, 224, 31, 1973.
92. Gordon, R. J., Viscosity considerations in the determination of regional circulatory adjustments to anemia, *Circ. Res.*, 31, 811, 1972.
93. Meiselman, H. J., Frasher, W. G., and Wayland, H., In vivo hemorheology employing outflow viscometric techniques, *Biorheology*, 10, 361, 1973.
94. Ing, T. S., Chen, W. T., Daugirdas, J. T., Geis, W. P., and Hano, J. E., The old and the new in ultrafiltration techniques, *Artif. Organs*, 3, 185, 1979.
95. Daugirdas, J. T., Ing, T. S., Chen, W. T., Vestal, R. E., Izzo, J. L., Jr., Hano, J. E., and Liu, P., Ultrafiltration hemodynamics in conscious, uremic dogs: effect of using a bicarbonate-buffered, plasma-resembling dialysate, *Trans. Am. Soc. Artif. Intern. Organs*, 28, 528, 1982.
96. Daugirdas, J. T., Ing, T. S., Chen, W. T., Vestal, R. E., Hano, J. E., and Al-Kudsi, R. R., Ultrafiltration hemodynamics in conscious, uremic dogs: effect of a decreasing plasma urea level, *Mineral Electrolyte Metab.*, in press.
97. Daugirdas, J. T., Purandare, V. V., Ing, T. S., Chen, W. T., Popli, S., Hano, J. E., and Klok, M. A., Ultrafiltration hemodynamics in an animal model: effect of a decreasing plasma sodium level, *Trans. Am. Soc. Artif. Intern. Organs*, 30, 603, 1984.
98. Razma, P. E., Daugirdas, J. T., Ing, T. S., and Izzo, J. L., Jr., Pre-existing hyponatremia is associated with a reduced total peripheral resistance response during isolated ultrafiltration (Abstr.), 17th Annu. Meet. Am. Soc. Nephrol., December 1984, 72A.
99. Daugirdas, J. T., Ing, T. S., Klok, M. A., Vestal, R. E., Hano, J. E., and Razma, P. E., Ultrafiltration hemodynamics in conscious dogs, Effect of extracorporeal blood temperature, *Trans. Am. Soc. Artif. Intern. Organs*, 29, 688, 1983.
100. Davidson, W. D., Rorke, S. J., Guo, L. S. S., and Morin, R. J., Comparison of acetate-1-^{14}C metabolism in uremic and nonuremic dogs, *Am. J. Clin. Nutr.*, 31, 1897, 1978.
101. Nawab, Z. M., Daugirdas, J. T., and Ing, I. L., Hemodialysis hemodynamics in a non-anesthetized animal model: effect of acetate (Abstr.) 17th Annu. Meet. Am. Soc. Nephrol., December 1984, 70A.
102. Daugirdas, J. T., Nawab, Z. M., and Ing, T. S., Acute hypotension during acetate dialysis in chemically sympathectomized dogs (Abstr.) American Society for Artificial Internal Organs, Boca Raton, Fla., 1985.
103. Gonzalez, F. M., Kirkendol, P. L., Pearson, J. E., and Reisin, E., The cardiovascular and metabolic effects of mixtures of acetate and succinate: a potential improvement in dialysate solutions, *Kidney Int.*, 26, 561, 1984.
104. Daugirdas, J. T., Nawab, Z. M., Ing, T. S., Leehey, D. J., Yang, M. C. M., Pang, P. K. T., and Klok, M. A., Studies of the vasorelaxant effect of acetate, *Trans. Am. Soc. Artif. Intern. Organs*, 30, 65, 1984.
105. Nawab, Z. M., Daugirdas, J. T., Ing, T. S., Leehey, D. J., Reid, R. W., and Klok, M. A., Calcium-complexing versus vasorelaxant effect of acetate, lactate, and other bases, *Trans. Am. Soc. Artif. Intern. Organs*, 30, 184, 1984.
106. Daugirdas, J. T., Nawab, Z. M., Leehey, D. J., and Ing, T. S., Vasorelaxant effects of pyruvate, lactate, alanine, and other intermediary metabolites (Abstr.) 17th Annu. Meet. Am. Soc. Nephrol., December 1984, 61A.
107. Walker, R. J., Lindsay, R. M., Peters, S. D., Sibbald, W. J., and Linton, A. L., A sheep model to study cardiopulmonary effects of blood dialyzer interactions, *ASAIO J.*, 6, 123, 1983.
108. Daugirdas, J. T., Nawab, Z. M., Leehey, D. J., and Ing, T. S., Lactate-buffered hemodialysis (Abstr.) American Society for Artificial Internal Organs, Boca Raton, Fla., 1985.
109. Nawab, Z. M., Daugirdas, J. T., and Ing, T. S., Lactate/acetate buffer-base cocktail for hemodialysis (Abstr.) 32nd Congr. Eur. Dial. Transpl. Assoc., June 1985.
110. Kelly, T. F., Jr., Morris, C. G., Jr., Crawford, E. S., Espada, R., and Howell, J. R., Perforation of the pulmonary artery with Swan-Ganz catheters, *Ann. Surg.*, 193, 686, 1981.
111. Michie, D. D., Wombolt, D. G., Carretta, R. F., Zencka, A. E., and Egan, J. D., Adverse reactions associated with administration of tricarbocyanine dye (Cardio-Green) to uremic patients, *J. Allergy Clin. Immunol.*, 48, 235, 1971.
112. Iseki, K., Onoyama, K., Fujimi, S., and Omae, T., Shock caused by indocyanine green dye in chronic hemodialysis patients, *Clin. Nephrol.*, 14, 210, 1980.

CURRENT STATUS OF PERITONEAL DIALYSIS IN VETERINARY MEDICINE AND DEVELOPMENT OF CAPD IN THE DOG AS A MODEL FOR DIALYSIS-RELATED DISORDERS IN MAN

Jerry A. Thornhill

INTRODUCTION

As reviewed in detail elsewhere, dialysis can be defined as clearance of solutes from one solution to another across a semipermeable membrane.[1] Dialysis of plasma may be performed by circulating blood out of the body through an artificial kidney to interface with fluid across semipermeable processed cellulose membranes (hemodialysis) or by circulating fluid in and out of the peritoneal cavity (peritoneal dialysis). In peritoneal dialysis, it is the peritoneum that invests most of the abdominal organs and lines the inner surface of the abdominal wall, that is the semipermeable exchange surface. Dialysate is the fluid into which solutes are transferred. It is delivered to the abdominal cavity through a peritoneal catheter, allowed to remain for a prescribed period, then removed. Dialysate is designed to be a physiological solution, though it contains varying concentrations of dextrose (1.5, 2.5, or 4.25%) for osmotic fluid removal. Dialysate does not contain urea, creatinine, or other uremic wastes, so these substances are cleared from the plasma into the dialysate over concentration gradients.

Since its introduction to human medicine in 1976,[2] continuous ambulatory peritoneal dialysis (CAPD), utilizing dialysate solution processed in plastic bags, has occupied an important place in the management of patients with end-stage renal disease and has been regarded as a valid alternative to hemodialysis in many cases.[3-12] However, in clinical veterinary medicine, due in part to little experience in development of dialysis techniques in the canine animal research model, peritoneal dialysis has not played a significant role as a therapeutic modality for acute or chronic end-stage renal failure patients.[13-21] Recently, though, three separate anephric dog studies have revealed the technique of CAPD to provide quality life support,[22,23,77] and the clinical application of CAPD to veterinary patients has been reviewed[24] and has been shown to provide successful life support in dogs with oliguric acute renal failure disorders.[25,26]

In the ensuing discussion, the principle of peritoneal dialysis will be reviewed and the technique of CAPD in the research and clinic dog, from catheter development, catheter placement, and bag exchange protocol, will be examined. The subject of peritonitis, which threatens to reduce the acceptability of the technique in the dog, and remains an ever present source of frustration in human patients, will be explored, and the use of the dog as an animal model for investigations into new horizons in peritoneal dialysis, such as volume, kinetics, advancements in alternative osmotic agents, solution improvements, sorbent based dialysate regeneration systems, access catheter developments, connectology improvements, further searches for uremic toxins, and in-line filter systems curbing peritonitis and protein losses, will be viewed and hopefully appreciated.

PRINCIPLE OF PERITONEAL DIALYSIS

Dynamics of Peritoneal Clearance

Peritoneal dialysis, as a system, involves the interrelationships of plasma, peritoneal membrane, and dialysate. In order for dialysis to be clinically efficient, mass transfer (diffusion) of uremic wastes across the peritoneum from plasma to dialysate must occur. Mass transfer, in turn, is dependent on the pore size of the peritoneum, molecular size of the solutes involves, volume and flow rate of dialysate instilled, temperature of the dialysate, and dialysate composition.[27-34]

Anatomically, the dialyzable membrane isolating the peritoneal cavity (theoretically the parietal peritoneum but splanchnic peritoneum of the intestine also serves as a dialyzable surface) is composed of capillary endothelial cells, capillary basement membrane, and mesothelial cells.[35] Transport of solutes from peritoneal capillary blood to dialysate within the peritoneal cavity occurs by diffusion and ultrafiltration over concentration gradients through the intercellular channels (pores) between the cells of the endothelium and mesothelium.[36,37]

The anatomic size of the intercellular channels is one of the major determinants of permeability of the peritoneum to certain solutes. The larger the diameter of the channel, the greater the molecular weight of substances that will diffuse through. The basement membrane apparently only impedes very large molecules from penetrating the peritoneal membrane barrier.[35] Intercellular channels may be enlarged by hypertonic dialysate which causes the endothelial and mesothelial cells to shrink, allowing for greater diffusion of solute from plasma to dialysate.[38] However, volume depletion and hyperglycemia can limit prolonged use of this type of solution. Vasodilators, such as nitroprusside or isoproterenol, which increase blood supply to the peritoneum thereby increasing ultrafiltration, may be added intravenously or intraperitoneally.[39-41] However, although these drugs increase urea and creatinine clearance, the unfavorable trade-off of systemic hypotension outweights their use. A safer vasodilator, theophyllin, was experimentally used in rabbits, but was found to not significantly increase peritoneal solute clearance.[42] To date, albeit many attemps have been made to increase peritoneal clearance, addition of drugs to the dialysate to improve peritoneal membrane permeability has been of little clinical value.

Urea (mol wt 60 daltons) and potassium (mol wt 19 daltons) are small molecules and diffuse more rapidly through the peritoneum than do creatinine (mol wt 113 daltons) or phosphate (mol wt 96 daltons). In 1 hr, equilibrium fron that fluid exchange is 98% complete for urea and potassium and from 60 to 80% complete for creatinine and phosphate.[19] However, in each hourly exchange only a portion of the total solute desired is removed. Peritoneal dialysis, therefore, is cummulative and may require up to 1 1/2 days (36 straight exchanges) to adequately dialyze a patient out of an acute uremic crisis.

The effect of dialysate volume on the kinetics of peritoneal dialysis has been in a state of flux. It has been established that small volumes of dialysate efficiently dialyze a patient as well as large volumes.[43] However, more recent investigation has revealed that the abdomen should be relatively full (not distended) for optimal clearance of solutes.[6,44] Contact of dialysate with the peritoneum is important for utilization of membrane surface area. However, dialysate overfill compounds the already low flow state that exists in peritoneal dialysis and decreases solute diffusion as pressure on mesothelial cells closes the diameter of intercellular channels.

Failure to adjust the temperature of the dialysate can affect the intercellular pore size of the peritoneum until the fluid is warmed by body heat. Dialysate warmed to a temperature of 37°C produces a 35% greater clearance of urea and creatinine than a dialysate temperature of 20°C.[45]

The dialysate flow rate is another factor that affects peritoneal clearance. The rate of removal of solute during peritoneal dialysis depends on both the volume of fluid instilled and the length of time it remains in the peritoneal cavity. Peritoneal dialysis can be improved by increasing dialysate volume and decreasing the dwell time, i.e., the time the dialysate remains in the peritoneal cavity.[30,46] Fluid delivered to the peritoneal cavity ultimately establishes equilibrium with extracellular fluid; however, osmotic equilibration (due to ultrafiltration of water) between plasma and dialysate occurs early in the exchange procedure, and once osmotic equilibrium between plasma and dialysate is approached, ultrafiltration slows significantly and dialysate begins to be very slowly reabsorbed.[47] During ultrafiltration, urea diffusion accompanies the bulk flow of water in amounts proportional to its concentration in plasma water. In contrast, creatinine, because of its higher molecular weight, lags in proportional diffusion-and continues to be cleared throughout the half-hour or hour of dialysate dwell time.

Because of the diffusion lag for creatinine and possibly other uremic wastes, Popovich and Moncrief,[2,4] in the late 1970s, instituted a program of continuous ambulatory peritoneal dialysis (CAPD), in which human patients were supported by maintenance peritoneal dialysis following intense half-hourly or hourly dialysis schedules. In CAPD, flow rate is decreased to 2 l/5- to 6-hr exchange. Dialysate remains in the peritoneal cavity at all times (24 hr a day, 7 days a week); this favors continued clearance of creatinine and other metabolic indicators of uremia, especially middle molecules (substances of 500 to 3000 mol wt).

Investigators have recently addressed the efficacy of long dwell exchanges in CAPD to allow equilibrium of small molecular weight solutes between plasma and dialysate, so that clearances of these solutes are essentially equal to drainage volumes in each exchange period.[48-51] Solute concentration gradients across the peritoneum from plasma to dialysate must, however, be preserved for optimal therapeutic dialysis (i.e., once in equilibrium mass transfer ceases). Nonetheless, a favorable trade-off for small solute equilibrium is the continued establishment of gradients for middle molecular weight compound clearance.

Dialysate Composition

In 1973, a workshop on peritoneal dialysis solutions was held by nephrologists, manufacturers, and the FDA, to establish guidelines for consistency in fluid preparation.[52] Based on recommendations from that workshop, peritoneal dialysis solutions are currently prepared in three strengths of dextrose: 1.5% (350 mOsm/l) which is moderately hypertonic to normal plasma, but is adequate for routine clinical use in uremic patients, 2.5%, and 4.25% (490 mOsm/l) which is reserved for oliguric renal failure patients with attendant volume expansion obligating extracellular fluid volume depletion. Since all solutions ultrafiltrate from patients more water than sodium, resulting in hypernatremia, the workshop recommended lowering the concentration of sodium in the dialysate to 132 mq/l favoring a diffusion gradient for plasma sodium across the peritoneum. No changes in the concentration of chloride (102 mq/l), magnesium (1.5 mq/l), or calcium (3.5 mq/l) were recommended.

Lactate or acetate in concentrations of 35 and 45 mq/l, respectively, were added as alkalinizing agents. They are rapidly abosorbed from dialysate across the peritoneum, and in patients with normal liver function, plasma levels should not exceed normal value. Both lactate and acetate solutions provide dilatation of the visceral peritoneum (lactate less than acetate) which may enhance solute clearance, however, their metabolism has been linked to aceleration of cholesterol and fatty acid synthesis.[53-55] Unfavorable trade-offs such as these have led nephrologists to search for better buffer systems, such as sorbent-based dialysate regeneration systems which manufacture bicarbonate (from urease splitting of urea) and return the buffer to the patient.[56,57]

Calculated Peritoneal Clearance

Clearance is defined as the rate (per minute) of solute removal from the plasma by an organ. A commonly used peritoneal clearance (P_{Cl}) formula for describing solute removal by perionteal dialysis is as follows:[58]

$$P_{Cl} = \frac{(C_D)(Q_D)}{C_P}$$

where C_D = the concentration of solute in the dialysate after exchange, Q_D = the flow rate of the dialysis fluid (total drainage volume ÷ duration of the exchange period), and C_P = the concentration of solute in plasma at the midpoint of the exchange. For example, if the patient is a 20-kg dog in renal failure with a C_P (serum creatinine) of 8.2 mg/dℓ, a C_D (dialysate creatinine) of 5.6 mg/dℓ, and a Q_D (dialysate flow rate) of 1 ℓ/hr, or 16.6 mℓ/minute,

$$P_{Cl} = \frac{(5.6 \text{ mg/d}\ell)(16.6 \text{ m}\ell/\text{min})}{8.2 \text{ mg/d}\ell}$$

$$P_{Cl} = 11.3 \text{ m}\ell/\text{min}$$

If the patient is on continuous peritoneal dialysis and is hypothetically anephric, the clearance for 1 week is (11.3 mℓ/min) (60 min/hr) (24 hr/day) (7 days/week) = 113,904 mℓ/week. The normal creatinine clearance for a dog is 3.7 mℓ min/kg.[59] The normal creatinine clearance for a 20-kg dog is therefore 74 mℓ/min. Clearance for such a dog for 1 week is (74 mℓ/min) (60 min/hr) (24 hr/day) (7 days/week) = 745,920 mℓ/ week. Thus, in this anephric dog model, continuous peritoneal dialysis theoretically provides approximately one sixth of the patient's normal function (113,904 ÷ 745,920 × 100 − 15.27%) and certainly enough to give life support. If the patient is not anephric and has partial renal function, then that patient's own endogenous creatinine clearance augments the creatinine clearance provided by peritoneal dialysis and improves the efficiency of the procedure.

Middle Molecules

Middle molecules are compounds of 300 to 1500 mol wt which are present in abnormal levels in dialyzed patients and are considered responsible for some of the clinical manifestations of uremia.[60,61] Except for certain hormones, i.e., parathyroid hormone (PTH) and antidiuretic hormone (ADH) the identity of these compounds and their origin are currently not well understood. Middle molecules may be one, or a mixture, of four groups of substances:

1. Normal waste products that accumulate during uremia
2. Normal hormones that achieve abnormally high levels in uremia
3. Abnormal peptides that accumulate as a result of disturbed cellular metabolism in uremia
4. Products of cellular breakdown[62]

Hormones, such as aldosterone, angiotensin, ADH, and PTH, that regulate water and electrolyte balance, have molecular weights within the middle molecular range. Elevation of these middle molecules could occur as a trade-off in the maintenance of altered sodium, potassium, phosphate, and total body water metabolism that occurs in renal insufficiency. A classic trade-off is secondary hyperparathyroidism for the maintenance of phosphate metabolism.[63] Phosphate retention in renal insufficiency produces reciprocal lowering of ionized calcium, which stimulates PTH secretion. Release of PTH influences renal excretion of phosphate and absorption of calcium from bone, intestine, and kidney, restoring the normal calcium to phosphorus ratio. This restoration, however, is not without a cost. Prolonged secretion of PTH has been shown to contribute to the acidosis, anemia, osteomalacia, pruritus, and neurological dysfunctions that accompany end-stage kidney disease.[64]

Peritoneal dialysis clears middle molecules more efficiently than hemodialysis does.[30,65-67] The hollow fiber artifical kidney has a large membrane surface area with many pores, but a small mean-pore radius, especially designed for clearance of small molecules (urea, creatinine, sodium, potassium, and phosphorus). Clearance of substances such as vitamin B_{12} (1355 mol wt) is only about one fourth that of creatinine. On the other hand, the peritoneal membrane has fewer pores, but a large mean-pore radius, which allows the clearance of larger molecular weight substances. There is still much to learn about middle molecules, their identification, and origin, but the importance they play in the biochemical consequences of renal failure appears to be accepted.

PERITONEAL ACCESS

The development of the Tenckhoff catheter in 1968 has allowed chronic peritoneal dialysis in man to be performed without requiring repeated puncture of the peritoneum.[68] Success of this access device in clinical medicine has been measured by its long-term use in chronic peritoneal dialysis for life support of patients with end-stage kidney disease.[69-71] The catheter is a soft silastic tube fenestrated at the distal end and furnished with two Dacron® velour cuffs into which fibrous tissue grows, walling off the catheter and preventing bacteria migration along the surface into the abdominal cavity. One cuff is placed preperitoneally, anchoring the catheter, and the second is placed subcutaneously at the catheter's site of exit from the skin (Figure 1)*.[72]

Tenckhoff catheters have been placed in dogs at Purdue University, but all failed within a week because of outflow obstruction (inability to recover dialysate from the peritoneal cavity because of blockage of the catheter with omentum, fat, or intestine).[73,74] Outflow obstruction also plagued earlier attempts at peritoneal dialysis in the dog when large-bore needles or straight-tube catheters were used for the peritoneal access.[13-18]

In 1972, at the University of California, Davis, development of the Parker straight-tube catheter and method for peritoneal access in the dog showed promise and contributed significantly to improving the technique of peritoneal dialysis in the dog.[19,20] Investigators at Purdue, however, were interested in developing a catheter that would not float free in the abdominal cavity, that would have negligible outflow obstruction, and that could withstand the rigors of peritoneal dialysis for weeks or months without removal or replacement; the catheter that was developed as a result of their efforts, the column disk catheter[2] (Figure 2), has shown excellent outflow characteristics.[73,74]

The catheter disk portion is 5.1 cm in diameter and 0.8 cm thick and consists of silicone sheets separated by numerous pillars, each 0.5 cm tall. Silastic tubing (O.D. 0.49 cm; I.D. 0.26 cm) enters the disk on one side to open into the space between the sheets (Figure 3). As with the Tenckhoff catheter, two Dacron® velour cuffs, one next to the base of the catheter and one near the free end, are placed preperitoneally and subcutaneously. The catheter is designed to be secured against the abdominal wall in either the right or left caudal abdominal quadrant (Figure 4).

The disk catheter must be placed surgically into the abdomen (Figure 5A and B). Through a paramedial incision in the skin and grid separation of muscle layers, the peritoneum is exposed and severed. The foldable disk is squeezed and placed through

* Figures for this chapter appear at the end of the text.

the surgical opening into the peritoneal cavity. The disk is pulled firmly against the parietal peritoneum which is then closed over the base of the catheter. Muscle layers are sewn tightly around the first Dacron® velour cuff, but sutures need not enter the cuff for a good seal. The free end of the catheter is tunneled subcutaneously to exit through a skin stab incision at a point beyond the second Dacron® velour cuff, which is buried. Heparinized saline (20 to 30 mℓ) is flushed through the catheter to assure patency and seal (Figure 6A to G). A titanium connector is fitted into the free end of the catheter and attached to the ever-lock end of Transfer Set* tubing (Figures 7 and 8). To suppress bacterial migration into the fitting, a povidone-iodine dressing followed by sterile gauze bandage (supplied by a CAPD™ Prep Kit*) is wrapped around the titanium-tube coupling (Figures 9 to 11). The spike end of the Transfer Set tubing, facilitated by a dialysate port clamp*, is introduced into the solution bag** (Figures 12A and B). This disconnect coupling is also wrapped with the povidone-iodine/gauze dressing (Figures 13 to 15). Dialysate to which has been added 1000 units per 2 ℓ of heparin and 250 mg/2 ℓ of cephalothin is immediately instilled at time of surgery (Figure 16).[75] If slight leakage around the catheter into the s.c. tissue occurs, a mild break-in period, with less volume of solution (with proportional mixes of heparin and cephalothin) administered per exchange, should be practiced for the next 48- to 78-hr period.[76] Once the catheter is sealed in place, a stepped up program of continuous ambulatory peritoneal dialysis (CAPD) with full volume exchanges can be practiced (i.e., 20 to 40 mℓ/kg in the dog).

CONTINUOUS AMBULATORY PERITONEAL DIALYSIS (CAPD)

As previously reported in this manuscript, three anephric research dogs utilizing the column disk catheter have been supported by CAPD for 56, 54, and 72 days, respectively.[22,23,77] At no time during these studies did catheter malfunction of outflow obstruction occur, and no catheter replacement surgery was required.

To fully understand the application of CAPD in the dog, one must recapitulate the events leading to the introduction of the technique in humans from which it is extrapolated. The success of the Tenckhoff catheter in the late 1960s as a chronic maintenance indwelling peritoneal access device led to renewed interest in peritoneal dialysis in man. At first, automated machines were developed which proportion concentrated dialysate with sterile deionized water, warm the solution, and control the delivery, dwell, and recovery of the fluid to and from the abdominal cavity.[7,46,78,79] The automated technique is used as a continuous form of dialysis. It is performed intermittently, as maintenance hemodialysis, 3 or 4 days a week for 10- to 14-hr sessions. The activity of the patients is limited during this time, but their comfort and mobility is greater than with hemodialysis.

In man, CAPD is a technique whereby the peritoneum serves as a dialysis membrane 24 hr a day, 7 days a week, rather than intermittently.[2] Premixed dialysate is processed in 2-ℓ collapsible plastic bags** (also available in 250-, 500-, and 1000-mℓ quantities). Transfer Set tubing* allows for a closed system to exist, so that the patient's Tenckhoff catheter is connected to a bag at all times. In CAPD, the patient, using aseptic technique, connects a fresh 2-ℓ bag of dialysate to the tubing to start a new exchange (Figure 17). A total of 6 to 10 min is required during inflow to fill the peritoneal cavity. When it is full, the patient clamps the tubing, folds the bag, wraps the tubing around the bag, and places the empty bag and tubing against the abdomen under an elastic wrap or into a cloth pouch which fits in a shirt or blouse or is suspended from a belt

* Travenol Laboratories, Inc., Deerfield, Ill. 60015.

** Dianeal® 137 with 1.5% dextrose — Viaflex; from Travenol Laboratories, Inc.

at the waist. At the end of the dwell time, which usually lasts 4 hr during the day and 8 hr during the night, the patient removes the empty plastic bag from the pouch, lowers it to the floor, unclamps the tubing, and lets the solution drain. Gravity drainage during outflow usually takes about 20 min. Bags of dialysate are changed with each new inflow; one set of tubing can be used for up to 1 month.

The theoretical and practical implications of CAPD have been well documented.[4,8,9,11,80] It is the most effective dialysis for middle molecule removal, primarily because it is utilized five to ten times longer than intermittent dialysis schedules. When a patient is dialyzed out of uremic crisis (by intense hourly peritoneal dialysis exchanges by bag or automated machines), but remains dialysis dependent, CAPD can provide maintenance support with adequate clearance of urea, creatinine, and middle molecules and acceptable equilibration of body fluid and electrolytes. Studies following patients with end-stage kidney disease treated by CAPD have revealed that uremic signs and symptoms that accompany renal failure are rectified by this chronic form of dialysis.[5,10,81]

Technique of CAPD in the Dog

As with the closed system used in man, dialysate can be delivered to and from the peritoneal cavity of an animal using the same collapsible bag for each inflow and outflow fluid-exchange cycle (Figure 18). The technique of CAPD is demonstrated in the following dialysis-support study of an anephric research dog.

Following a 4- to 6-hr dialysate dwell period, the dog is brought into the dialysis exchange room (should be an area in a hospital or research laboratory where there is little pedestrian traffic flow) and prepared for the dialysate outflow segment of the exchange cycle (Figure 19). The storage pocket of the protective harness* is unzipped and the viaflex bag is retrieved, unrolled, and straightened (Figures 20A and B). The bag is suspended from a cart serving as a dialysis table, and outflow is initiated after tubing and catheter clamps are released (Figure 21). At a 1-m hydrostatic head, gravity outflow of 2 *l* should take approximately 5 to 7 min (Figure 22). At the end of the outflow, both catheter and tubing clamps are reapplied.

Changing the Transfer Set tubing from the old bag (most recent outflow) to the new bag becomes the most critical time in the CAPD exchange schedule, for a break in technique can predispose the animal to bacterial contamination and peritonitis. The used dialysate bag is situated on the dialysis table to the right side with the new bag to the left. Dialysate port clamps are applied to the inflow/outflow ports of each bag. The protective sleeve over the port of the fresh bag is left intact (Figure 23). If medications are to be added (at Purdue, 1000 units of heparin and 250 mg of cephalothin per 2 *l* are added to each exchange bag for the first 5 days on dialysis),[75] they are instilled at this time through the medication port of the fresh bag after a preparation of the port with Betadine®. Protective gloves can be worn, but do not make the wearer sterile, as strict attention must be paid to prevention of the tubing spike from contacting other than the protected lips of the dialysate inflow/outflow ports. A mask covering the face of all dialysis personnel in the immediate area should, however, be worn.

By grasping the port clamp of the used bag, the tubing spike is swiftly pulled from the bag. The protective sleeve is removed from the dialysate port of the new bag, and facilitated by grasping its port clamp, the new bag is immediately spiked (Figure 24A and B). This procedure should be accomplished in swift motion, so that the tubing spike has little exposure to room air. The port clamp on the new bag is removed and a providone/gauze wrap as previously discussed is applied (Figure 25A, B, and C).

* Alice King Chatham Medical Arts, Los Angeles, Calif. 90043.

The next cycle is initiated by instilling fresh dialysate to the dog. After suspending the solution bag, the catheter and tubing clamps are released and flow begins. At a 1-m hydrostatic head above the animal's abdomen, inflow should take approximately 5 min (Figure 26). After the inflow cycle, clamps are reapplied, bag is rolled up, and tubing is wrapped around the bag which is placed back in the pocket of the harness (Figure 27). If a harness is not used, the tube-wrapped bag can be placed against the animal's side or back and secured with an elastic bandage, as shown in another research dog pictured (Figure 28). Note the free ambulatory nature of this animal on CAPD.

Peritoneal dialysis in an anephric research dog (or a severely compromised, induced renal failure subject) and spontaneously uremic canine patients using the CAPD technique should be repeated hourly until an acceptable level of azotemia, i.e., a level of waste product accumulation that the patient can tolerate, is reached. A dialysis progress chart should be established and fluid exchanges should be recorded (Table 1). Dialysate outflow on the first two exchanges will not match inflow because the abdominal cavity is being primed, but volume exchanges should closely match outflow for inflow thereafter. During hourly cycles, dialysate inflow (of 5 min) should be followed by an abdominal dwell time of approximately 45 min and then by approximately 5 to 10 min of outflow.

Once a patient is in compensation following intense hourly dialysis, but remains dialysis dependent, a less intense, "spread-out" schedule of CAPD can be practiced. Dialysate can be allowed to remain in the abdominal cavity from 4 to 6 hr during the day and for up to 8 hr during the night before it is exchanged. It is reasonable to assume that in a selected number of cases, owners of pets on dialysis can be trained in CAPD and can carry out this spread-out schedule at home, thereby diminishing costly hospitalization fees.

Indications for CAPD

The biochemical consequences of uremia have been well documented in primate and animal species.[1,64,83-86] In man, dialysis-dependent patients with acute or chronic renal failure have received successful life support from peritoneal dialysis,[3,9,10,80,82,87-89] and although there is a dramatic difference in the lengths of time on dialysis among these patients (i.e., patients with acute oliguric renal failure may require dialysis for 1 to 3 weeks until partial or total renal function returns, whereas the chronic end-stage renal patient may require maintenance dialysis for life), there is little doubt that the peritoneal technique of dialysis is chosen nearly as frequently as hemodialysis and is no longer considered a second choice.[90,91]

In the dog and cat, peritoneal dialysis is indicated for life support when renal function is insufficient to excrete nonprotein nitrogenous wastes and conservative management has failed to control the signs and symptoms of uremia (Figure 29). Since dialysate has become available in smaller quantity bags (250, 500 and 1000 ml), acute peritoneal dialysis and chronic maintenance dialysis using the CAPD technique is now applicable to small animal species of any body size.

PERITONITIS ASSOCIATED WITH PERITONEAL DIALYSIS

One of the major complications accompanying peritoneal dialysis is peritonitis,[92-96] and although the incidence was higher when the peritoneum was repeatedly punctured for dialysis, infection of the peritoneal membrane is still recognized as one of the leading causes for peritoneal dialysis failure in patients with chronic indwelling catheters.[97-100]

Table 1
FLOW CHART OF FLUID VOLUME EXCHANGES*

Exchange No.	Time started	Inflow vol. (m*l*)	Outflow vol. (m*l*)	Δm*l*
1	10:00 a.m.	2000	1700	−300
2	11:00 a.m.	2000	1800	−200
3	12:00 noon	2000	2105	+105
4	1:00 p.m.	2000	2010	± 10
5	2:00 p.m.	2000	1985	− 15
6	3:00 p.m.	2000	2000	0
7	4:00 p.m.	2000	2000	0
8	4:00 p.m.	2000	1990	− 10
9	6:00 p.m.	2000	2005	+ 5

* Exchanges were recorded during a period of peritoneal dialysis on a dog using 2-*l* dialysate bags.

From Thornhill, J. A., Peritoneal dialysis in the dog and cat: an update, *Compendium on Continuing Education for the Practicing Veterinarian*, Vol. 3, 1981, 20. With permission.

Detection of Peritonitis

The diagnosis of peritonitis has been extensively reviewed.[75,101] Peritonitis implies peritoneal inflammation. Basically, in order to diagnose its presence, two of the following three criteria in any combination should be present:

1. Cloudy fluid with inflammatory cells — Observation for turbid drainage and follow-up cellular evaluation clinically has become the most important factor in the diagnosis of peritonitis. Turbid drainage is due to increases in dialysate WBC count. In the absence of peritonitis, drainage WBC counts average well below 50 cells per cubic millimeter, with mononuclear cells predominating. In the presence of peritonitis, turbidity becomes grossly obvious when WBC counts exceed 100 to 200 cells per cubic millimeter, with the majority of cells in the dialysate being neutrophils.
2. Presence of organisms in Gram stain or in culture — Following first detection of cloudy dialysate drainage, the bag is prepared for collection of some of its fluid content for bacteriologic examination. A clamp is placed over the dialysate port of the bag to prevent leakage. The medication port of the bag and the top of the culture bottle, following removal of its cap, are then cleaned with povidone-iodine and allowed to dry. With a sterile syringe and needle, the bag is penetrated through the medication port and 15 to 20 m*l* of dialysate is withdrawn. The needle on the syringe is then replaced and harvested fluid is injected into the culture bottle, allowing vacuum inside the bottle to withdraw 4 to 5 m*l* from the syringe. Both the syringe with its remaining fluid and the culture bottle are delivered to the laboratory for Gram stain, cell counts, and culture and sensitivity tests.
3. Clinical signs of peritoneal inflammation — In the dog, clinical signs of peritonitis generally develop when dialysate WBC counts exceed 200 cells per cubic millimeter. Typically, dogs become depressed and anorectic. Peripheral WBC counts increase with a left shift, but fever may or may not be present depending on the level of azotemia. Abdominal rebound tenderness is initially mild; however, if

therapy is not instituted or is unsuccessful, patients repeatedly resist ballottement of the abdomen. As peritonitis advances, diarrhea with or without vomiting ensues, patients become recumbent with reluctance to move, the abdomen is extremely tense and tight on palpation, and respirations are frequent and shallow.

Treatment of Peritonitis

There are two approaches to treatment of peritonitis. It can be treated when it occurs or prophylactic measures can be initiated ahead of time to prevent it. Prophylaxis becomes extremely important if there is a break in sterile technique during changing of dialysate bags or if a dog traumatizes its catheter or fluid line, thereby exposing and contaminating the luminal surfaces.

Presence of Peritonitis

With the early detection of turbid dialysate drainage, the clinician should suspect peritonitis. Dialysate cell counts must be instituted promptly with aliquots of solution submitted for Gram stain and culture examination as discussed previously. If the dialysate WBC count is greater than 100 cells per cubic millimeter, antibiotic and heparin should be added to the solution bags immediately. Just prior to each exchange, cephalothin (250 mg/2-*ℓ* bag) and heparin (1000 units per 2-*ℓ* bag) should be aseptically added to the dialysate bag through the medication port. Cephalothin is a form of cephalosporin. Not all patients will receive 2-*ℓ* exchange cycles, because smaller dogs may be on 250-, 500-, or 1000-m*ℓ* exchanges. Therefore, antibiotic and heparin dosage should be proportioned with 2-*ℓ* bag doses as reference.

If the research dog or patient improves, treatment should be continued for 10 days. If dialysate turbidity does not decrease or signs do not diminish over the next 24 hr, the Gram stain is repeated and initial cultures are evaluated for causative organisms. Therapy at this point is governed by results of these findings.

If Gram-negative organisms are revealed on the initial Gram stain or after isolation and the identification with culture separation techniques, aminoglycoside therapy with gentamicin or tobramycin is recommended.[102,103] An initial i.m. loading dose of 4 mg/kg[104,105] is administered to the dog followed proportionally by addition of 10 mg/2-*ℓ* bag exchange for 10 days. Kinetic studies on bidirectional fluxes of gentamicin across the peritoneum in dialysis have shown that there is rapid absorption of the drug from peritoneal fluid into the blood compartment, thereby maintaining therapeutic serum concentrations but providing negligible exchange from the blood compartment to dialysate.[106] The latter was supported by a study at Purdue of an anephric dog which was maintained on peritoneal dialysis. A least square linear regression plot of gentamicin concentrations in blood following an initial i.m. injection revealed persistence of therapeutic concentrations of greater than 4 μg/m*ℓ* for approximately 18 hr (Figure 30). There was negligible peritoneal clearance of gentamicin and a half-life of 75 hr, which greatly exceeded the half-life of 1.26 hr observed in intact dogs. If after initial therapy with cephalothin only Gram-negative organisms are found and clinical response has been poor, the cephalothin should be discontinued and aminoglycoside therapy should be begun. However, if multiple organisms require both aminoglycoside and cephalosporin therapy, each safely can be mixed with the other in the bag, but should not be mixed in the same syringe before instillation. Heparin is inactivated by aminoglycosides and, therefore, addition of the anticoagulant is contraindicated.

Although cephalosporin and aminoglycoside antibiotics have been chosen specifically to treat patients with peritonitis at Purdue, culture- and antibiotic-sensitivity findings may dictate use of other antimicrobials. Extrapolated from work in man (until further experience in the dog mandates change), these drugs can be added to dialysate in the following concentrations: penicillin G (50,000 units per liter), ampicillin (50 mg/ℓ), cloxacillin (100 mg/ℓ), ticarcillin (100 mg/ℓ), vancomycin (30 mg/ℓ), amikacin (50 mg/ℓ), clindamycin (50 mg/ℓ), and sulfadiazine to trimethoprim (25:5 mg/ℓ). Heparin should not be mixed with penicillin or vancomycin.

For fungal peritonitis, i.v. amphotericin B with addition of the drug to dialysate (5 mg/ℓ) is recommended. Nonetheless, the prognosis is poor as human patients rarely improve without termination of peritoneal dialysis and transfer to hemodialysis due to residence of the fungus organisms within the lumen of the catheter.[107] In the anephric research dog or clinical dialysis-dependent canine patient, this is devastating as a majority of the peritoneal patients cannot be transferred to hemodialysis due to body size and equipment restrictions.

Prophylaxis

Strict dialysis technique must be followed by all personnel executing fluid bag exchanges. The dialysis station ideally should be in an isolated room with little extraneous traffic flow. Masks should be worn and dialysate tubing and bags should have all disconnect coupling ends overlaid with a povidone-iodine gauze pad followed by sterile gauze wrapping provided in a dialysis preparation kit. Transfer of the dialysate line spike from the spent dialysate bag to the new bag must be swift with immediate wrap of the povidone-iodine gauze combination around the coupling after its completion.

To assist in early prophylaxis, cephalothin (250 mg/2ℓ) and heparin (1000 units per 2ℓ) are added to the bag exchanges for the first 5 days following catheter implantation and initiation of dialysis. Thereafter, observation for dialysate turbidity and the foregoing principles discussed for detection of peritonitis are applied.

If there is a break in sterile technique or if the peritoneal catheter, line, or any of the disconnect couplings are traumatized by the dog, thereby allowing luminal surfaces to be exposed, a technique of saline, saline plus iodine flush of the abdomen, described in man, should be instituted immediately.[108,109] Procedure for the saline, saline plus iodine wash is performed as follows:

1. Initially and then continuing once a day thereafter for 5 days, following the removal of dialysate, isotonic saline (the same quantity that is used for dialysate exchanges, up to 1 ℓ) is instilled and immediately removed without dwell time.
2. Next, isotonic saline, to which 0.2 mℓ of 2% solution of iodine USP per liter has been added, is instilled and allowed to remain in the abdominal cavity for 4 min (the length of time required for the small quantity of dextrose that is left following the saline flush to convert iodine to iodide).
3. The saline plus iodine solution is then removed and followed with the prescribed quantity of dialysate to initiate another dialysis cycle.

Pharmacologically, iodine reacts with water to form hypoiodous acid: $I_2 + H_2O \rightleftharpoons HIO + H^+ + I$. Both I_2 and HIO are microbicidal (at low pH levels, elemental iodine prevails; at high pH levels, I_2 converted to hypoiodous acid prevails). Residual dextrose from the just-completed dialysis exchange needs to be removed from the abdominal cavity with a saline flush prior to instillation of the saline plus iodine solution. In the presence of dextrose, iodine is converted to inactive iodide (although iodide may enhance the polymorphonuclear leukocyte and monocyte microbiocidal capacity via activation of their hydrogen peroxide-myeloperoxidase-halide system).

Trials with the saline, saline plus iodine wash of the abdominal cavity, in research dogs and canine patients receiving peritoneal dialysis at Purdue University, have shown that the technique will prophylactically prevent peritonitis, after break in technique without the concomitant use of antibiotics. If peritoneal infection does occur, the amount of 2% iodine used per liter of saline solutin can be doubled from 0.2 to 0.4 m*ℓ* or the frequency of flushes with 0.2 m*ℓ* iodine can be increased to 2 times a day. If control of the peritonitis is not provided by iodine within 24 hr, the procedure is abandoned and cephalothin and heparin are added to the bag.

A noninfectious, peritonitis will appear periodically in dogs on dialysis. It is distinguished by a slightly turbid exchange, negative Gram stain and culture results, and cell counts ranging between 200 to 600 cells with characteristically large populations of eosinophils. In these episodes, dexamethasone added (10 ml/2 *ℓ*) to exchange bags generally will clear the effluent solution within 3 days. Therapy is extended for 3 to 5 days beyond clinical improvement.

ADVANCEMENTS IN PERITONEAL DIALYSIS — THE FUTURE

Albeit successful, peritoneal dialysis in man is attendant with certain problems. Peritonitis remains a major impediment to otherwise quality amelioration of signs and symptoms of uremia by CAPD. Axiomatically, improvements in the methodology of the exchange cycle are needed to circumvent human error.[57,110] Automated connectology has produced a sterile connecting patient line device (SCK') which thermally cuts and seals old line to new line precluding spiking solution bags at time of exchange. Purdue is currently investigating the efficacy of the device in curbing peritonitis in clinic and research dogs on CAPD.

Another problem area in CAPD is the trade-off of glucose absorption from dextrose additions to solution bags. Hyperglycemia in diabetics, potential predisposition to obesity, and aggravation of hyperlipidemia have been attributed to absorption of glucose from the peritoneal cavity.[111-115] As discussed previously, dextrose is added to dialysate for ultrafiltration of extracellular fluid from the patient across the peritoneum to the dialysis solution. For this purpose, glucose is an effective osmotic agent, is inexpensive, and is readily available. It presumably has a negligible detrimental effect on the peritoneal membrane and its absorption into the bloodstream stimulates secretion of insulin which may, by its anabolic effect, promote positive nitrogen balance.[116]

However, in addition to the above endocrine considerations, to prevent caramelization of glucose, hydrochloric acid must be added which lowers the pH of the dialysate (to less than 5.5) and may be responsible for inflow pain experienced in some patients. Furthermore, investigators have shown that the low pH inhibits polymorphonuclear phagocytosis and hampers intracellular killing of bacteria.[117] Since most patients respond to antibiotic therapy, the peritoneal membrane must not be irreversibly damaged by polymorphonuclear cell suppression. Nonetheless, more investigative work is needed in evaluating the efficacy of the glucose agent as an osmotic additive to peritoneal dialysate solutions.

Alternative osmotic agents heretofore proposed as substitutes of glucose have included sorbitol, polyanions (such as dextran, xylitol, fractose), and amino acids.[116] Sorbitol was used in the early 1970s as a substitute for glucose, but was found to predispose human patients to CNS disorders (i.e., convulsion and coma) due to hyperosmolar states.[118,119] Dextran sodium sulfate was studied in vitro as an osmotic agent and in a simulated CAPD system.[120,121] However, the compound as a dialysate additive has not been studied in an animal model, but would be expected to be relatively nontoxic since the molecule should be metabolized if absorbed. Xylitol was thought to show promise as an osmotic agent, especially in diabetic patients, due to its metabolism being independent of insulin.[122] However, patients in this study, as well as in a previously reported study, developed hyperuricemia and lactic acidosis, precluding use of xylitol, expecially in individuals with volume overload who need higher concentrations of an osmotic agent and more frequent exchanges.[123]

Addition of fructose to dialysate in the early 1970s revealed favorable studies in diabetic patients as compared to equivalent concentrations of dextrose.[124] Blood glucose levels showed better control with fructose and ultrafiltration was slightly improved. Based on these findings alone, fructose deserves further evaluation as an osmotic agent, especially in the canine model on CAPD. There is some suggestion, however, that fructose may promote hypertriglyceridemia over glucose.[125] However, again, this needs investigation.

Amino acids have been shown to be promising agents for ultrafiltration as they increase the oncotic pressure of the dialysate producing fluid draw.[126] Additionally, due to loss of protein in CAPD patients,[127] amino acids fluxing across the peritoneum to plasma might serve to correct existing amino acid deficits, promote positive nitrogen balance, and correct gross hypoproteinemia if present.[128,129] In a preliminary study performed on dogs undergoing CAPD at Purdue, protein losses were favorably reduced and bacteria were curbed from entering the peritoneal cavity, by use of a protein/microorganism rejection in-line 0.22-μm filter* from Millipore.[130]

Sorbent-based dialysate regeneration delivery systems should be readily available and clinically applicable in the near future. The efficacy of sorbent compounds, such as urease, zirconium phosphate, and charcoal, for the removal of uremic wastes and maintenance of chemical balance, has been well documented.[56,131-135] Use of these and other compounds as functional systems to purify small quantities of recycled peritoneal fluid back to the patients, therefore providing life support, needs further sophistication and refinement. Nevertheless, a couple of strong attributes accompany the use of sorbent-based machines regenerating dialysate. First, sorbent systems containing urease generate bicarbonate back to the patient,[56] precluding the need for lactate or acetate additions as alkalinizing agents. Second, glucose can be added to the sorbent recipe commensurate on the patient's needs, since ultrafiltration can be controlled by fluctuations in negative pressure application in the regenerating dialyzer.

The Canine Animal Model

Many of the above paragraphs in this section have dealt with continuing problems present in today's system of CAPD and the lack of an adequate animal dialysis testing model for investigation and elucidation of the unsettled issues. Clearly, the reader should recognize the ease of application of CAPD to the dog and how this model could facilitate harvesting of much of the in vivo data needed to support investigators' hypotheses concerning the points of peritoneal dialysis in question. A recent study did support the use of the canine model, as ultrafiltration kinetic data for dextrose and ^{14}C-dextran were obtained using anephric dogs supported by CAPD.[136]

With all the interest in the efficacy of i.p. administration of insulin to "end-stage" diabetics on CAPD,[137,138] a canine model with laboratory-induced insulin-dependent diabetes showed marked improvement of electron microscopical membranous glomerulopathy and massive proteinuria following i.p. administration of insulin,[139,140] and so, with experimentally induced renal disease and/or failure, or with the large population of dogs having spontaneously occurring acute and chronic renal failure disorders, the species appears to be an excellent reservoir of information needed to better understand like disorders in man. In the research or clinic dog with a dialysis-dependent state, then, each animal should be viewed as a learning experience, greatly increasing our understanding of the identity of uremic toxins, the optimization of dialysis prescriptions and choice of dialysate solutions and concentrations used, the efficacy of drug therapy for peritonitis if present, and the overall improvements needed to make the therapeutic modality of CAPD as lifelike as possible.

* Millipak 100 GLCA, Millipore Corporation, Bedford, Mass. 01730.

SUMMARY

In human medicine, peritoneal dialysis has grown over the past 6 years to become a valid alternative to hemodialysis in the management of patients with end-stage renal disease. It may, in fact, be the treatment of choice in chronic renal failure involving children, diabetics, and elderly patients. The growth in peritoneal dialysis has been attributed to the development of a functionally reliable catheter, better control of peritonitis and the evolvement of a new dialysis method, and continuous ambulatory peritoneal dialysis (CAPD), whereby the peritoneum serves as a dialysis membrane 24 hr a day, 7 days a week. CAPD represents a closed system where dialysate stored in a plastic bag is instilled into the peritoneal cavity, allowed to remain for a prescribed period, then gravity drained back into the same bag which is discarded, completing the cycle. A fresh bag of dialysate initiates the next cycle. In between exchanges, during dialysate dwell time, the empty bag is rolled up with its tubing, which remains attached to the peritoneal catheter, and wrapped to the individual's side, allowing complete freedom of patient movement during the procedure.

In veterinary medicine, with the most recent development of a new peritoneal access catheter at Purdue University, which efficiently drains the abdominal cavity of the dog, CAPD has shown to be an effective therapeutic modality for short- and long-term support of clinical patients with renal disease and research dogs with compromised renal function. Because of the incidence of renal disease and failure in the dog, the species becomes a natural model for studying the long-term metabolic effects of peritoneal dialysis support, and, whether addressing advancements in peritoneal dialysis or attacking consequential problems attendent with conventional methods, such as peritonitis, an animal model, such as the dog, is needed to generate the comprehensible in vivo data.

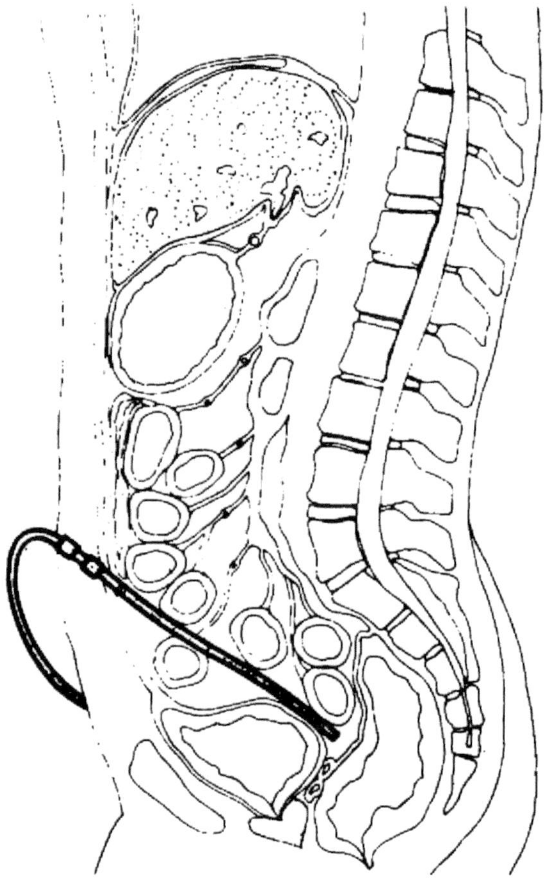

FIGURE 1. Drawing of placement of the Tenckhoff catheter in the lower abdominal quadrant, adjacent to the bladder, in man. Dacron,® velour cuffs around the tube can be seen at the abdominal muscle layer and in the s.c. tissue before the catheter exits through the skin. (Courtesy of Physio-Control Corp., Redmond, Wash. 98052.)

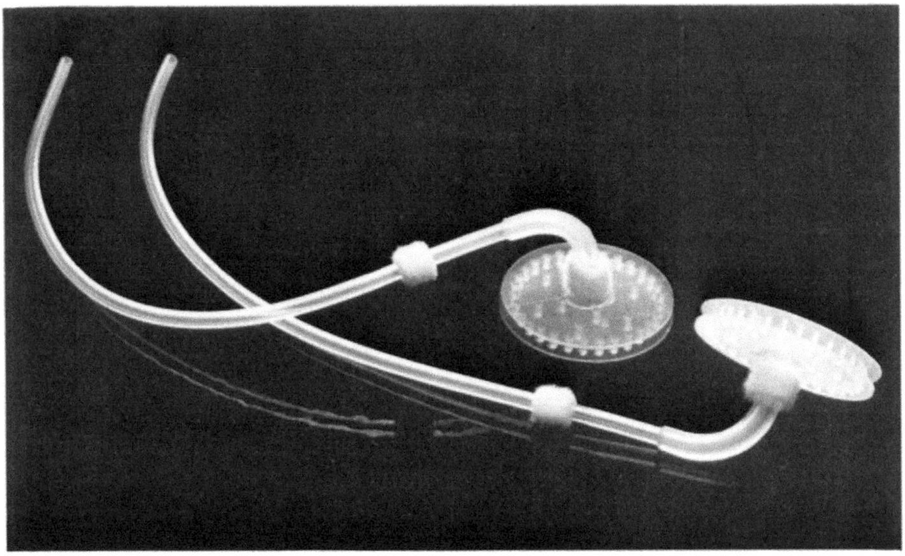

FIGURE 2. Silastic injection molded Column Disc Catheter (Life Cath)™ with double-layered disk containing 40 separative pillars and two Dacron® velour cuffs positioned, one adjacent to the disk and the other on the tubing. (Courtesy of Physio-Control Corp.)

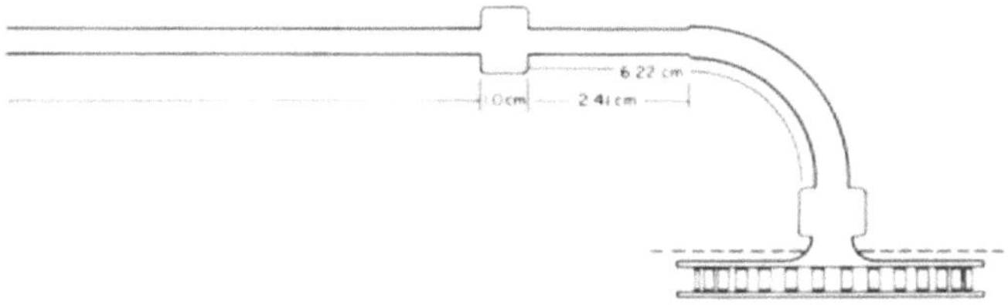

FIGURE 3. Drawing of the catheter in cross-section with dimensions. Dotted line represents position of the disk against the peritoneal membrane. Diameter of disk, 5.1 cm; columns, 40 total. (Courtesy of Physio-Control Corp.)

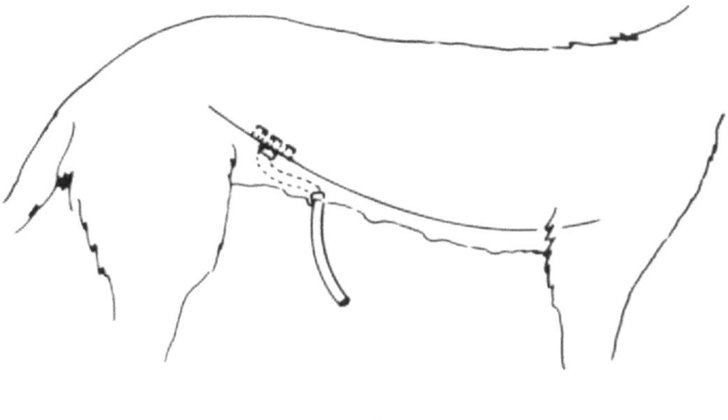

A

FIGURE 4. Diagrams (A and B) illustrating caudal paramedial position of the Column Disc Catheter.

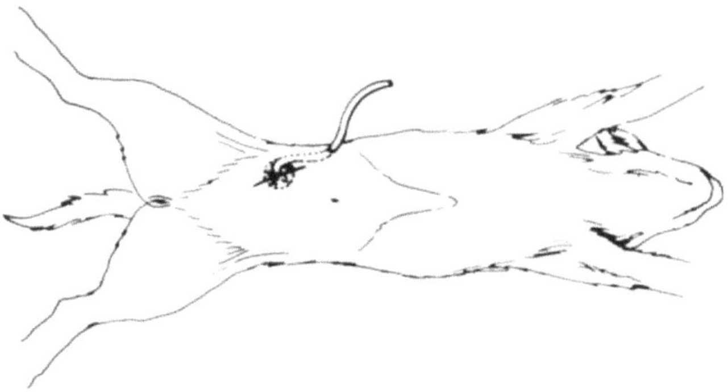

FIGURE 4B

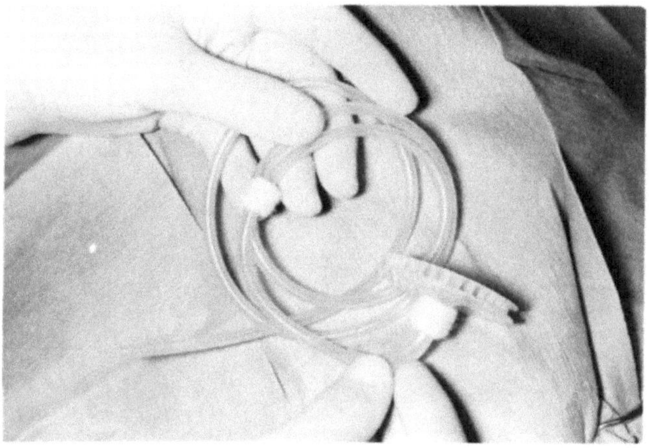

A

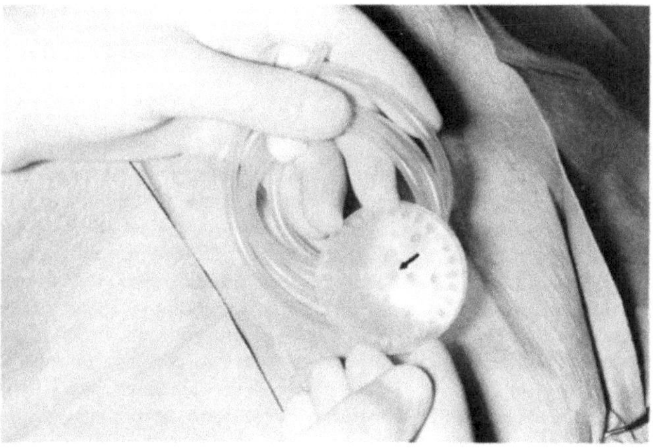

B

FIGURE 5. (A) Column Disc Catheter before surgical implanatation. To clear air bubbles trapped within the Dacron ® velour cuffs, assuring better seal in tissue, the catheter is presoaked in sterile saline at this time. (B) Arrow points to the entry and exit fluid port of the catheter located in the sheet of the disk next to the tubing.

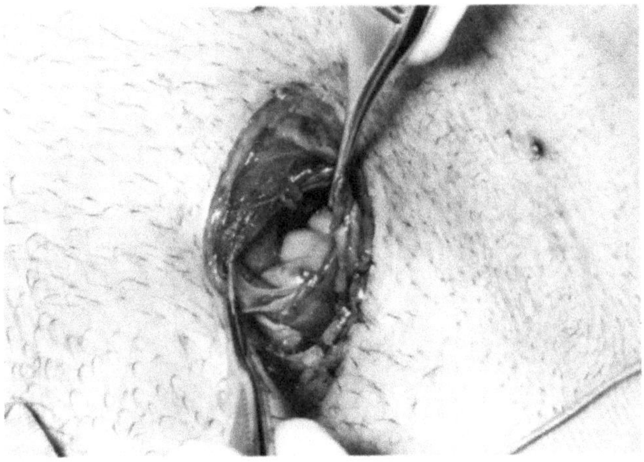

A

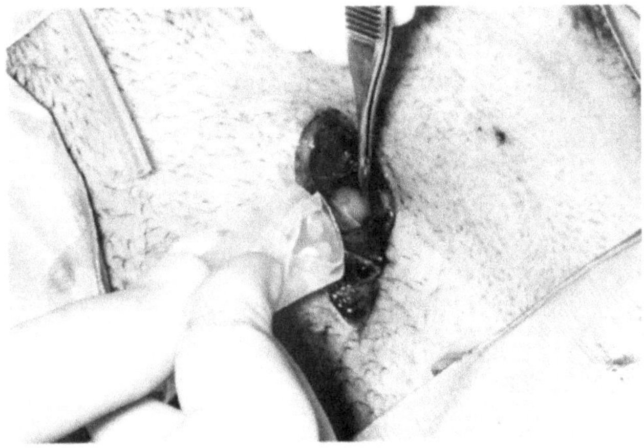

B

FIGURE 6. Surgical placement of the Column Disc Catheter. (A) Following a paramedial skin incision in the caudal abdomen and grid separation of muscle layers, the peritoneum is identified and severed; (B) disk portion of the catheter folded and introduced into the abdominal cavity; (C) disk pulled snugly against the peritoneal membrane; (D) closure of remaining muscle layers and adipose tissue firmly around the first Dacron® velour cuff; (E) heparinized saline flush to test bidirectional patency of the system and to confirm absence of fluid leak after closure of peritoneum and adjacent muscle layer around the base of the catheter; (F) free end of catheter pulled through the s.c. tissue with a set of Doyen clamps; (G) catheter exiting through a stab skin incision beyond the site of the second cuff, which is buried. Again, the catheter is flushed with heparinized saline to check bidirectional patency. A Beta-Cap clamp (Quinton Instrument Company, Seattle, Wash.) is applied to the catheter to control flow (arrow).

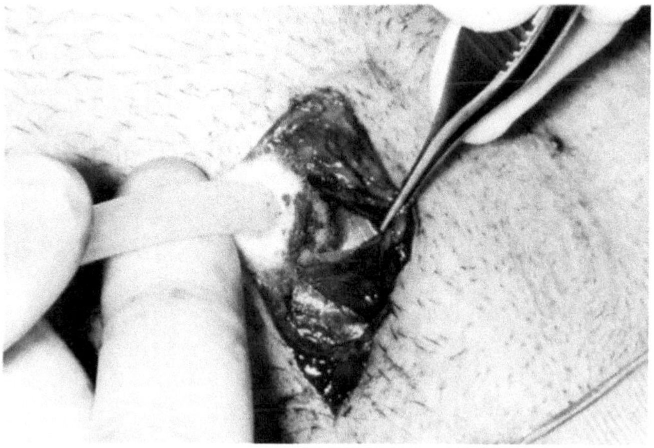

FIGURE 6C

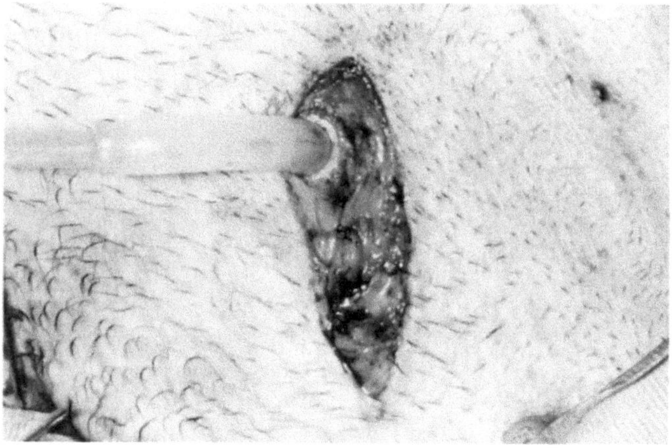

FIGURE 6D

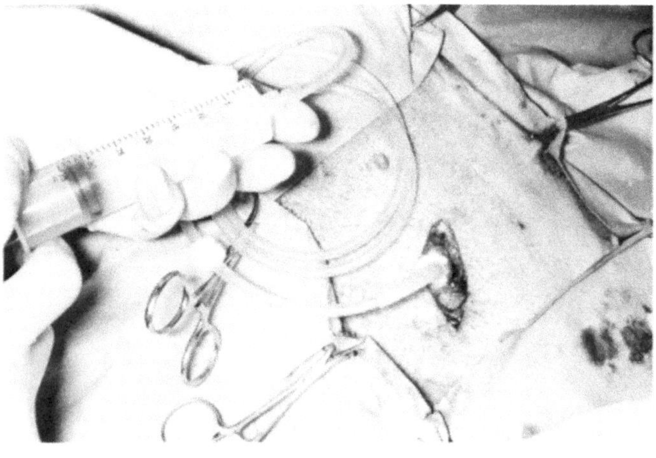

FIGURE 6E

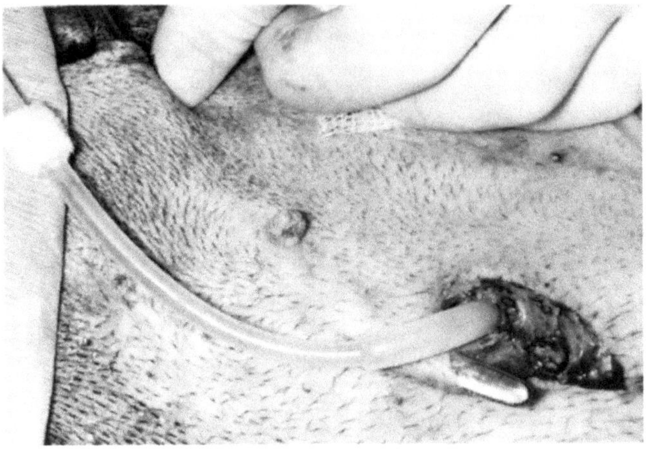

FIGURE 6F

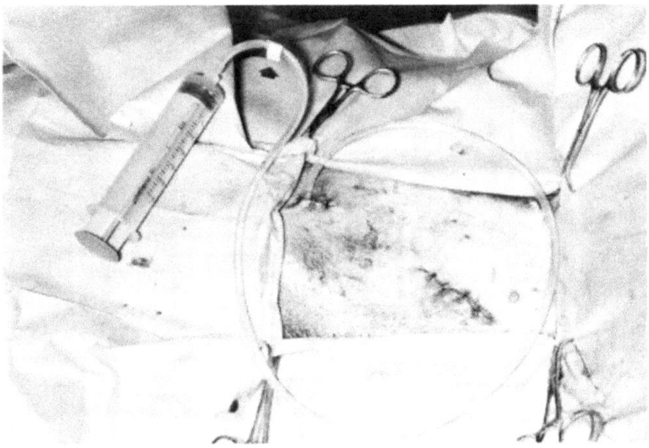

FIGURE 6G

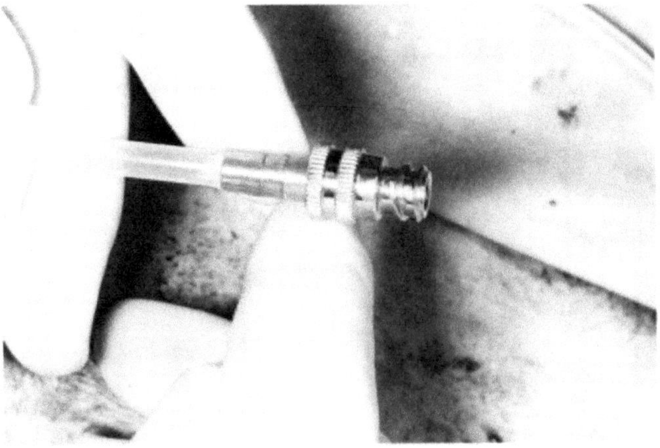

FIGURE 7. Titanium connector fitted into distal end of catheter.

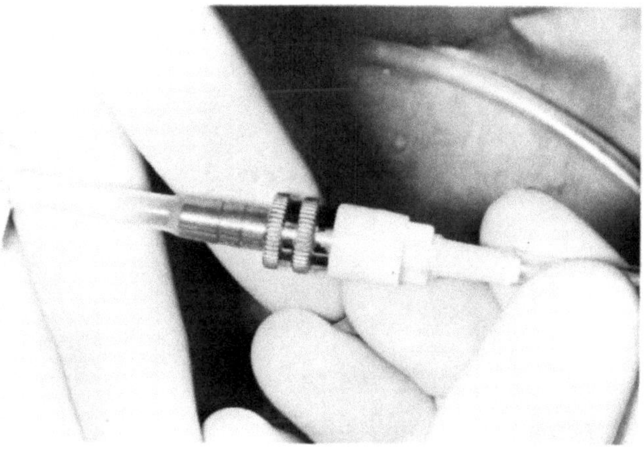

FIGURE 8. Attachment of Transfer-set tubing (Luer-lock end) to titanium connector.

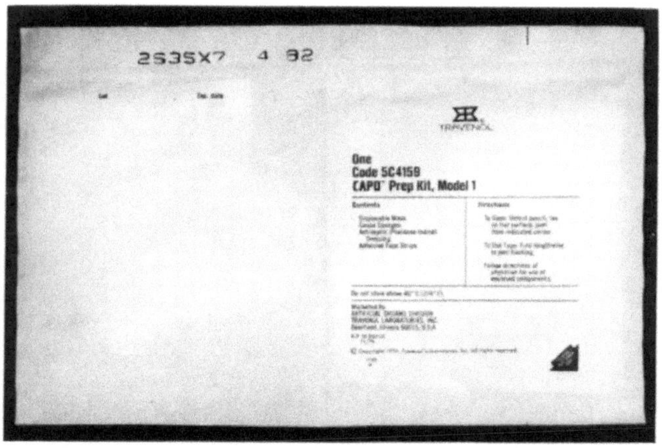

A

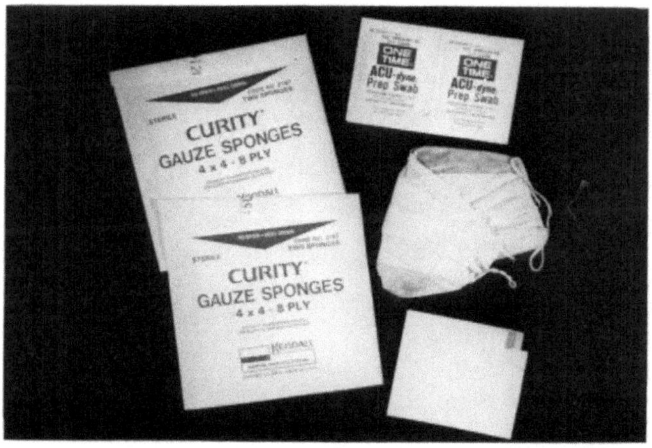

B

FIGURE 9. CAPD Prep Kit (A and B).

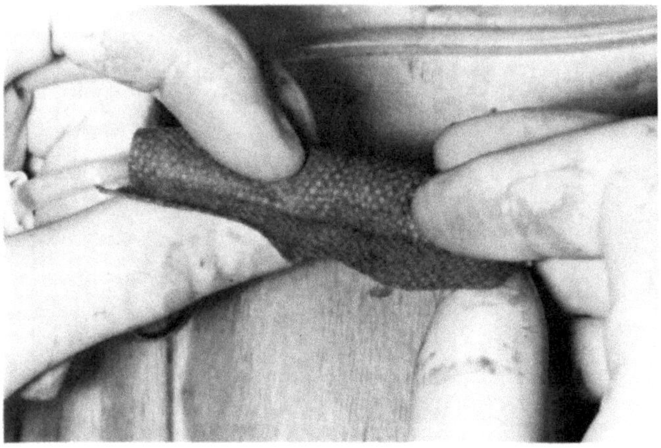

FIGURE 10. Povidone-iodine dressing from CAPD® Prep Kit applied to titanium-tubing connection.

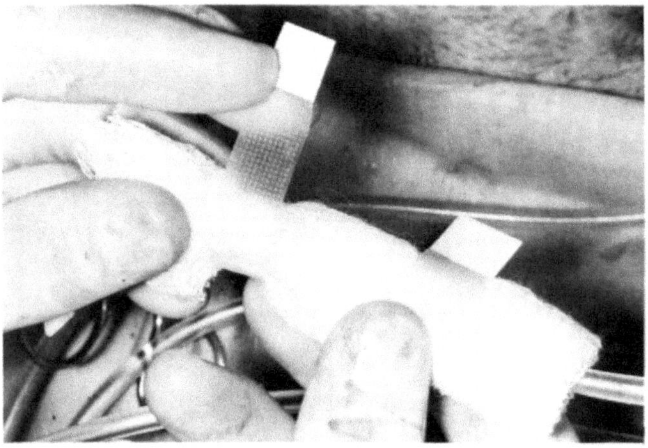

FIGURE 11. Sterile gauze and adhesive tape strips from Prep Kit applied to previous iodine dressing.

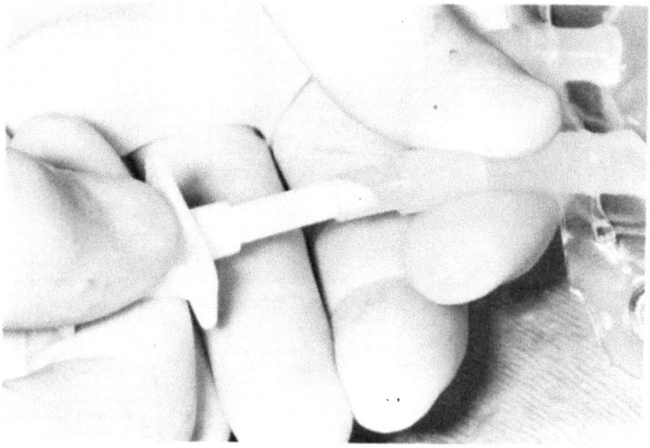

A

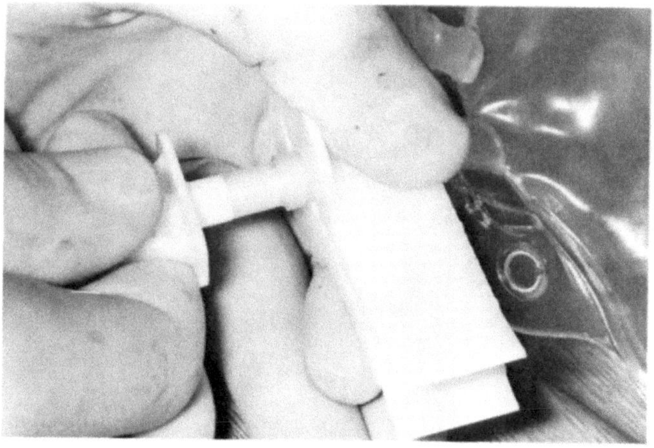

B

FIGURE 12. (A) Attachment of Transfer-Set tubing (spike end) to dialysate bag, facilitated in lower picture (B) by placement of a port clamp around dialysate inflow/outflow port (Travenol Laboratories, Inc.).

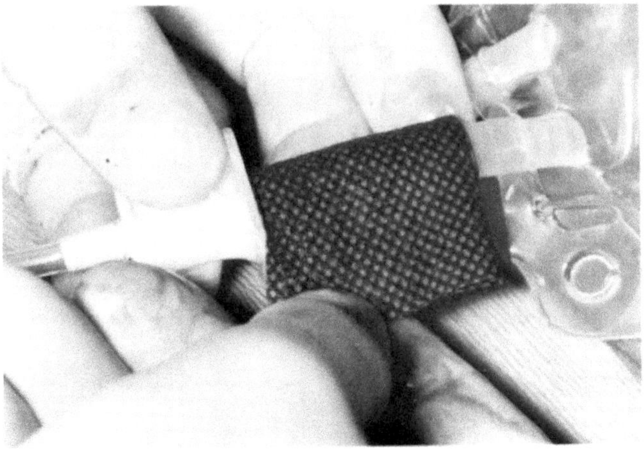

FIGURE 13. Povidone-iodine dressing from Prep Kit wrapped around connection site of spike and dialysate port of bag.

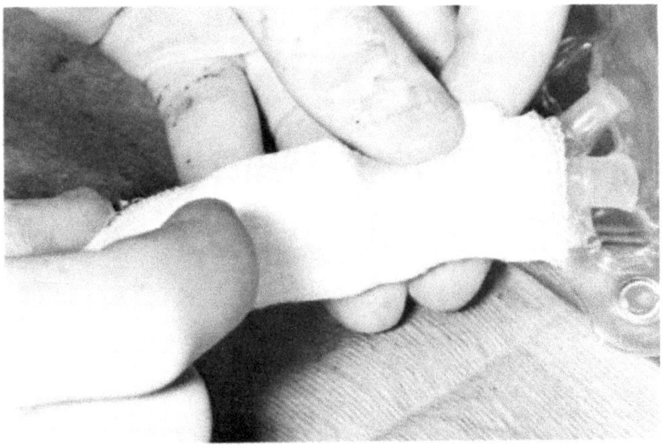

FIGURE 14. Sterile gauze wrap of iodine dressing.

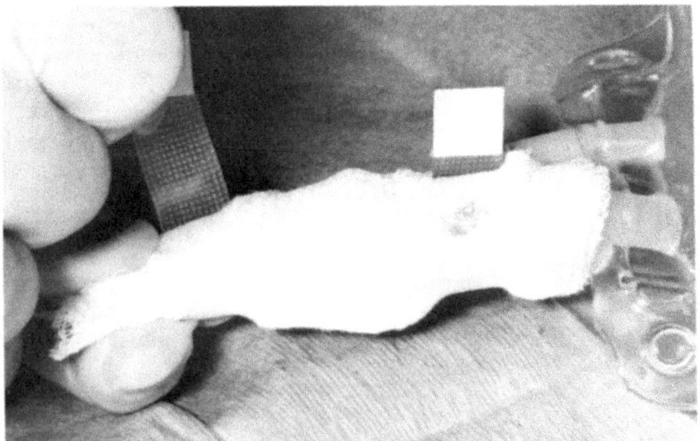

FIGURE 15. Adhesive tape application to gauze wrap.

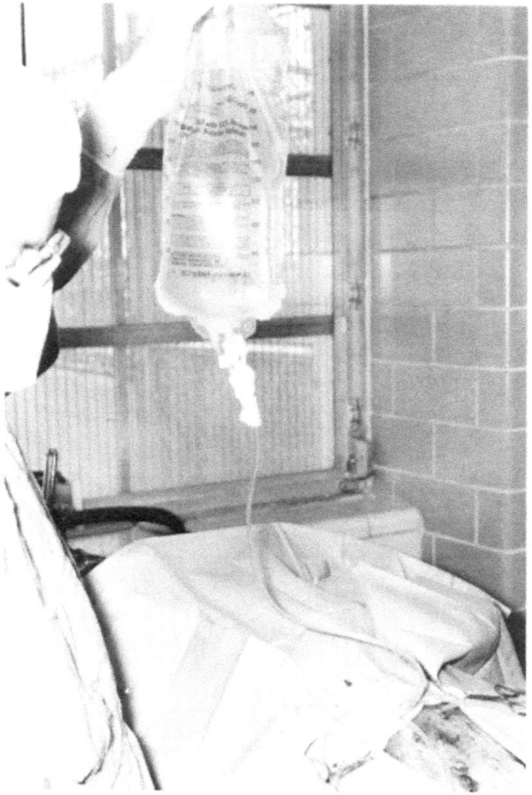

FIGURE 16. Instillation of dialysate, to which 250 mg of cephalothin and 1000 units per 2ℓ bag of heparin has been added, to patient immediately postsurgical implantation of catheter.

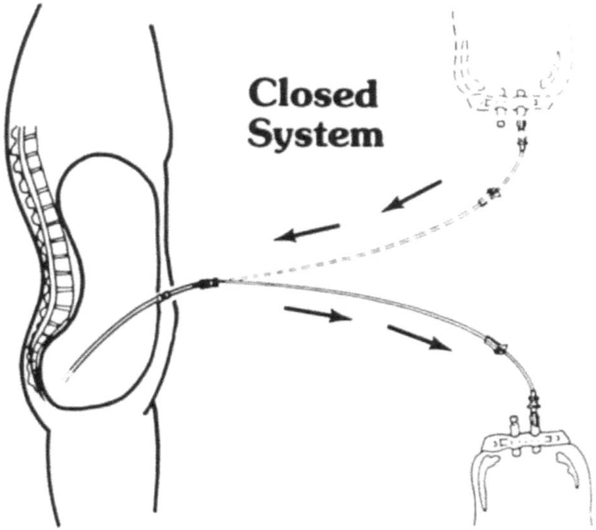

FIGURE 17. Closed system of CAPD in man utilizing dialysate in bags and tubing, which is changed infrequently, connected to a Tenckhoff catheter. Following dialysis solution inflow, bag and tubing (clamped) are rolled up and stored against the abdomen under a wrap or in a pouch suspended from the waist. At time of dialysate outflow, bag and tubing are unrolled, clamp on tubing is released, and bag is lowered to the floor for gravity drainage. (Courtesy of Travenol Laboratories.)

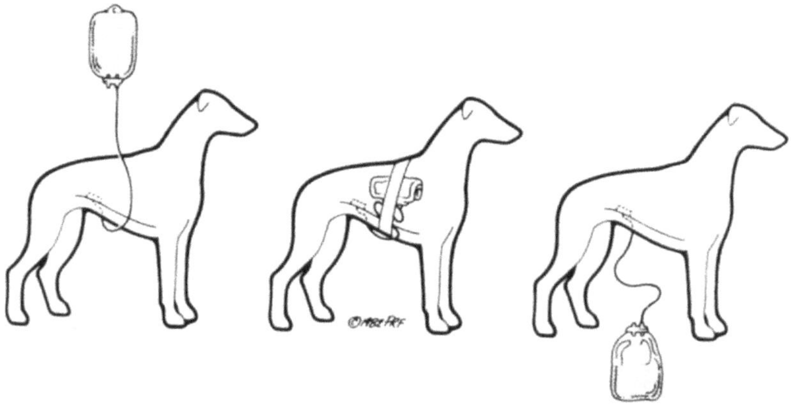

FIGURE 18. Diagrams (left to right) illustrating a complete CAPD cycle of dialysate inflow, dwell, and outflow, in the dog.

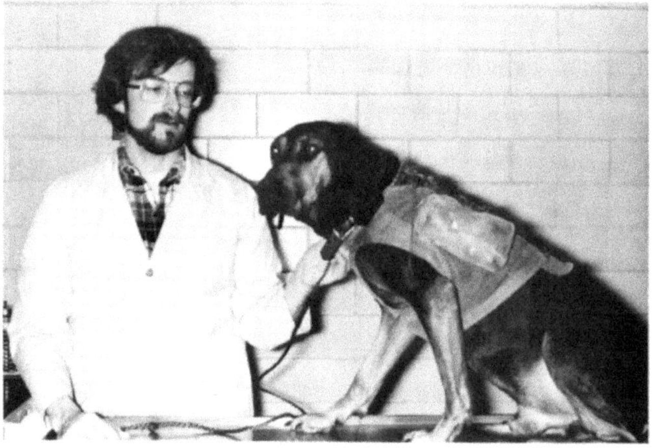

FIGURE 19. Patient at end of CAPD cycle, following dialysate dwell, being prepared for dialysate outflow. The harness system shown greatly facilitates protection and storage of dialysate bag and tubing.

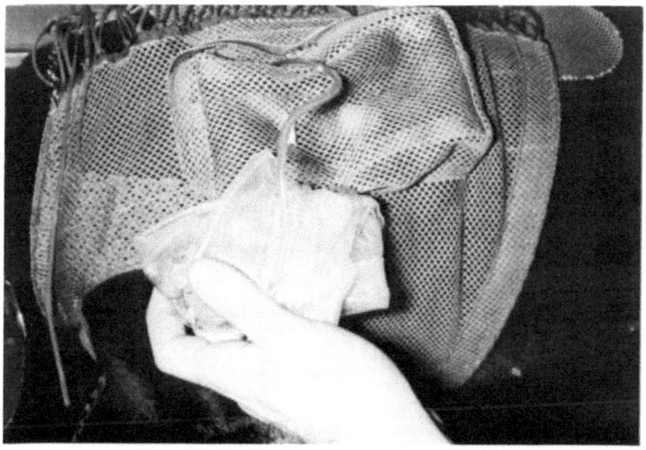

A

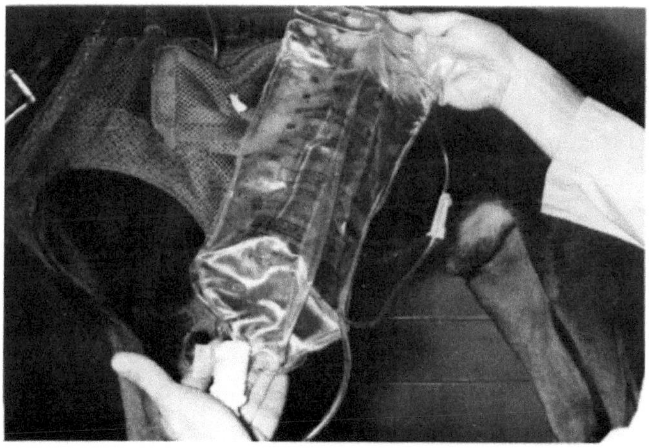

B

FIGURE 20. Storage pocket of harness is unzipped (A) and rolled-up bag is removed and straightened (B).

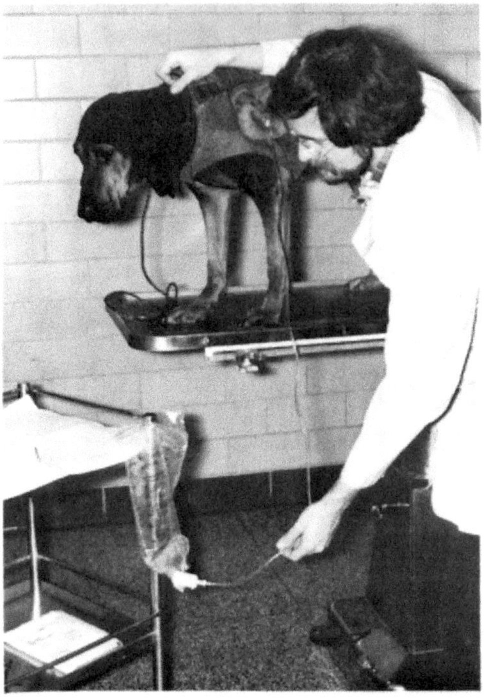

FIGURE 21. Dialysate outflow initiated by gravity drainage following release of Beta-cap clamp on catheter and roller clamp on tubing.

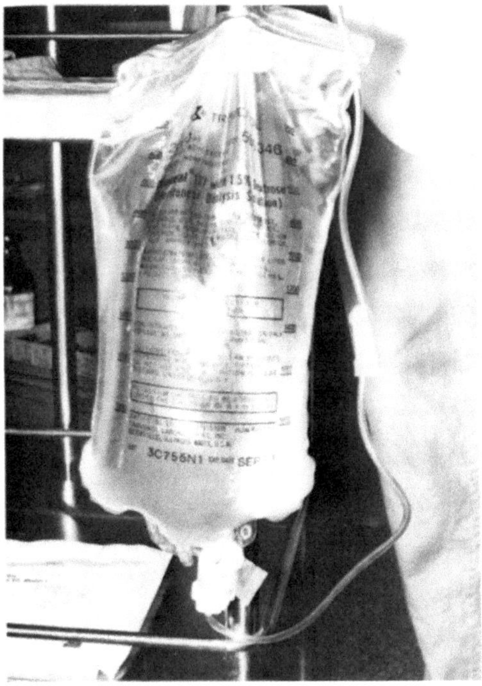

FIGURE 22. Filled dialysate bag following patient drainage.

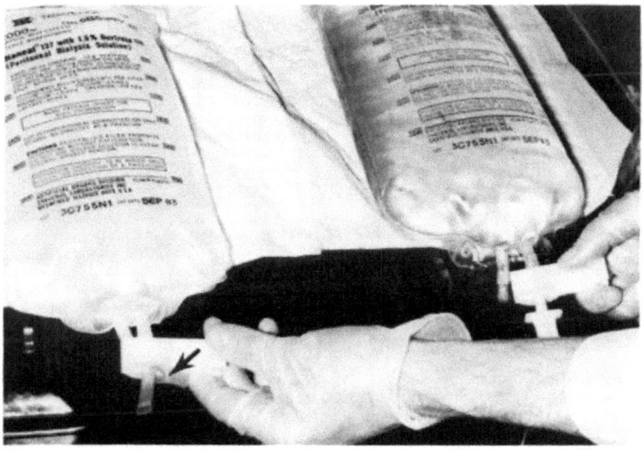

FIGURE 23. Used dialysate bag is placed on cart to right with new bag situated to its immediate left. Tape, gauze, and iodine dressing are removed from spike connection of used bag, but protective sleeve over dialysate inlet/outlet port of new bag is left in place (arrow). Port clamps are applied to inlet/outlet arms of both bags.

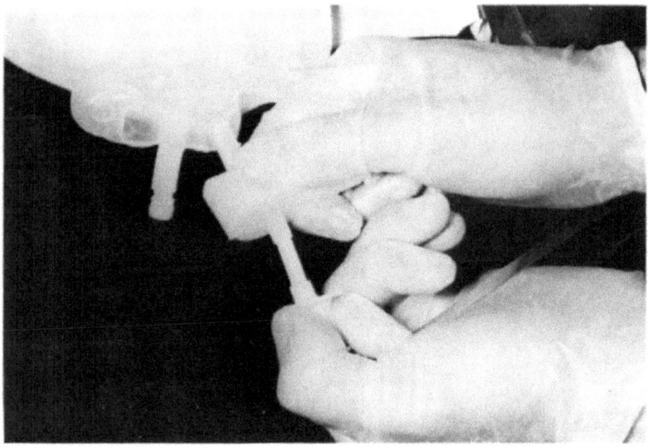

A

FIGURE 24. Exchange of Transfer-Set tubing from old bag (A) to new bag (B) is swiftly accomplished. Immediately after spike is removed from used bag, protective sleeve is pulled from dialysate port of fresh bag and spike is introduced.

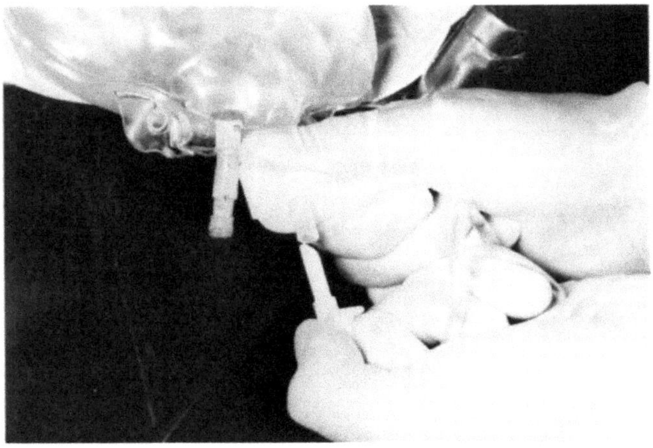

FIGURE 24B

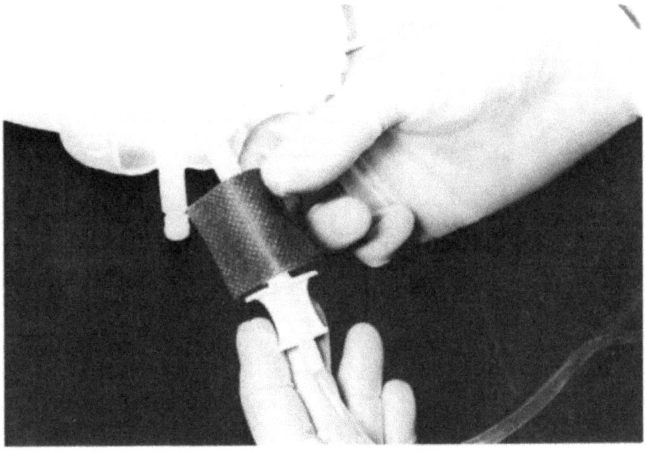

A

FIGURE 25. New spike connection is again wrapped with povidone-iodine dressing from a CAPD® Kit; followed (B) a gauze; and (C) adhesive tape.

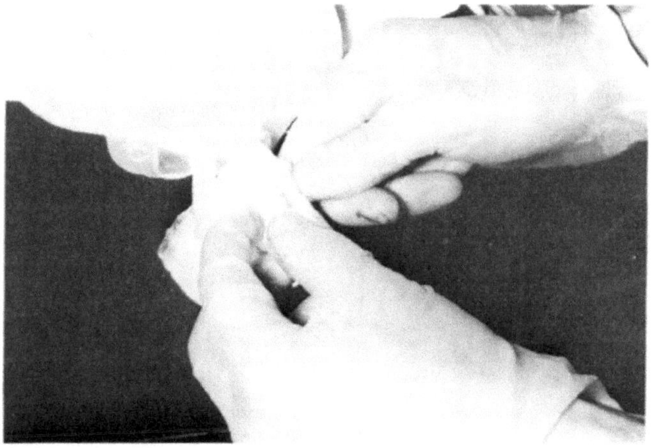

FIGURE 25B

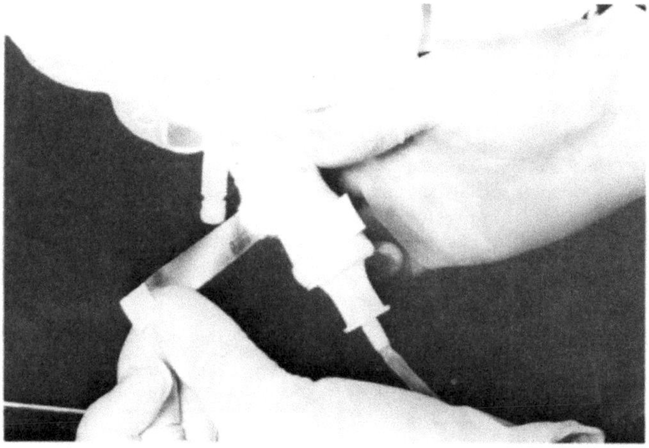

FIGURE 25C

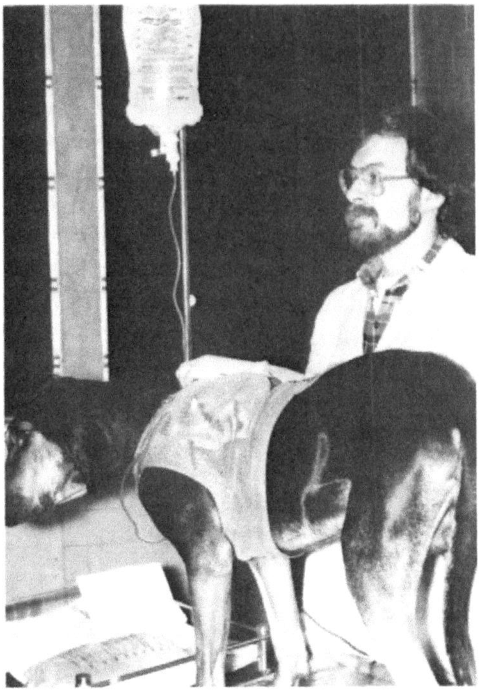

FIGURE 26. Initiation of the next CAPD exchange cycle with instillation of dialysate to the patient from previously prepared fresh bag.

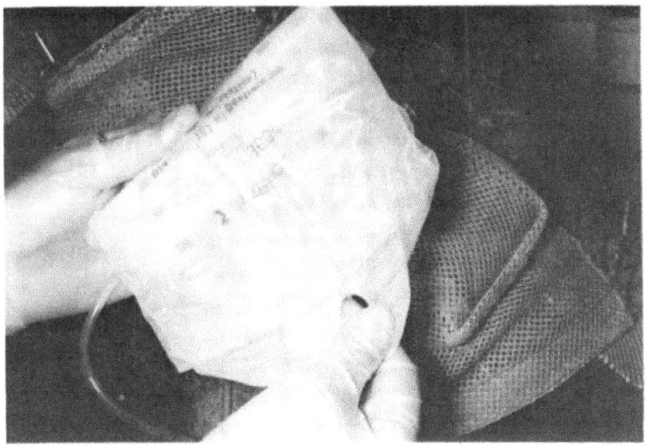

FIGURE 27. Bag rolled up after evacuation of solution in preparation for placement inside the pocket of the harness.

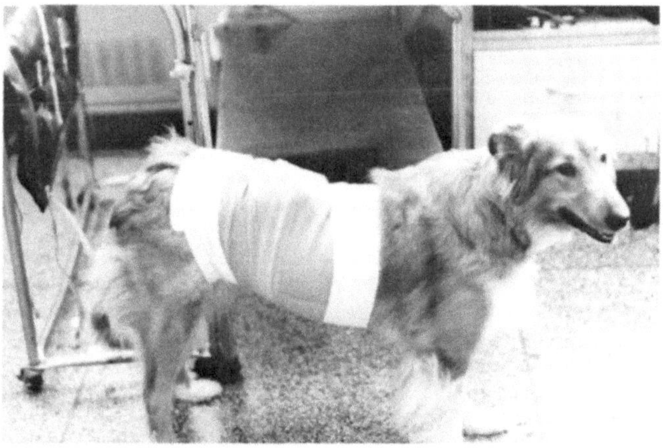

FIGURE 28. Collie research dog on CAPD with empty bag placed against the back under a protective elastic bandage. Ambulatory nature of procedure during the dialysate dwell period is revealed.

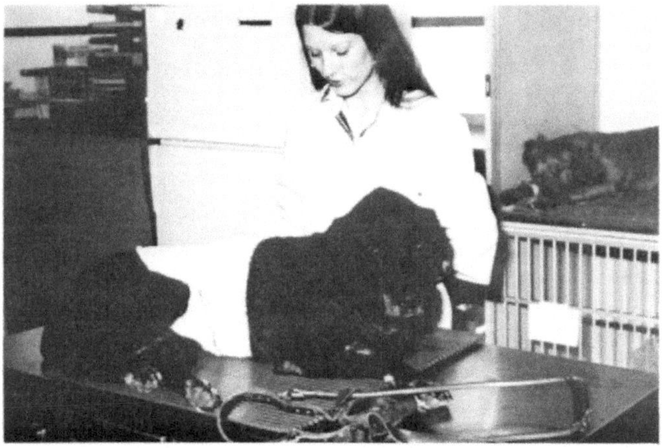

FIGURE 29. Clinical patient, a Seeing Eye dog, on CAPD at time of exchange.

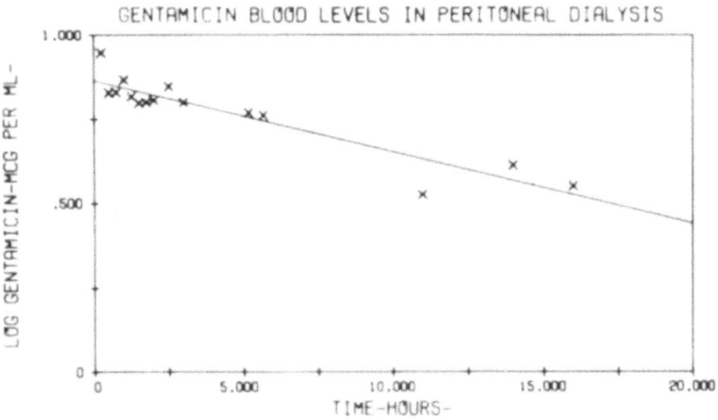

FIGURE 30. Least squares linear regression plot of an initial i.m. injection of gentamicin in the anephric dog.

REFERENCES

1. Osborne, C. A., Low, D. G., and Finco, D. R., Reversible versus irreversible renal disease in the dog, *J. Am. Vet. Med. Assoc.*, 155, 2062, 1969.
2. Popovich, R. P., Moncrief, J. W., Decherd, J. B., Bomar, J. B., and Pyle, W. K., The definition of a novel portable/wearable equilibrium dialysis technique, *Trans. Am. Soc. Artif. Intern. Organs*, 5 (Abstr.), 64, 1976.
3. Blumenkrantz, M. J., Maintenance peritoneal dialysis as an alternative for the patient with end-stage renal failure, *Clin. Dig.*, 6, 1, 1977.
4. Moncrief, J. W., Nolph, K. D., Rubin, J., and Popovich, R. P., Additional experience with continuous ambulatory peritoneal dialysis (CAPD), *Trans. Am. Soc. Artif. Intern. Organs*, 24, 476, 1978.
5. Nolph, K. D., Popovich, R. P., and Moncrief, J. W., Theoretical and practical implications of continuous ambulatory peritoneal dialysis, *Nephron*, 21, 117, 1978.
6. Oreopoulos, D. G., Chronic peritoneal dialysis, *Clin. Nephrol.*, 9, 165, 1978.
7. Orepoulos, D. G., Renewed interest in chronic peritoneal dialysis, *Kidney Int.*, 13, S117, 1978.
8. Popovich, R. P., Moncrief, J. W., Nolph, K. D., Ghods, A. J., Twardowski, J., and Pyle, W. K., Continuous ambulatory peritoneal dialysis, *Ann. Intern. Med.*, 88, 449, 1978.
9. Robson, M. D. and Oreopoulos, D. G., Continuous ambulatory peritoneal dialysis, a revolution in the treatment of chronic renal failure, *Dial. Transplant.*, 7, 999, 1978.
10. Zappacosta, A. R., Peritoneal dialysis for chronic renal replacement therapy, *Today's Clinician*, 2, 39, 1978.
11. Moncrief, J. W., Continuous ambulatory peritoneal dialysis, *Dial. Transplant.*, 8, 1077, 1979.
12. Sorrels, A. J., Continuous ambulatory peritoneal dialysis, *Am. J. Nursing*, 79, 1400, 1979.
13. Grollman, A., Turner, L. B., and McLean, J. A., Intermittent peritoneal lavage in nephrectomized dogs and its application to the human being, *Arch. Intern. Med.*, 87, 379, 1951.
14. Houch, R. C., Problems in maintenance of chronic bilaterally nephrectomized dog, *Am. J. Physiol.*, 176, 175, 1954.
15. Kirk, R. W., Peritoneal lavage in uremia in dogs, *J. Am. Vet. Med. Assoc.*, 131,101,1957.
16. Jackson, R. F., The use of peritoneal dialysis in the treatment of uremia in dogs, *Vet. Rec.*, 76, 1481, 1964.
17. Roesler, A. R., Carper, H. A., and Smith, L. K., Peritoneal dialysis in acute renal failure, Report of a case, *Am. J. Vet. Clin. Pathol.*, 1, 37, 1967.
18. Tabotabo, L. L., Escudero, S. H., III, and DeVera, L. T., A study on the efficacy of peritoneal dialysis in induced acute uremia in dogs, *Philipp. J. Vet. Med.*, 9, 1, 1970.
19. Parker, H. R., Gourley, I. M., and Bell, R. L., Current developments in peritoneal and hemodialysis, Gaines 22nd Vet. Symp., Stillwater, Okla., 1972, 3.
20. Gourley, I. M. and Parker, H. R., Peritoneal dialysis, in *Current Veterinary Therapy*, Vol. 6, Kirk, R. W., Ed., W. B. Saunders, Philadelphia, 1977, 1144.
21. Parker, H. R., Current status of peritoneal dialysis, in *Current Veterinary Therapy*, Vol. 7, Kirk, R. W., Ed., W. B. Saunders, Philadelphia, 1980, 1106.
22. Simmons, E. E., Lockard, B. S., Moncrief, J. W., Hiatt, M. P., McCollough, W. S., and Popovich, R. P., Experience with continuous ambulatory peritoneal dialysis and maintenance of a surgically anephric dog, *Southwest. Vet.*, 33, 129, 1980.
23. Thornhill, J. A., Hartman, J., Boon, G. D., Riviere, J. E., Jacobs, D., and Ash, S. R., Support of an anephric dog for 54 days with ambulatory peritoneal dialysis and a new peritoneal catheter design, *Am. J. Vet. Res.*, 45, 1156, 1984.
24. Thornhill, J. A., Peritoneal dialysis in the dog and cat: an update, *Compend. Cont. Educ. Pract. Vet.*, 3, 20, 1981.
25. Thornhill, J. A., Ash, S. R., Dhein, C. R., Polzin, D. J., and Osborne, C. A., Peritoneal dialysis with the Purdue column disc catheter, *Minn. Vet.*, 20, 27, 1980.
26. Thornhill, J. A., Continuous peritoneal dialysis support (CAPD) in a dog with oliguric actue renal failure induced by ethylene glycol intoxication, *J. Am. Vet. Med. Assoc.*, in review.
27. Brown, P. and Kolph, K. D., Chemical measurements of insulin concentrations in peritoneal dialysis solution, *Clin. Chim. Acta*, 76, 103, 1977.
28. Nolph, K. D, Ghods, A. J., Brown, P., VanStone, J. C., Miller, F. N., Wiegmann, D. L., and Harris, P. D., Peritoneal dialysis efficiency, *Dial. Transplant.*, 6, 52, 1977.
29. Boen, S. T., Haagsma-Schouten, W. A. G., and Birnie, R. J., Long-term peritoneal dialysis and a peritoneal dialysis index, *Dial. Transplant.*, 7, 377, 1978.
30. Nolph, K. D., Popovich, R. P., Ghods, A. J., and Twardowski, Z., Determinants of low clearances of small solutes during peritoneal dialysis, *Kidney Int.*, 13, 117, 1978.
31. Rubin, J., Nolph, K. A., Arfania, D., Brown, P., and Prowant, B., Follow-up of peritoneal clearances in patients undergoing continuous ambulatory peritoneal dialysis, *Kidney Int.*, 16, 619, 1979.

32. Gutman, R. A., Toward enhancement of peritoneal clearance, *Dial. Transplant.*, 8, 1072, 1979.
33. Rubin, J., Nolph, K., Arfania, D., Brown, P., Moore, H., and Rust, P., Influence of patient characteristics on peritoneal clearances, *Nephron*, 27, 118, 1981.
34. Gokal, R., Fryer, R., Prescott, R., Petty, R., Ward, M. D., and Kerr, D. N. S., Plasma and dialysate correlations for small solutes after overnight dwell in CAPD patients, *Dial. Transplant.*, 11, 406, 1982.
35. Barbour, G. L., The kinetics of peritoneal dialysis: a review, *Dial. Transplant.*, 8, 1055, 1979.
36. Maher, J. F., Pharmacology of peritoneal dialysis and permeability of the membrane, *Dial. Transplant.*, 9, 197, 1980.
37. Nolph, K. D., Miller, F. N., Pyle, W. K., Popovich, R. P., and Sorbin, M. I., An hypothesis to explain the ultrafiltration characteristics of peritoneal dialysis, *Kidney Int.*, 20, 543, 1981.
38. Nolph, K. D., Hano, J. E., and Teschan, P. E., Peritoneal sodium transport during hypertonic peritoneal dialysis, *Ann. Intern. Med.*, 70, 931, 1969.
39. Gutman, R. A., Nixon, W. P., McRae, R. G., et al., Effect of intraperitoneal and intravenous vasoactive amines on peritoneal dialysis: study in anephric dogs, *Trans. Am. Soc. Artif. Intern. Organs*, 22, 570, 1976.
40. Miller, F. N., Wiegman, D. L., Joshua, I. G., Nolph, K. D., and Rubin, J., Effects of vasodilators and peritoneal dialysis solution on microcirculation of the rat cecum, *Proc. Soc. Exp. Biol. Med.*, 161, 605, 1979.
41. Nolph, K. D., Ghods, A. J., Brown, P. A., and Twardowski, Z. J., Effects of intraperitoneal nitroprusside on peritoneal clearances in man with variations of dose, frequency of administration and dwell time, *Nephron*, 24 114, 1979.
42. Maher, J. F., Cassetta, M., Shea, C., and Hohnadel, D. C., Peritoneal dialysis in rabbits, *Nephron*, 20, 18, 1978.
43. Goldschmidt, Z. H., Pote, H. H., Katz, M. A., and Shear, L., Effect of dialysate volume on peritoneal dialysis kinetics, *Kidney Int.*, 5, 240, 1974.
44. Robson, M., Oreopoulos, D. G., Izatt, S., Ogilve, R., Rapoport, A., and DeVeber, G. A., Influence of exchange volume and dialysate flow rate on solute clearance in peritoneal dialysis, *Kidney Int.*, 14, 486, 1978.
45. Gross, M. and McDonald, H. P., Jr., Effect of dialysate temperature and flow rate on peritoneal clearance, *JAMA*, 202, 363, 1967.
46. Vaamonde, C. A. and Perez, G. O., Perionteal dialysis today, *Kidney*, 10, 31, 1977.
47. Rubin, J., Nolph, K. D., Popovich, R. P., Moncrief, J. W., and Prowant, B., Drainage volumes during continuous ambulatory peritoneal dialysis, *Am. Soc. Artif. Intern. Organs J.*, 2, 54, 1979.
48. Nolph, K. D., Twardowski, Z. J., Popovich, R. P., and Rubin, J., Equilibration of peritoneal dialysis solutions during long-dwell exchanges, *J. Lab. Clin. Med.*, 93, 124, 1979.
49. Twardowski, Z. and Janicka, L., Three exchanges with a 2.5-liter volume for continuous ambulatory peritoneal dialysis, *Kidney Int.*, 20, 281, 1981.
50. Twardowski, Z., Majdan, K. M., Janicka, L., Bochenska-Nowacha, E., Sokolowska, G., Gutka, A., and Zbikowska, A., Kinetics of continuous ambulatory peritoneal dialysis (CAPD) with four exchange per day, *Clin. Nephrol.*, 15, 119, 1981.
51. Waugh, W. H., Retrospective appraisal of daily equilibrium peritoneal dialysis and a presage for maintenance dialysis, *Dial. Transplant.*, 11, 712, 1982.
52. Vidt, D. G., Recommendations on choice of peritoneal dialysis solutions, *Ann. Intern. Med.*, 78, 144, 1973.
53. Rubin, J., Nolph, K. D., Arfaania, D., Wiegman, D. L., Miller, F. N., and Harris, P. D., Comparison of the effects of lactate and acetate on clinical peritoneal clearances, *Clin. Nephrol.*, 12, 145, 1979.
54. Biasioli, S., Feriani, M., Chiaramonte, S., Fabris, A., Pisani, E., Ronco, C., and LaGreca, G., Buffers for peritoneal dialysis — sodium acetate. I, *Contemp. Dial.*, 3, 16, 1983.
55. Biasioli, S., Feriani, M., Chiaramonte, S., Fabris, A., Pisani, E., Ronco, C., and LaGreca, G., Buffers for peritoneal dialysis — sodium acetate. II, *Contemp. Dial.*, 3, 40, 1983.
56. Ash, S. R., Barile, R. G., Wilcox, P. G., Wright, D.L., Thornhill, J. A., Dhein, C. R., Kessler, D. P., and Wang, N-H., L., The sorbent suspension reciprocating dialyzer: a device with minimal sorbent saturation, *Am. Soc. Artif. Intern. Organs J.*, 1, 28, 1981.
57. Rife, M., Nolph, K. D., Oreopoulos, D. G., Twardowski, Z. J., Blumenkrantz, M., Maher, J. F., and Diaz-Buxo, J. A., Peritoneal dialysis — what's ahead?, *Dial. Transplant.*, 11, 663, 1982.
58. Ku, K., Anderson, R., and Schoenfeld, P., Kinetic modeling of urea in peritoneal dialysis, *Dial. Transplant.*, 12, 374, 1983.
59. Bovee, K. C. and Joyce, T., Clinical evaluation of glomerular function: 24-hour creatinine clearance in dogs, *J. Am. Vet. Med. Assoc.*, 174, 488, 1979.

60. Bergström, J. and Furst, P., Uremic middle molecules, *Clin. Nephrol.*, 5, 143, 1976.
61. Gotch, F. A., A quantitative evaluation of small and middle molecule toxicity in therapy of uremia, *Dial. Transplant.*, 9, 183, 1980.
62. Kjellstrand, C. M., The clinical significance of middle molecules, *Dial. Transplant.*, 8, 860, 1979.
63. Slatopolsky, E., Caglar, S., Pennell, J. P., Taggart, D. D., Catebury, J. M., Reiss, E., and Bricker, N. S., On the pathogenesis of hyperparathyroidism in chronic experimental renal insufficiency in the dog, *J. Clin. Invest.*, 50, 492, 1971.
64. Massry, S. G. and Goldstein, D. A., Role of parathyroid hormone in uremic toxicity, *Kidney Int.*, 13, S39, 1978.
65. Brown, E. A., Kliger, A. S., and Finkelstein, F. O., Peritoneal dialysis clearances. A practical approach to the measurement of small- and middle-molecular clearances, *Nephron*, 21, 310, 1978.
66. Nolph, K. D., Short dialysis, middle molecules, and uremia, *Ann. Intern. Med.*, 86, 93, 1977.
67. Sargent, J. A., Kinetic modeling in guidance of dialysis therapy, *Dial. Transplant.*, 8, 1101, 1979.
68. Tenckhoff, H. and Schecter, H., A bacteriologically safe peritoneal access device, *Trans. Am. Soc. Artif. Intern. Organs*, 14, 181, 1968.
69. Brewer, T. E., Caldwell, F. T., Patterson, R. M., and Flanigan, W. J., Indwelling peritoneal (Tenckhoff) dialysis catheter. Experience with 24 patients, *JAMA*, 219, 1011, 1972.
70. Tenckhoff, H., Peritoneal dialysis, catheter implantation, *Dial. Transplant.*, 1, 18, 1972.
71. Devine, H., Oreopoulos, D. G., Izatt, S., Mathews, R., and DeVeber, G. A., The permanent Tenckhoff catheter for chronic peritoneal dialysis, *Can. Med. Assoc. J.*, 113, 219, 1975.
72. Tenckhoff, H., *Chronic Peritoneal Dialysis. A Manual for Patients, Dialysis Personnel and Physicians*, University of Washington School of Medicine, Seattle, 1974.
73. Ash, S. R., Johnson, H., Hartman, J., Granger, J., Koszuta, J., Sell, L., and Thornhill, J. A., The column disc peritoneal catheter. A peritoneal access device with improved drainage, *Am. Soc. Artif. Intern. Organs*, 3, 109, 1980.
74. Thronhill, J. A., Dhein, C. R., Johnson, H., and Ash, S. R., Drainage characteristics of the column disc catheter: a new chronic peritoneal access catheter, *Proc. Dial. Transplant Forum*, 10, 119, 1980.
75. Thornhill, J. A. and Riviere, J. E., Peritonitis associated with peritoneal dialysis: diagnosis and treatment, *J. Am. Vet. Med. Assoc.*, 182, 721, 1983.
76. Ash, S. R. and Carr, D. J., Peritoneal access devices: biocompatibility and hydraulic function, in *Clinical Dialysis*, Nissenson, A., Fine, R., and Gentile, D., Eds., Appleton-Century-Crofts, East Norwalk, Conn., 1984, 221.
77. Thornhill, J. A., Knab, W. R., and Johnson, R. C., Continuous ambulatory peritoneal dialysis (CAPD) support of an anephric dog, *Abstr. 63rd Annu. Conf. Res. Workers Anim. Dis.*, Chicago, November 1982, 35.
78. Tenckhoff, H., Peritoneal dialysis today: a new look, *Nephron*, 12, 420, 1974.
79. Giordano, C., DeSanto, N. G., Papa, A., Capodicasa, G., Cirillo, D., Capasso, G., Quarto, E., Gallo, B., and Sanna, G., Short daily peritoneal dialysis, *Kidney Int.*, 7, S425, 1975.
80. Randerson, D. H. and Farrell, P. C., Clinical assessment of CAPD, *Dial. Transplant.*, 10, 389. 1981.
81. Oreopoulos, D. G., Khanna, R., McCready, W., Katirtzoglous, A., and Vas, S., Continuous ambulatory peritoneal dialysis in Canada, *Dial. Transplant.*, 9, 224, 1980.
82. Abramowicz, M., Ed., Home peritoneal dialysis for end-stage renal disease, *Med. Lett. Drugs Ther.* 21, 69, 1979.
83. Osborne, C. A., Low, D. G., and Finco, D. R., Reversible versus irreversible renal disease in the dog, *J. Am. Vet. Med. Assoc.*, 155, 2062, 1969.
84. Bovee, K. C., The uremic syndrome: patient evaluation and treatment, *Compend. Cont. Educ. Pract. Vet.*, 1, 279, 1979.
85. Lynn, K. L., Neale, T. J., Bailey, R. R., and Little, P. J., Oliguria and its sequelae, *N. Z. Med. J.*, 86, 563, 1977.
86. Cohen, B. D., Uremia: pillar of salt or column of nitrogen?, *Dial. Transplant.*, 9, 535, 1980.
87. Palmer, R. A., Newell, J. E., Gray, E. G., and Ouinton, W. E., Treatment of chronic renal failure by prolonged peritoneal dialysis, *N. Engl. J. Med.*, 274, 248, 1966.
88. Bartecchi, C. E., When should peritoneal dialysis be considered in the elderly patients?, *Geriatrics*, 30, 47, 1975.
89. Valk, T. W., Swartz, R. D., and Hsu, C. H., Peritoneal dialysis in acute renal failure: analysis of outcome and complications, *Dial. Transplant.*, 9, 48, 1980.
90. Hall, J. W., Johnson, W. J., Maher, T. F., and Hunt, J. C., Immediate and long-term prognosis in acute renal failure, *Ann. Intern. Med.*, 73, 515, 1970.
91. Giordano, C. and DeSanto, N. G., Peritoneal dialysis: Cinderella or princess?, *Int. J. Artif. Organs*, 1, 3, 1978.
92. Ribot, S., Jacobs, J. G., Frankel, H. J., and Bernstein, A., Complications of peritoneal dialysis, *Am. J. Med. Sci.*, 246, 505, 1966.

93. Miller, R. B. and Tassistro, C. R., Peritoneal dialysis, *N. Engl. J. Med.*, 281, 945, 1969.
94. Gault, M. H., Ferguson, E. L., and Corbin, R. P., Fluid and electrolyte complications of peritoneal dialysis. Choice of dialysis solutions, *Ann. Intern. Med.*, 75, 253, 1971.
95. Fenton, S., Wu, G., Cattran, D., Wadgymar, A., and Allen, A. F., Clinical aspects of peritonitis in patients on CAPD, *Peritoneal Dial. Bull.*, 1, 54, 1981.
96. William, C. C., Cattran, D., Fenton, S. S., Khanna, R., Manuel, A., Saiphuo, C., Oreopoulos, D. G., and Roscoe, J., Peritonitis in CAPD — three years' experience in Toronto, *Peritoneal Dial. Bull.*, 1, 57, 1981.
97. Black, H. R., Finkelstein, F. O., and Lee, R. V., The treatment of peritonitis in patients with chronic indwelling catheters, *Trans. Am. Soc. Artif. Intern. Organs*, 20, 115, 1974.
98. Rubin, J., Oreopoulos, D. G., Liu, T. T., Mathews, R., and DeVeber, G. A., Management of peritonitis and bowel perforation during chronic peritoneal dialysis, *Nephron*, 16, 220, 1976.
99. Golper, T. A., Bennett, W. M., and Jones, S. R., Peritonitis associated with chronic peritoneal dialysis. A diagnostic and therapeutic approach, *Dial. Transplant.*, 7, 1173, 1978.
100. Chan, L. K. and Oliver, D. O., Simple method for early detection of peritonitis in patients on continuous ambulatory peritoneal dialysis, *Lancet*, 2, 1336, 1979.
101. Vas, S. I., Microbiological aspects of peritonitis, *Peritoneal Dial. Bull.*, 1, 511, 1981.
102. Smithivas, T., Hyams, P. J., Matalon, R., Simberkoff, M. S., and Rahal, J. J., Jr., The use of gentamicin in peritoneal dialysis. Pharmacologic results, *J. Infect. Dis.*, 124, 577, 1971.
103. Golper, T. A., Drugs in peritoneal dialysis, *Dial. Transplant.*, 8, 41, 1979.
104. Szwed, J. J., Luft, F. C., Black, H. R., Elliott, R. A., and Kleit, S. A., Comparison of the distribution of tobramycin and gentamicin in body fluids of dogs, *Antimicrob. Agents Chemother.*, 5, 444, 1974.
105. Baggot, J. D., *Principles of Drug Disposition in Domestic Animals: The Basis of Veterinary Clinical Pharmacology*, W. B. Saunders, Philadelphia, 1977, 152.
106. Somani, P., Shapiro, R. S., Stokard, H., and Higgins, J. t., Unidirectional absorption of gentamicin from the peritoneum during continuous ambulatory peritoneal dialysis, *Clin. Pharmacol. Ther.*, 32, 113, 1982.
107. Arfania, D., Everett, E. D., Nolph, K. D., and Rubin, J., Uncommon causes of peritonitis in patients undergoing peritoneal dialysis, *Arch. Intern. Med.*, 141, 61, 1981.
108. Stephen, R. L., Kablitz, C., Kitahara, M., Nelson, J. A., Duffy, D. P., and Kolff, W. J., Peritoneal dialysis: peritonitis: saline-iodine flush, *Dial. Transplant.*, 8, 584, 1979.
109. Everett, E. D., Tarka, E., Knolph, K. D., and Morris, J. S., Saline-iodine flush in peritoneal dialysis: in vitro effects, *Dial. Transplant.*, 12, 233, 1983.
110. Ash, S. R., Wimberly, A. L., and Mertz, S. L., Peritoneal dialysis for acute and chronic renal failure: an update, *Hosp. Pract.*, 18, 179, 1983.
111. Cattran, D. C., Fenton, S. S. A., Wilson, D. R., and Steiner, G., Defective triglyceride removal in lipemia associated with peritoneal dialysis and hemodialysis, *Ann. Intern. Med.*, 85, 29, 1976.
112. Roncari, D. A. K., Breckenridge, W. C., Khanna, R., and Oreopoulos, D. G., Rise in high-density lipoprotein-cholesterol in some patients treated with CAPD, *Peritoneal Dial. Bull.*, 1, 136, 1981.
113. Grodstein, G. P., Blumenkrantz, M. J., Kopple, J. D., Moran, J. K., and Coburn, J. W., Glucose absorption during continuous ambulatory peritoneal dialysis, *Kidney Int.*, 19, 564, 1981.
114. Hanna, A. K., Control of blood glucose in diabetics on CAPD — the importance of blood glucose control in the prevention of diabetic complications, *Peritoneal Dial. Bull.*, 2, 522, 1982.
115. Khanna, R., Breckenridge, C., Roncari, D., Digenis, G., and Oreopoulos, D. G., Lipid abnormalities in patients undergoing continuous ambulatory peritoneal dialysis, *Peritoneal Dial. Bull.*, 3, 513, 1983.
116. Wu, G., Osmotic agents for peritoneal dialysis solution, *Peritoneal Dial. Bull.*, 2, 151, 1982.
117. Vas, S. I., Suwe, A., and Weatherhead, J., Natural defense mechanisms of the peritoneum: the effect of peritoneal dialysis fluid on polymorphonuclear cells, in *Peritoneal Dialysis*, Atkins, R. C., Thomson, N. M., and Farrell, P. C., Eds., Churchill Livingstone, London, 1981, 41.
118. Rasib, R. M., Moros, J. G., Kramer, M. S., and Rosenbaum, J. L., Hyperosmolar coma complicating peritoneal dialysis with sorbitol dialysate, *Am. Intern. Med.*, 73, 993, 1970.
119. Bischel, M. D. and Barbour, B., Peritoneal dialysis with sorbitol versus dextrose dialysate, *Nephron*, 12, 449, 1974.
120. Rubin, J., Nolph, K. D., and McGary, T. J., Osmotic ultrafiltration with dextran sodium sulfate; potential for use in peritoneal dialysis, *J. Dial.*, 3, 251, 1979.
121. McGary, T. J., Nolph, K. D., and Kartinos, J. N., Polyanions as osmotic agents in a simulated in vitro model of peritoneal dialysis, *Trans. Am. Soc. Artif. Intern. Organs*, 27, 314, 1981.
122. Bazzato, G., Coli, U., Landini, S., Francasso, A., Morachiello, P., Righetto, F., and Scanferla, F., Xylitol and low dosages of insulin: new perspectives for diabetic uremic patients on CAPD, *Peritoneal Dial. Bull.*, 2, 161, 1982.

123. Thomas, D. W., Edwards, J. B., Gilligan, J. E., Lawrence, J. R., and Edwards, R. G., Complications following intravenous administration of solutions containing xylitol, *Med. J. Aust.*, 1, 1238, 1972.
124. Rasib, M. R., Kramer, M. S., Manchande, R., Lazaro, N., and Rosenbaum, J. L., Peritoneal dialysis with fructose dialysate, *Ann. Intern. Med.*, 79, 511, 1973.
125. MacDonald, I., Effects of dietary carbohydrates on serum lipids, *Prog. Biochem. Pharmacol.*, 10, 216, 1973.
126. Williams, P. F., Marliss, E. B., Anderson, G. H., Oren, A., Stein, A. N., Khanna, R., Pettit, J., Brandes, L., Rodella, H., Mupas, L., Dombros, N., and Oreopoulos, D. G., Amino acid absorption following intraperitoneal administration in CAPD patients, *Peritoneal Dial. Bull.*, 2, 124, 1982.
127. Dombros, N., Oren, A., Marliss, E. B., Anderson, G. H., Stein, A. N., Khanna, R. Petitt, J., Brandes, L., Rodella, H., Leibel, B. S., and Oreopoulos, D. G., Plasma amino acid profiles and amino acid losses in patients undergoing CAPD, *Peritoneal Dial. Bull.*, 2, 27, 1982.
128. Kobayashi, K., Manji, T., Hiramatsu, S., Maeda, K., and Uemura, J., Nitrogen metabolism in patients on peritoneal dialysis, *Contrib. Nephrol.*, 17, 93, 1979.
129. Oreopoulos, D. G., Marliss, E., Anderson, G. H., Oren, A., Dombros, N., Williams, P., Khanna, R., Rodella, H., and Brandes, L., Nutritional aspects of CAPD and the potential use of amino acid containing dialysis solutions, *Peritoneal Dial. Bull.*, 3, 510, 1983.
130. Dhein, C. D., Thornhill, J. A., Chiosi, C., and Ash, S. R., Protein and bacterial rejecting filters for peritoneal dialysis, *Abstr. 13th Annu. Meet.*, American Society of Nephrology, Washington, D.C., November 1980, 38A.
131. Zinsser, H. H. and Cohen, A. B., The capacity of resins to simulate chemical processing by the kidney, *Trans. Am. Soc. Artif. Intern. Organs*, 1, 19, 1955.
132. Zatzidis, H., A convenient hemoperfusion apparatus over charcoal for the treatment of endogenous and exogenous intoxicants. I. Its use as an effective artificial kidney, *Proc. Eur. Dial. Transplant Assoc.*, 1, 83, 1964.
133. Gordon, A., Popovitzer, M., Greenbaum, M., Morantz, L., McArthur, M., DePalma, J. R., and Maxwell, M. H., Zirconium phosphate — a potentially useful absorbent in the treatment of chronic uremia, *Proc. Eur. Dial. Transplant Assoc.*, 5, 86, 1968.
134. Gordon, A., Gral, T., DePalma, J. R., Greenbaum, M. A., Marantz, L. B., McArthur, M. J., and Maxwell, M. H., A sorbent-based low volume dialysate system: preliminary studies in human subjects, *Proc. Eur. Dial. Transplant Assoc.*, 7, 63, 1970.
135. Ward, R. A., Klein, E., and Wathen, R. L., Sorbent regenerated dialysis delivery systems, *Peritoneal Dial. Bull.*, 3, 541, 1983.
136. Johnson, R. C., Block, F., Knab, W., and Martis, L., A model for study of the kinetics of continuous ambulatory peritoneal dialysis (CAPD), Abstr. 29th Annual Meeting, The American Society for Artificial Internal Organs, Toronto, April 1983, 52.
137. Zingg, W., Shirriff, J. M., and Lerbel, B. S., Control of blood glucose in diabetics on CAPD — the importance of blood glucose control in the prevention of diabetic complications, *Peritoneal Dial.Bull.*, 2, 524, 1982.
138. Roscoe, J. M., Practices of insulin administration in CAPD, *Peritoneal Dial. Bull.*, 2, 527, 1982.
139. Janle-Swain, E., Thornhill, J. A., Carter, J. M., Hinsman, E., Jackson, H. D., and Ash, S. R., Case study of a diabetic dog with chronic membranous glomerulopathy treated with continuous intraperitoneal insulin infusion, *Am J. Vet. Res.*, 43, 2044, 1982.
140. Janle-Swain, E., Thornhill, J. A., Carter, J. M., Hinsman, E., Jackson, H., Swain, S. L., and Ash, S. R., Improvements of kidney function with continuous intraperitoneal insulin infusion (CIPII) in severe diabetic glomerulopathy, *Diabetic Nephropathy*, 2, 16, 1983.

DEVELOPMENT OF A CANINE MODEL FOR CONTINUOUS AMBULATORY PERITONEAL DIALYSIS*

Jack W. Moncrief, Robert P. Popovich, Carol Mullins-Blackson, Alison Booth Harter, A. J. Sorrels, Leslee Angell, W. K. Pyle, Mary Hiatt, and Everett E. Simmons

INTRODUCTION

Peritoneal dialysis has become increasingly utilized for the treatment of acute and chronic renal failure in humans. This increase in popularity has been primarily due to the development of continuous ambulatory peritoneal dialysis (CAPD.) [1,2] CAPD was developed in 1975 and has taken its place as an alternative form of dialysis therapy for patients with end-stage renal disease. This increased utilization of peritoneal dialysis has stimulated a need for basic research in the transport characteristics of the peritoneal membrane. This research has benefited from the development of a dog animal model for peritoneal dialysis.[3] The dog model allows the testing of computer models for ultrafiltration and small and large molecule transport. It is useful to evaluate changes in transport associated with inflammation and drug application, as well as the study of the vascular supply and surface area of the peritoneum.

The measurement of fundamental transport characteristics of the peritoneal membrane in adult dogs and the capability of sustaining a surgically anephric dog model has allowed these same parameters to be utilized in determining the feasibility of peritoneal dialysis in newborn puppies.[4] This information has been used in developing parameters and characteristics of peritoneal dialysis in a newborn human infant.

DEVELOPMENT OF ANIMAL MODEL

Chronic peritoneal access has been the major impediment to successful maintenance of dogs in peritoneal dialysis. At the Acorn Research Laboratory, Austin, Texas, we unsuccessfully attempted peritoneal access utilizing the previously reported Parker technique.[5] Catheters and techniques that were tried but found to be inadequate because of rapid obstruction include intermittent peritoneal puncture, the Tenckhoff catheter, and the Toronto Western catheter. The development of the Purdue column disk catheter by Ash and Thornhill[6] at Purdue University (West Lafayette, Ind.) in 1978 revolutionized the access device to the peritoneal cavity for our dog peritoneal dialysis animal model. We have modified the Purdue catheter by changing the i.c. (intracutaneous) and s.c. portion to produce a straight catheter utilizing the intra-abdominal portion of the column disk and the i.c. portion of the Gore-tex® (W. L. Gore and Associates, Newark, Del.) peritoneal dialysis catheter (Figure 1).

SURGICAL TECHNIQUE

The development of the surgical technique for implantation of the straight modified column disk catheter required that the catheter exit through a stab wound independent of the initial skin incision. This is accomplished by retraction of the skin of the abdominal wall immediately prior to the incision. Once the incision is made through to the

* This study was supported in part through a grant from Baxter-Travenol Laboratories, Inc.

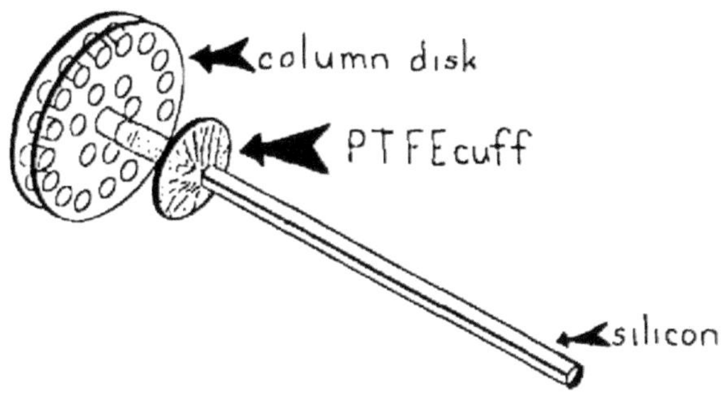

FIGURE 1. Modified Purdue catheter.

abdominal cavity, the disk segment of the catheter is placed in the abdominal cavity. The retraction of the skin is then released and the skin incision moves laterally. A small stab wound can then be made directly over the column disk segment and the catheter can be brought out through the small stab wound. The small incision in the s.c. tissue, not including the skin, is made to allow the small polytetrafluorethylene (PTFE) s.c. disk cuff to be placed in the tissue at the catheter exit site. The original skin incision, which is lateral to the exit site, is then sutured. The exit site should be small enough so that no suturing is required.

POSTOPERATIVE CATHETER IRRIGATION

Immediately after the catheter insertion and while the field is still sterile, a catheter adapter, of either plastic or titanium, is inserted into the external end of the modified Purdue catheter. An extension tubing is then connected to the adapter and a small irrigation bag of Dianeal® (Baxter-Travenol) containing heparin (1000 units) is used to irrigate the abdominal cavity to remove fibrin and blood left from the surgery. If the irrigation remains clear, the catheter can be maintained by irrigation with 250 to 500 m*l* of Dianeal® and heparin twice daily. More frequent irrigation will be required if the dialysate is bloody or fibrinous. Tobramycin (1 mg/kg of body weight) is given intramuscularly approximately 1 hr prior to surgery, and then 1 to 1.5 mg/kg in 2 divided doses is given for 3 days in the postoperative period. A continuously closed system is used except when irrigation bag exchanges are made. An abdominal binder, the Bettybandage (Figure 2), and a disposable Buster collar (Figure 3) are used to prevent the animal from damaging or obtaining access to the peritoneal catheter exit site. Within 24 hr the animal is ambulatory and eating. If no leakage from the catheter occurs within 48 hr, 500 to 1000 m*l* of dialysis solution may be instilled in the peritoneal cavity and changed twice daily to ensure catheter patency.

USE OF THE ANIMAL MODEL

The peritoneal dialysis animal model has been used to develop techniques for measuring transport of various size water-soluble molecules across the peritoneal membrane. Nondiffusable, radio tagged dextrans (70,000 mol wt) have been utilized to measure ultrafiltration rates in the peritoneum of the normal animal during controlled episodes of peritonitis and to evaluate the effect of drugs given both systemically and intraperitoneally (Appendix 1). Since 1978, 41 peritoneal access devices have been placed in adult mongrel dogs and maintained from 2 to 13 weeks. Six of these animals had bilateral nephrectomies and were maintained on CAPD for up to 90 days. Seven other animals had peritoneal access catheters: three were removed because of dialysate

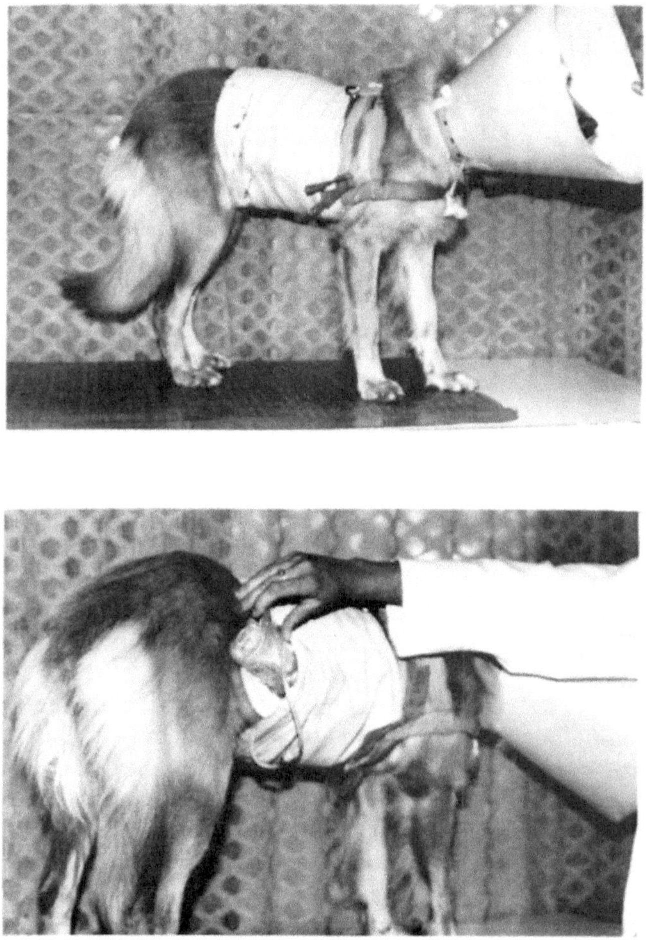

FIGURE 2. Bettybandage.

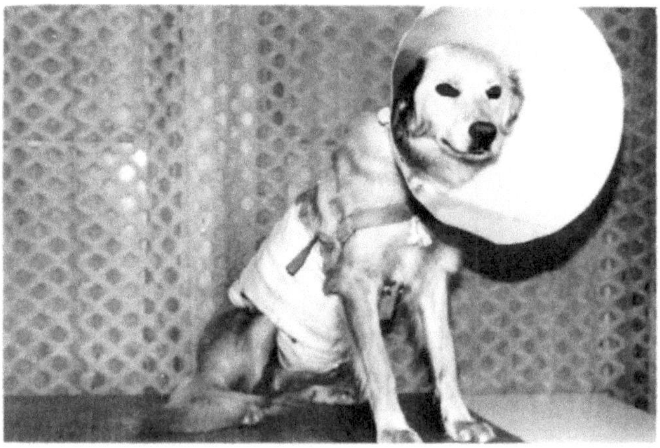

FIGURE 3. Buster collar.

leakage and infection and four were removed because of obstruction secondary to omental wrap. To prevent this latter complication, we are presently evaluating the necessity for partial omentectomy at the time of catheter insertion. Partial omentectomies have been performed on eight animals. None of these animals has had omental wrap, but several have had increased fibrin and bloody dialysate for 7 to 10 days postoperatively, thus causing a delay in the experimentation as well as increasing project costs.

MAINTENANCE OF THE UREMIC ANIMAL ON CAPD

The development of an adequate access to the peritoneal cavity for long-term dialysis in dogs has allowed research protocols to determine techniques for maintenance of uremic animals for both experimental and clinical purposes. To prove the validity and practicality of long- and short-term maintenance of uremic dogs, a bilaterally nephrectomized model was developed. The nephrectomy was performed to eliminate the possibility that residual renal function would serve any role in the successful maintenance of the uremic animal.

A general anesthesia and endotracheal intubation was performed 1 week post catheter insertion. Bilateral, paracostal flank incisions and surgical removal of both kidneys were carried out in a single procedure. The peritoneal membrane of the canine is somewhat different from that of the human. The peritoneal membrane of the dog consists of a very thin, fibrinous, less distinct tissue layer. The bilateral nephrectomy, therefore, could not be performed without partial entry to the peritoneal cavity. Bloody and fibrinous dialysate can be expected postoperatively, and s.c. wound leakage into the flank incisions is common. Tight surgical closure and small-volume dialysis every 2 hr during the day and every 4 hr during the night for the first 24-36 hr was required to prevent excessive leakage of dialysate into the flank incisions. Rapid development of azotemia and hyperkalemia will occur if adequate dialysis in the postoperative period is not carried out. Within 24 hr postbilateral nephrectomy, 6 exchanges per day with 50 mℓ/kg of body weight per exchange and a low-protein and low-potassium diet will prevent dangerous hyperkalemia and severe azotemia. The size of the peritoneal cavity in the dog is disproportionately large as compared to the human, and in most cases 100 mℓ/kg of body weight per exchange can easily be infused after a good seal from the bilateral nephrectomy has occurred. This exchange volume performed five times per day has proved adequate to maintain bilaterally nephrectomized dogs in good nutritional status (4 to 9 g of protein per kilogram of body weight per day) for prolonged periods.

GROWTH RESULTS IN PUPPY MODEL

To demonstrate the adequacy of the dialysis and to evaluate the cause of growth retardation in children undergoing hemodialysis, a protocol was designed to maintain bilaterally nephrectomized puppies utilizing the animal model described. This included forced feedings to assure sufficient nutrition.

Four mongrel puppies from the same litter were entered into the study after weaning at 5 weeks of age. Peritoneal catheters were inserted into three of the animals; one animal was maintained as a control. Excellent catheter function was obtained after 1 week of irrigation at which time a bilateral nephrectomy was performed on two of the animals. The third animal underwent a sham nephrectomy with general anesthesia and bilateral flank incisions as well as kidney mobilization, but no nephrectomy was performed. All four animals were then maintained on an identical diet. The diet was adjusted to produce total consumption of all food offered to the control and sham-

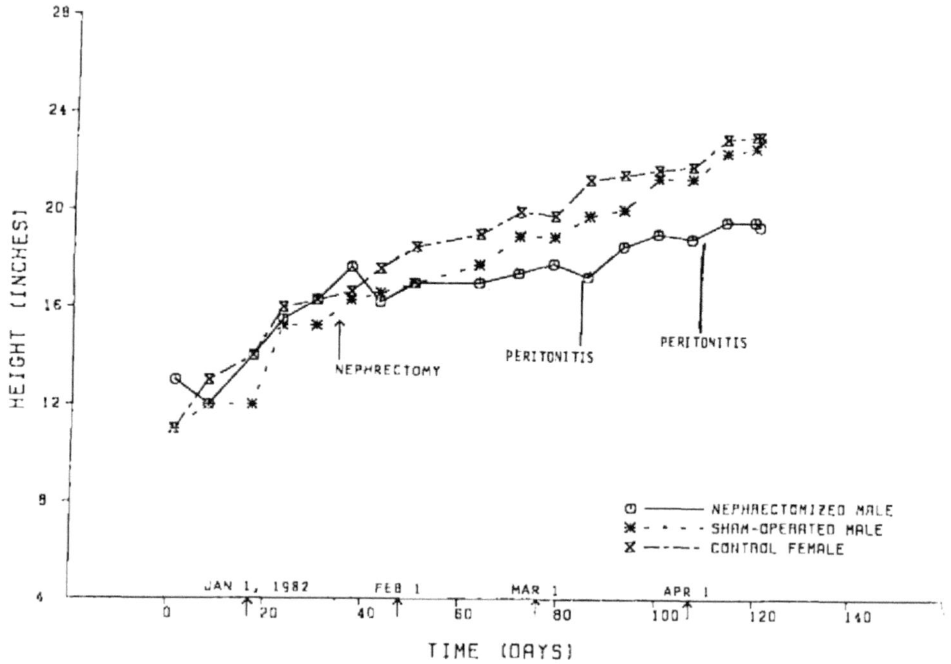

FIGURE 4. Canine growth curves — height. Litter-mates born 10/21/81.

operated animals. Protein intake was 9.2 g/kg of body weight per day.[7] The nephrectomized animals were unable to ingest this high quantity of protein, and all uningested food was administered through an intermittently placed oral feeding tube. Animals were maintained in a separate outdoor dog kennel, and no special restrictions were placed on the nephrectomized animals. Blood urea nitrogen (BUN) and creatinine rose rapidly in the nephrectomized animals to approximately 100 and 12 mg %, respectively. Mild acidosis developed and was controlled by the addition of sodium bicarbonate to the dietary regimen of the nephrectomized animals.

Growth and development of the animals was measured by activity observation, skin fold measurement, body length and height, and head circumference. Within 2 weeks of the bilateral nephrectomizations, severe anemia developed in both test animals and blood transfusions were required. One of the animals developed a severe transfusion reaction and died in what appeared to be shock lung without evidence of fluid overload. Adjustments in the technique of cross-matching were successful in preventing a similar reaction in the remaining nephrectomized test animal[8] (Appendix 2). During the remaining 3 months of the growth trial, seven blood transfusions were required, but no further allergic reactions were manifested. After each blood transfusion, the test animal was vigorous and active and played with the two control animals. No obvious visual difference in the growth rate or head circumference between the control and sham-operated animals and the nephrectomized animal could be detected though there was a slight delay in the linear growth of the nephrectomized animal with measurement. During the 3-month period, two episodes of peritonitis occurred and each of these episodes was associated with a slight delay in the growth of the nephrectomized animal. Following the peritonitis epidsodes, an increasing rate of growth recurred (Figure 4). There was no difference in the growth rate of the circumference of the head (Figure 5) and only a slight retardation in weight gain (Figure 6).

RESULTS

Unfortunately, the death of one of the nephrectomized animals due to a transfusion

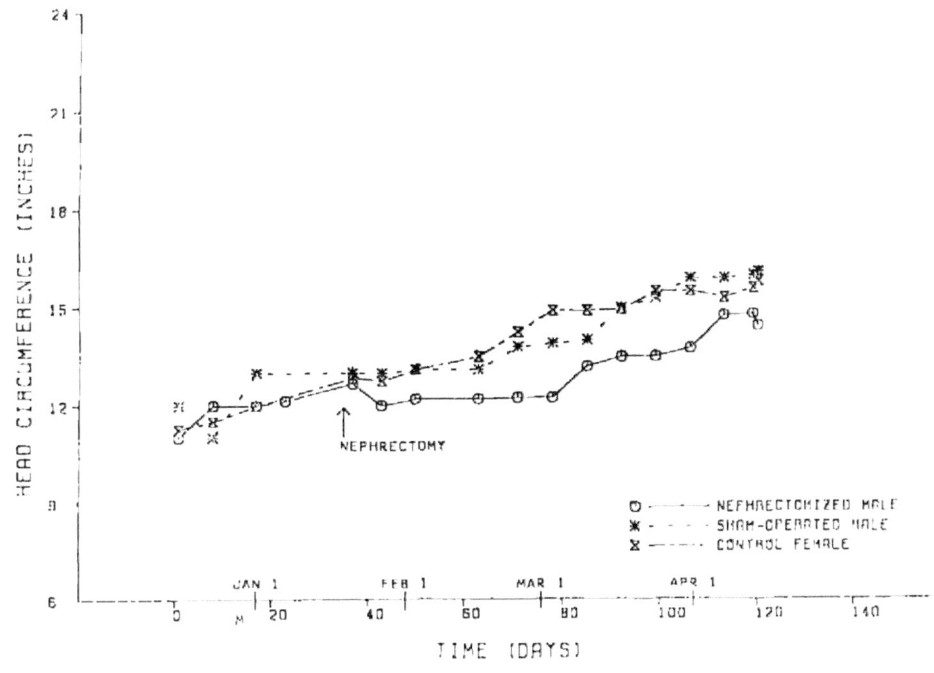

FIGURE 5. Canine growth curves — head circumference. Litter-mates born 10/21/81.

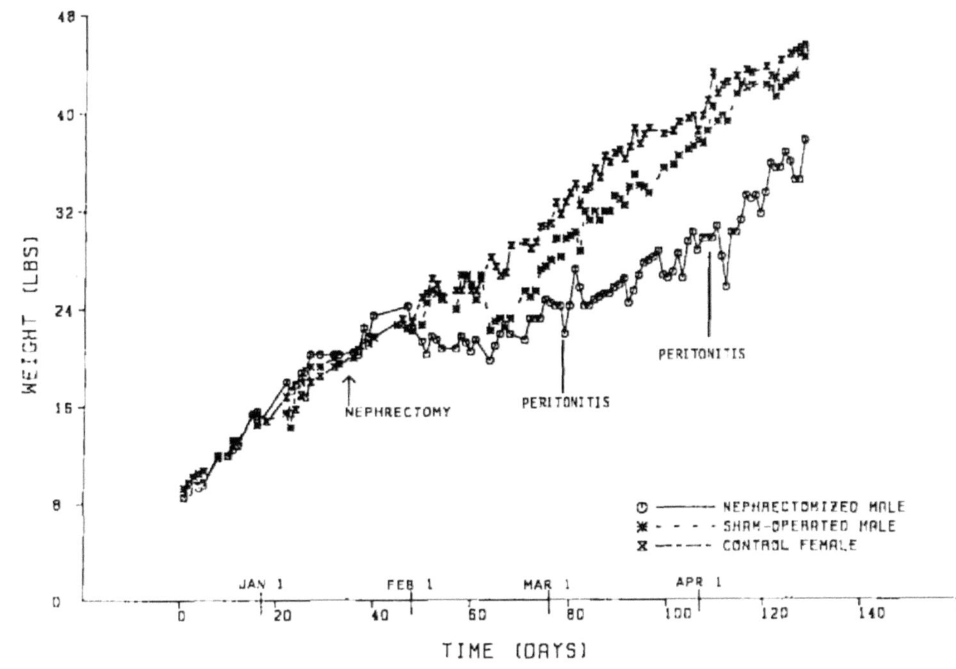

FIGURE 6. Canine growth curves — weight. Litter-mates born 10/21/81.

reaction made evaluation of the growth rate and the effect of CAPD on growth in newborn puppies somewhat difficult. There was, however, minimum growth retardation in an infant dog maintained for 3 months with CAPD and forced nutrition. Two episodes of peritonitis during the test period appeared to be temporarily related to a decreased growth rate. Our results would suggest that:

1. Chronic maintenance of anephric animals on peritoneal dialysis is possible and practical.
2. Forced feedings to prevent malnutrition in infants and small children undergoing dialysis may prevent growth retardation.
3. High-protein diet, maintenance of normal serum bicarbonate level, control of phosphorus and calcium metabolisms, and adequate dialysis appears to improve and allow near normal growth in our newborn puppy peritoneal dialysis models.

Improved rates of growth with forced feeding in newborn infants undergoing CAPD for uremia has been confirmed and recently reported.[9]

SUMMARY

CAPD has improved the technique and possibility of management of patients with end-stage renal disease and is increasing in both popularity and application. A dog model has been developed which has allowed experimental development of techniques for measuring solute transport, osmotically controlled ultrafiltration rate, and systemic and topical (i.p.) drug effect. Use of this model in newborn puppies has allowed evaluation of growth and development in anephric newborn puppies, suggesting that growth retardation in newborn infants undergoing dialysis is primarily related to nutrition.

Modification of the Purdue catheter with a straight tunnel, utilizing a Z-skin closure technique has allowed long-term adequate peritoneal access. This development has made CAPD practical for both experimental and veterinary purposes in maintaining animals in temporary renal failure.

APPENDIX 1

CONTINUOUS AMBULATORY PERITONEAL DIALYSIS

Mass Transfer Evaluation Protocol

The purpose of this procedure is to obtain timed blood and dialysate samples and to measure the concentration levels of selected substances in these samples. This data will be analyzed to determine peritoneal mass transfer parameters.

Prior to Exchange of Interest

1. Obtain informed consent (in writing) from patient or guardian.
2. Complete the general information section of the attached data report form or otherwise record this information.
3. Place bag of fresh solution on work area and apply port wrapping: place Betadine®-soaked gauze pad around injection port; wrap with sterile 4 × 4 gauze pad, and tape in place.

Exchange Procedure

1. Dialysate in the peritoneal cavity should be drained as completely as possible.

2. Any exogenous solutes to be added to the dialysis solution should be carefully measured and aseptically added to the fresh dialysate bag. The solution should be mixed well by repeated inversion of the bag.
3. The fresh solution bag should be connected to the patient and infusion initiated. Note the time required for infusion. No solution should remain in the bag.
4. Approximately 2 min *before* completion of infusion, an initial blood sample sufficient for analysis must be taken. The time should be carefully noted and recorded.
5. As soon as infusion of the fresh solution is complete, the initial dialysate sample (D1) should be taken in the manner described below. The elapsed time from the completion of infusion must be recorded (actual time, not necessarily the ideal time indicated). The volume of the samples should be kept to the minimum necessry for accurate analysis.
6. Additional dialysate samples should be obtained and noted as above at 30, 60, 90, 120, 180, 240, 360, and 480 min (measured from the completion of infusion).
7. Additional blood samples should be obtained and noted as above at 180 min and immediately after initiation of drainage.
8. Record the drained volume.

Dialysate Sampling Procedure

1. Unroll the dialysate bag and place in the drain position.
2. Release the tubing clamp and allow approximately 50 m*l* of dialysate to flow into the bag. Close the tubing clamp. Record the time that the clamp is closed as the "actual sample time".
3. Place the bag on a suitable surface and remove the injection port wrapping.
4. Aseptically withdraw a sufficient volume of dialysate for analysis from the bag via the injection port.
5. Raise the bag and reinfuse any dialysate remaining in the bag into the patient.
6. Wrap the injection port with a fresh wrapping as described above.
7. Transfer the sample to a clean, suitably labeled container.

Blood Sampling Procedure

Blood samples are obtained in the normal manner. Note and record the sample time when about half of the blood sample has been collected. Label the container appropriately.

APPENDIX 2

Blood Transfusions

During the initial canine blood transfusions given at this lab, several animals became ill during or after a transfusion. On three occasions, the animals died in 48 to 72 hr following transfusions. In this study, one of the nephrectomized animals died within 72 hr post-blood transfusion. Documented information along with personal references stated that most canines do not require more than one transfusion in their lifetime. Also, dogs do not have naturally occurring erythrocyte isoantibodies so that reactions are not expected with the first transfusion. In situations requiring more than one transfusion, there is often inadequate laboratory facilities to accommodate more sophisticated analysis.

Cross-match

During this study, cross-matches were done before and after each transfusion. The usual cell suspension of 2 to 3% normal saline, autologous serum, and Coombs antiserum were used. Incubation temperatures included 30°C, room temperature, and 4°C. After several transfusion reactions, it was found that some canine antibodies were temperature specific and reactions were only seen at refrigerated temperatures.

Blood Typing

Dogs have three significant erythrocyte antigens which cause problems after multiple transfusions. For this reason numberous donors were required during this 6-month study. Because canine blood typing was not available, cross-matching was the only method utilized for preventing erythrocyte incompatibility.

REFERENCES

1. Moncrief, J. W., Continuous ambulatory peritoneal dialysis, *Dial. Transplant.*, 7(8), 809, 1978.
2. Moncrief, J. W., Popovich, R. P., Nolph, K. D., Rubin, J., Robson, M., Dombros, N., de Veber, G., and Orepoulos, D., Clinical experience with continuous ambulatory peritoneal dialysis, *ASAIO J.*, 2(8), 114, 1979.
3. Simmons, E. E., Lockard, D. T., Moncrief, J. W., Hiatt, M. P., McCollough, W. S., and Popovich, R. P., Experience with continuous ambulatory peritoneal dialysis in the maintenance of a surgically anephric dog, *Southwest. Vet.*, 33(2) 1980.
4. Popovoch, R. P., Hiatt, M. P., Pyle, W. K., Simmons, E. E., and Moncrief, J. W., Clinical experience with a dog model to investigate continuous ambulatory peritoneal dialysis, in Proc. 8th Int. Cong. Nephrol., Greece, June 1981.
5. Parker, H. R., Gourley, I. M., and Bell, R. L., Current developments in peritoneal and hemo dialysis, in Proc. 22nd Vet. Symp., Stillwater, Okla., September 1972, 3.
6. Ash, S. R., Johnson, H., Hartman, J., Granger, J., Koszuta, J., Sill, L., Dhein, C., Blevins, W., and Thornhill, J. A., The column disc peritoneal catheter: a peritoneal access device with improved drainage, *ASAIO J.*, 3(3) 109, 1980.
7. Subcommittee on Dog Nutrition, Committee on Animal Nutrition, Board on Agriculture and Renewable Resources, and the National Research Council, Nutrient Requirements of Dogs, Number 8, Revised 1974.
8. Greene, C. E., Practical considerations of blood transfusion therapy, in AAHA'S 47th Annu. Meet. Proc., 1980, 187.
9. Wilson, D. J., Conley, S., and Brewer, E., Adaptation to home dialysis: the use of CAPD and CCPD in infants and small children, in press.

EXPERIMENTAL MODELS OF ACUTE TUBULAR NECROSIS

George A. Tanner

INTRODUCTION

The topic of acute tubular necrosis (ATN) has been extensively discussed in review articles and symposia in the last decade.[43,70,71,78,100,107,115,117,119,121,129,133,164,202,203,221,237-239,241,243,247,293,310,334,335,350] The pathophysiological mechanisms involved in this disorder have received considerable attention and will not be detailed here. Although great progress has been made, much is still not known. This chapter will focus on some of the animal models that have been developed to study the mechanisms involved in ATN. We will describe the methods used to produce ATN in animals and highlight some areas where future work needs to be done.

ATN is a term used to designate the common type of potentially reversible acute renal failure (ARF) which follows shock, sepsis, trauma, transfusion reactions, or nephrotoxins.[47] The term is commonly used, but suggests that tubule cell death (necrosis) is an important feature of this disorder. The significance and extent of this morphological change are, however, disputed.[305] In man, extensive necrosis of tubular cells may be prominent in nephrotoxin-induced ATN, but is usually not observed in ischemia-induced ATN. The term ATN may overemphasize the importance of tubule cell damage. It also neglects possible glomerular changes and functional changes such as impaired renal hemodynamics. Despite these reservations, we will employ this widely used name to designate acute intrinsic renal failure due to an ischemic episode, heme pigments, or nephrotoxins.

ATN has also been called "lower nephron nephrosis, shock kidney, acute tubulo-interstitial nephropathy", and "vasomotor nephropathy". "Lower nephron nephrosis" is an older term, not often used today, that emphasizes pathological changes in the distal portions of the nephron.[205] Actually, changes in the proximal nephron are often more obvious and important.[251] "Shock kidney" suggests the frequent association between ATN and a preceding period of severe hypotension or shock.[341] It is important to distinguish "shock kidney" from "the kidney during shock". During shock, renal function is depressed, primarily as a result of reduced renal blood flow (RBF) and glomerular capillary pressure. This depression is usually promptly reversed when blood volume and blood pressure are reestablished. By contrast, with "shock kidney" renal failure persists for days to weeks after the circulation has been restored. The name "acute tubulo-interstitial nephropathy"[44] suggests the presence of tubular changes and interstitial inflammation as prominent histological alterations. This name, however, is too broad and includes many other diseases of the kidney, e.g., interstitial nephritis due to various drugs and pyelonephritis. "Vasomotor nephropathy" stresses an inadequate RBF as the explanation for the origin and maintenance of renal insufficiency.[238] A low RBF is not, however, essential to the established phase of ATN, and so this term is deficient.

ATN is characterized by a rapid decrease in renal function and a consequent progressive accumulation of nitrogenous wastes (e.g., urea, creatinine) in the blood (azotemia). The kidney abnormalities in human ATN are several. Most important, the glomerular filtration rate (GFR), estimated from the endogenous creatinine clearance or inulin clearance, is less than 5% of normal. A very low urine flow rate (oliguria), defined as a urine output less than 400 mℓ/day, is usually present during the early phases of ATN. It is now recognized, however, that many patients with ATN have normal or even elevated urine outputs.[3,344] Patients with this nonoliguric variety typically have fewer complications and an overall lower mortality than patients with oliguria. Urine sodium concentration is usually elevated above 25meq/*l*, in spite of a low GFR and hormonal stimuli for sodium reabsorption. This suggests damage to the tubular reabsorptive mechanism for sodium. Low urine/plasma concentration ratios for urea and creatinine (below 5 and 15, respectively) and a urine approximately iso-osmotic to plasma indicate an impaired ability to concentrate the urine. The urine frequently contains casts, epithelial cells, blood cells, and sometimes hemoglobin and myoglobin. Morphological changes in human ATN vary with the stage of the disease and to some extent with the underlying etiology. The basic lesions observed are necrosis and/or degeneration of the tubule epithelium and obstruction of tubule lumens by casts or debris.[101,251] In patients with ATN, the glomeruli are usually,[86,180] but not always,[66] normal in appearance. Excretory insufficiency and impairment of reabsorptive mechanisms lead to altered composition and volume of the body fluids in the patient with ATN.

The clinical course of ATN is quite variable. It is customary to divide the disorder into three phases: the initial, maintenance, and recovery phases. During the initial phase, events occur which lead to the development of ATN. Obviously, this is the time during which preventive measures are important. The maintenance phase comprises established renal failure; this may last for days to weeks, during which time GFR stays at a persistently low level. The recovery phase is signaled by a progressively increasing urine flow rate. At the beginning of this polyuric stage, the tubular cells show evidence of regeneration.[87] Renal function in patients continues to improve for 3 to 12 months after an episode of ATN. The seriousness of ATN is underscored by a mortality rate of about 60%, even in patients who are dialyzed frequently. Death in these patients usually results not from the disturbed renal function per se, but from the underlying condition, superimposed infection, or malnutrition.

The causes of ATN in man are diverse (Table 1). For simplicity, they may be divided into three categories: ischemia, heme pigments, and nephrotoxins. A decreased RBF (ischemia), such as might occur during shock, trauma, sepsis, surgery, or burns, is generally accepted as the major cause of ATN. The release of heme pigments, myoglobin and hemoglobin, into the circulation from damaged muscle or hemolyzed red cells may also contribute to the development of ATN. Nephrotoxins are agents which primarily damage kidney tubule cells. Such substances include certain antibiotics, iodinated contrast agents used in radiology, heavy metals (mercury, uranium, chromium, platinum, arsenic, bismuth, lead, and cadmium), organic solvents (carbon tetrachloride, chloroform), ethylene glycol, and pesticides. Because of the clinical importance of some potentially nephrotoxic antibiotics (e.g., gentamicin), this topic is covered in a separate chapter in this book. Even though a single insult may not be of sufficient magnitude to cause ATN, a combination of insults (e.g., ischemia plus heme pigments or a nephrotoxin plus renal ischemia) may cause life-threatening renal insufficiency.

Animal models have helped us to understand the nature of the underlying disorders in human ATN. Also, such studies are helpful in testing possible preventive measures and therapy. The obvious advantage of such studies is that they permit us to study ATN under well-defined conditions with techniques that cannot be used in man. The evolution of the disorder can be accurately described, and functional, structural, and biochemical changes can be correlated. A wide variety of models have been studied in an attempt to mimic conditions seen in human ATN (Table 2). Investigators disagree as to what is the "best" model. In reality, no single animal model would be completely satisfactory, because in man the etiology and development of ATN are diverse.

An ideal animal model should be easy to induce, consistent in its outcome, and bear

Table 1
FREQUENT CAUSES OF ACUTE TUBULAR NECROSIS IN MAN[a]

Ischemic disorders
- Major trauma
- Major surgery (e.g., aortic surgery)
- Cardiogenic shock
- Massive hemorrhage
- Septic shock
- Burns
- Thrombosis or embolization of major renal vessels
- Pregnancy-related disorders

Heme pigments
- Intravascular hemolysis (transfusion reaction, hemolysis due to toxins or immunological factors, malaria)
- Rhabdomyolysis and myoglobinuria (muscle trauma, muscle disease, and severe exercise)

Nephrotoxins
- Antibiotics (aminoglycosides, cephalosporins, tetracyclines, amphotericin B, polymyxins, and sulfonamides)
- Radiographic contrast agents
- Heavy metals (mercury, uranium, platinum, chromium, arsenic, and bismuth)
- Organic solvents (carbon tetrachloride)
- Ethylene glycol
- Analgesics (acetaminophen and glaphenine)
- Pesticides

[a] This list is not exhaustive.

a close resemblance to the clinical, biochemical, and histological picture seen in man. More specifically, an animal model for human ATN should show:

1. A marked depression of the apparent filtration rate in the presence of a normal arterial blood pressure
2. Development of severe uremia (elevated BUN and plasma creatinine concentrations)
3. Recovery of renal function to essentially normal values in surviving animals
4. An overall mortality of 10 to 60% (without dialysis)

Oliguria is not an essential feature in an animal model, since many patients with ATN are nonoliguric. Massive tubular cell necrosis usually does not occur in human ischemia-induced ATN and so should not be present in an ideal animal model. Use of small mammals is preferred to dogs, ruminants, etc. because of cost factors and a high disease mortality. Rats are inexpensive, easy to maintain, and require small quantities of supplies for experiments. Two approaches have been used to develop animal models. One is to cause ATN with an agent known or suspected to be important in the human disorder. An example of this approach is the administration of mercuric chloride or heme pigments. A second approach is to produce ATN by wholly artificial means, to analyze the factors contributing to the ATN, and to compare the results with the human disorder. An example of this approach is the glycerol model.

In comparing different animal studies, it is well to keep in mind the many variables that can influence the results of experiments. These include animal species, type of insult, hydration state of the animal, whether there has been prior renal failure,[155,246,354] and whether there is a combination of insults. The intensity of the precipitating agent is of obvious importance. In the case of nephrotoxic drugs, this will be affected by the

Table 2
EXPERIMENTAL MODELS OF ACUTE TUBULAR NECROSIS

	Ref.[a]
Ischemic	
Temporary renal artery occlusion	189
Intra-arterial norepinephrine	77
Glycerol	244
I.V. epinephrine	30
Heme pigment	
Glycerol	244
Myoglobin	52
Hemoglobin	176
Methemoglobin	214
Nephrotoxin	
Mercuric chloride	25
Uranyl nitrate (or uranyl acetate)	38
Dichromate	33
Cisplatin	62
Aminoglycoside antibiotics	See Chapter 6
Tubular obstruction	
Dextran 40	61
Globin	226
Uric acid	307
Folic acid	168
Glaphenine	136

[a] This is a key reference describing production of the model.

dose and route of administration. The duration and intensity of induced renal ischemia will clearly affect the outcome. The time of study after induction of ATN determines the results, and hence interpretation, of experiments. Many animal experiments involve complicating factors of anesthesia and extensive surgery.

Four factors have been proposed to explain the marked reduction in GFR that is characteristic of ATN (Figure 1). A *persistent decrease in RBF* has been considered to be an important mechanism explaining the decreased GFR during ATN in man.[162,163,256,264,265] It should be noted, however, that acute restoration of RBF to normal in man may not lead to any improvement in renal function during the established phase of ATN.[195,256,264] *Tubular obstruction* by casts and cell debris has been clearly demonstrated in patients dying from ATN[251] and in several animal models of ATN. Obstruction leads to an increase in tubular pressure and hence a reduction in GFR. Obstruction may also enhance back-leak of filtered substances across the tubule wall by elevating the intratubular pressure. It may contribute to a reduction in RBF by causing an increase in intrarenal pressure which compresses the blood vessels or by activating a nephron feedback mechanism.[5,118,119,322] Abnormal *tubular leakiness* to the substances which normally can be used to measure GFR (inulin, creatinine) will cause a reduction in the apparent GFR. Tubular leakiness has now been clearly demonstrated in patients who developed ATN after open heart surgery.[232] A *decrease in glomerular ultrafiltration coefficient* (K_f) has been found in some experimental models of ATN; morphological changes in glomeruli have also been observed. K_f is equal to the product

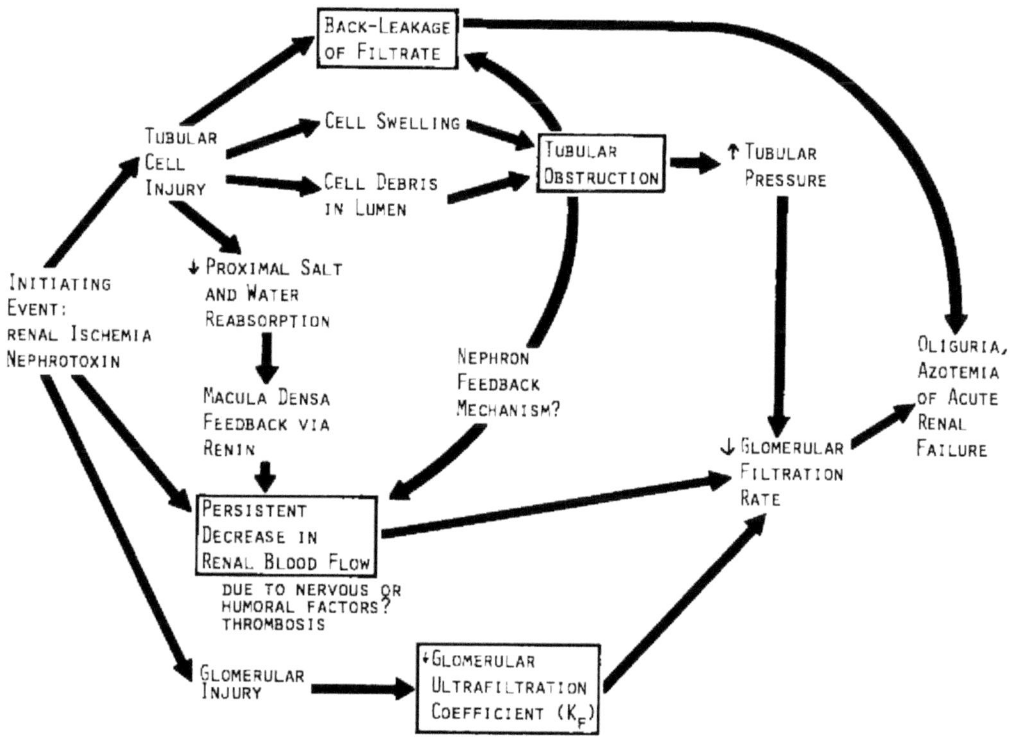

FIGURE 1. Schema of possible mechanisms involved in acute renal failure produced by ischemia or nephrotoxins.

of the total surface area and hydraulic permeability of the glomerular capillaries. It is calculated from the quantity of filtrate formed divided by the mean effective filtration pressure gradient across the glomerular capillary wall. The role of glomerular changes in human ATN is uncertain. The relative importance of these four factors in causing the impairment in GFR varies from model to model and may indeed vary with the course of the disease in a single subject. Each of these factors could conceivably account for the reduced clearances of substances by the kidney. Renal clearance is a measure of the efficiency of excretion (removal from the body) and is equal to the rate of urinary excretion divided by the plasma concentration. Renal clearances are usually expressed in terms of milliliters of plasma per minute. A substance that is freely filtered by the glomerulus and is not reabsorbed or secreted by the kidney tubules (e.g., inulin) has a clearance equal to the GFR. If, however, a substance is reabsorbed, then its clearance becomes less than the GFR.

In the past, investigators have often favored one or another of the above pathophysiological mechanisms, because one mechanism appeared to be of major importance in the particular model they studied. It is now clear that there is no single pathophysiological mechanism involved in ATN. Usually, multiple disturbances are present which may interact and reinforce each other. Even in a single model, it is difficult to assign to each mechanism its proper quantitative weight.

TECHNIQUES OF STUDY

A great variety of techniques have been used to assess renal changes in studies of ATN. The techniques vary in ease from simple measurements of urea nitrogen (BUN) or creatinine concentrations in blood plasma (or serum) to sophisticated in vivo micropuncture and in vitro microperfusion techniques. Morphological techniques and biochemical measurements are valuable, especially when correlated with functional measurements. Most laboratory studies have been done on rats, rabbits, and dogs; there is a paucity of studies using primates.

The severity of ATN has often been assessed by a simple determination of BUN. This measurement has its weaknesses because BUN is influenced by the rate of protein catabolism and by the rate of urine flow, both of which may be markedly altered in some renal failure models. Increased protein catabolism, which may be due to tissue destruction or infection, leads to an elevated BUN. A low urine flow rate is associated with enhanced back-diffusion of filtered urea from the tubules and hence reduced urea excretion and higher BUN levels. Plasma creatinine concentrations probably afford a more reliable index of changes in GFR. Both BUN and plasma creatinine are hyperbolic functions of GFR, so when GFR is low, striking changes in plasma concentrations of these substances can occur with very modest changes in GFR.[93] This last point should be kept in mind when evaluating reports that GFR was improved or worsened by an experimental maneuver. Recent experiments by Cabili and Charney[53] underscore the limitations of BUN in assessing GFR in ATN.

Clearance methods[301,315] provide important information about kidney function as a whole. Clearance methods can be used in unanesthetized animals, but often anesthesia and surgery are employed. As usually performed, these methods require the collection of urine; therefore, if urine flow is very low (oliguria or anuria), the clearance approach may not be feasible. As already mentioned, the inulin clearance will not afford a reliable measure of GFR if inulin is reabsorbed across damaged, leaky tubules. The clearance of *p*-aminohippurate (PAH), an organic anion avidly secreted by the renal proximal tubules, is normally close to the renal plasma flow. If the tubules are damaged, however, the extraction of PAH from the blood passing through the kidney will be reduced, and under these circumstances PAH clearance will greatly underestimate the true renal plasma flow. RBF is calculated by dividing the renal plasma flow by the plasmacrit (the fraction of whole blood volume occupied by plasma). RBF may also be measured using a number of other techniques: electromagnetic flow probe or accumulation or wash-out of various tracers (e.g., microspheres, radioactive inert gases). Measurements of intrarenal oxygen tension[252,353] may provide interesting and important information about tissue blood flow vs. metabolic rate.

Micropuncture techniques[39,200,360] in the intact, anesthetized animal (rat, dog) have been widely used to study renal functions in ATN. This technique permits one to measure tubular transport functions and nephron pressures and flows. This approach has been most valuable in defining the abnormalities present in ATN. In assessing data obtained with this approach, one should bear in mind that anesthesia and extensive surgery are required for these experiments. Animals with ATN may be less tolerant of such procedures than normal animals. Also, the procedures are difficult to perform, and technical problems encountered in experiments on normal kidneys are magnified in damaged kidneys. For example, it may be difficult to collect tubule fluid when flow rates are markedly decreased. The tubules may be fragile, making microinjection studies hazardous. Unlike the normal kidney, there is often a great heterogeneity in the appearance of surface nephrons in the damaged kidney. Some tubule segments may be widely dilated and others collapsed. It is difficult to obtain a truly representative sample of nephron function under these circumstances. Finally, only the most superficial nephrons can be studied with the micropuncture technique, and these nephrons may not be representative of the nephron population as a whole.

Recently, in vitro techniques have been used to study the function of isolated tubules [149-151] and glomeruli[54,282] in ATN. An advantage of this approach is that it is possible to study nephron segments which are inaccessible to in vivo micropuncture. Since tubules are difficult to dissect in mammalian species other than the rabbit, in vitro microperfusion studies have only been done in this species. This is unfortunate because relatively little work has been done in the intact rabbit, compared to numerous studies in rats and dogs. The problem of tubule selection with this technique must be formidable. This would be especially true in models of ATN in which scattered lesions are found. Short-term culture of isolated tubules[9] may permit the assessment of kidney tubule cell viability and nutritional requirements in different ATN models.

Morphological studies complement functional measurements. Nephron dissection techniques[88,251] and light and electron microscopy may reveal changes that have important functional consequences. Morphological techniques for studying the kidney have been described by Bulger and Dobyan.[45] Histochemical and biochemical studies are also valuable when correlated with functional and morphological data. In various models of ATN, biochemical measurements have provided information on the metabolic changes produced by injury and during recovery. Investigators have studied oxygen consumption by whole kidney,[22,252] kidney slices,[259,260,262,286] and isolated mitochondria,[346] tissue adenine nucleotide levels,[58,139,190] tissue enzyme activities,[8] utilization and production of metabolites,[8,259,260] and synthesis of nucleic acids.[80] The development of micromethods for measuring enzyme activities in single nephron segments[227,288] should permit, in the near future, a more complete description of the renal metabolic changes in ATN.

SPECIFIC ANIMAL MODELS

Ischemic Insults

A markedly reduced RBF, such as occurs during circulatory shock, is generally believed to be a major cause of ATN in man. Investigators have been unable to produce ATN in experimental animals by pure hemorrhagic shock.[57,97,327,341] Usually the animal dies from circulatory collapse or develops only a transient renal insufficiency (not ATN). This and other evidence suggests that the development of ATN after shock in patients is usually not due to blood loss alone, but results from a combination of insults, e.g., hemorrhage, trauma, and often sepsis.[327,350] To produce ATN in animals, investigators have had to turn to other methods of producing renal ischemia, such as renal artery occlusion, intra-arterial norepinephrine infusion, or i.m. glycerol injection. These methods of induction of ATN are artificial, but can produce reversible acute renal failure that shows many features similar to human ATN. There is good evidence that many of the pathophysiologic mechanisms which operate in these models also are important in the human disease.[231,251]

Renal Artery Occlusion

Complete, temporary renal artery occlusion produces a "pure" ischemic insult to the kidneys and, if sufficiently prolonged, causes ATN. Occlusion of the renal vein, together with the renal artery, appears to produce more severe renal damage.[233] In some studies, the entire renal pedicle (artery, vein, and ureter) is temporarily occluded to produce ATN. The latter procedure prevents blood flow to the kidney by way of the ureteral artery,[294] but may favor the development of hydronephrosis.

The usual procedure, as carried out in rats, is as follows.[6,181,323,324] The animal is anesthetized with pentobarbital (40 to 50 mg/kg body weight) injected intraperitoneally and the flank is shaved and prepared with iodine solution. One or both kidneys are exposed by small subcostal incisions. Prior to clamping the renal artery, heparin (100 to 500 units per kilogram) is given intravenously to reduce the risk of intrarenal thrombosis and infarction of the kidney. The artery is occluded by applying a smooth-surfaced tension clamp (Schwartz clip artery forceps, A. H. Thomas Co.) or bulldog clamp or by ligating or pulling on the artery with silk thread.[64,218,285] The kidney blanches and then turns dark purple with prolonged ischemia. Intravenously injected dyes (e.g., lissamine green) should not appear on the kidney surface. During the artery clamping, the kidney should be covered (e.g., with cellophane or by closing the abdomen) and kept at body temperature. Release of the renal artery obstruction is usually accompanied by a uniform reappearance of a pink color to the kidney, within a matter of seconds or after a few minutes. Typically, the kidney swells after release. The wound is closed by suturing the muscle layer and by using metal clips on the skin. Rectal temperature of the rat is maintained at 37 to 38°C with a heat lamp until the animal awakens from anesthesia. For these experiments, male rats are preferred because recovery from pentobarbital anesthesia is much slower in female rats.[161] An animal on which surgery has been performed should always be kept in an individual cage.

If the kidney is exposed to air during the period of artery occlusion, the outermost layer of tubules may survive due to diffusion of oxygen from the environment.[295] Mason et al.[217] recently measured intracellular electrolytes in rat kidneys subjected to 20 to 90 min of ischemia and found that the electrolyte composition of surface tubules was normal if the exposed kidney was maintained in air. It would seem preferable then to maintain the kidney in an atmosphere of moistened pure nitrogen[217,323] during artery occlusion, so that the surface tubules become as hypoxic as the deeper lying tubules. This may be especially important for micropuncture studies, since only surface tubules are studied.

The procedures for producing renal ischemia in the rabbit[14,49,151,235,294,303] and dog[1,134,148,356] are basically the same as in the rat. Sterile techniques should be used in performing chronic studies in these species.

The duration of ischemia determines the severity of the ensuing ATN. In the rat, Koletsky and Gustafson[189] found that bilateral occlusion of the renal artery and vein for 30 min produced only a transient reduction in renal function (as assessed by increase in BUN), slight morphological damage, and no mortality. A total of 1 hr of occlusion produced a greater increase in BUN, extensive tubular necrosis, a mortality of 13%, and 90 min or 2 hr of occlusion produced large increases in BUN and massive tubular necrosis. Mortality averaged 73% after 90 min and 100% after 2 hr of occlusion; all of the animals which died expired after 2 to 6 days. Roth et al[273] clamped one renal pedicle in rats and removed the contralateral kidney. They observed a mortality of 40% after 1 hr and 100% after 100 min of ischemia. Anderson et al.[2] clamped one renal artery and vein for 50 min and removed the contralateral kidney. A total of 20 to 30% of their rats died on the 2nd to 4th days after clamping. Schimmelpfennig and Hagemann[285] temporarily ligated one renal artery in rats for varying times and removed the contralateral kidney. Mortality after 30 min of renal ischemia was 0%, after 1 hr was 25%, and after 90 min was 67%. In the rabbit, Badenoch and Darmady[14] found that partial clamping of one renal artery with contralateral nephrectomy produced the following mortality results: 1 hr of ischemia, 20%; 90 min, 60%; and 2 hr, 100%. Dahlager[85] found that all rabbits survived 60 to 90 min of ischemia; 2 hr produced a 60% mortality and 3 hr produced a 90% mortality. In the dog, by contrast, all animals survived complete renal ischemia of 2-hr duration, and 4 hr of complete ischemia were necessary to produce death from uremia.[148] The greater susceptibility of the kidney of the rat to temporary ischemia probably reflects the higher metabolic rate in this small mammal.

The most frequently studied models have involved renal artery occlusion for 1 hr in rats, for 1 to 2 hr in rabbits, and 90 min to 2 hr in dogs. These durations of ischemia produce comparable impairments of renal function and usually a reversible ATN. Especially in studies in rats, investigators have often studied the function of one damaged kidney, while the other kidney is untouched. Such unilateral ischemia has the advantage that no mortality results. The situation, however, is quite unlike clinical ATN in that the composition and volume of the body fluids are not disturbed and the animal is quite healthy. The function of the damaged kidney is certainly influenced by the presence of its normal mate. For example, 1 day after unilateral renal artery occlusion for 1 hr in rats, the damaged kidney is oliguric. If both kidneys have been damaged by 1-hr bilateral artery occlusion, a nonoliguric renal failure is observed. Furthermore, proximal tubular pressures are higher when both renal arteries have been temporarily occluded.[324] The situation is similar to the differences observed after bilateral vs. unilateral ureteral occlusion.[357] The recovery of a damaged kidney after 1 hr of renal artery occlusion is inhibited by the presence of a contralateral normal kidney.[116] The normal kidney undergoes a compensatory hypertrophy and in effect suppresses its mate. Bilateral renal artery occlusion (or unilateral renal artery occlusion with contralateral nephrectomy) represents a more relevant model for the life-threatening ARF seen in patients than does the unilateral model with one normal kidney.

A total of 1 hr of renal artery occlusion in the rat produces a reduction in the measured GFR to less than 10% of normal. RBF is decreased to one half to three quarters of normal for several days after releasing the occlusion.[6,103,268,312] There is widespread tubular cell necrosis and cast formation (Figure 2).[261,275,342] The extensive necrosis, more than is usually seen in patients with postischemic ATN, is considered to be a weakness of this rat model. The course of recovery of renal function in the rat over a period of days to weeks has been examined in a number of studies.[104,116,120,318] There is general agreement that the reduction in GFR is due to tubular obstruction, tubular leakage, and persistent vasoconstriction[6,7,99,103,104,119,120,312,323-325,342]

In the rabbit and dog, comparable changes in renal function are produced by temporary renal ischemia. RBF is often reported to be normal[1,235,356] at a time when GFR is severely depressed. Tubular obstruction and leakiness contribute to the reduced filtration clearance.[151,174] In addition, in the dog kidney, the glomerular K_f is reduced and glomerular fenestrations are reduced in size and number, suggesting an important role for glomerular abnormalities.[282,356] Solez et al.[303] found that 1 hr of temporary renal artery and vein occlusion in the rabbit produced a degree of histological damage which is quite similar to that observed in patients with ATN, and therefore they recommend this model.

Norepinephrine Infusion

Direct infusion of norepinephrine, of sufficient dose and duration, into the renal artery will produce ATN. The first studies in the dog[75,188] used a dose of 0.75 μg/min/kg and a duration of 2 hr. This produced irreversible ATN.[77,290] When the infused kidney was examined after 8 weeks, it was found to be severely shrunken, its RBF was about 5% of normal, and oliguria was present. Norepinephrine infusion at 0.75 to 1 μg/min/kg for 40 min produces reversible ATN.[77,290] In uninephrectomized dogs, intrarenal artery infusion of norepinephrine for 40 min results in a mortality of 22%. GFR is markedly depressed for days after stopping the infusion, oliguria may or may not be present, and RBF is about half of normal; functional recovery occurs with time. The 40-min infusion model produces uniform proximal tubular necrosis, but less severe damage than is observed in the 2-hr model. No structural glomerular abnormalities are observed with the 40-min infusion, in contrast to observations after 2 hr of infusion.[75,76]

In anesthetized dogs, the norepinephrine is infused either by way of an angiographic catheter [75-77,188,290] or curved needle (21 to 25 gauge) placed directly into the renal artery.[17,48,255] The angiographic catheter (e.g., French 7 - 8, Cordis Corp., Miami, Fla.) is inserted by way of a femoral artery cutdown and is guided under fluoroscopic con-

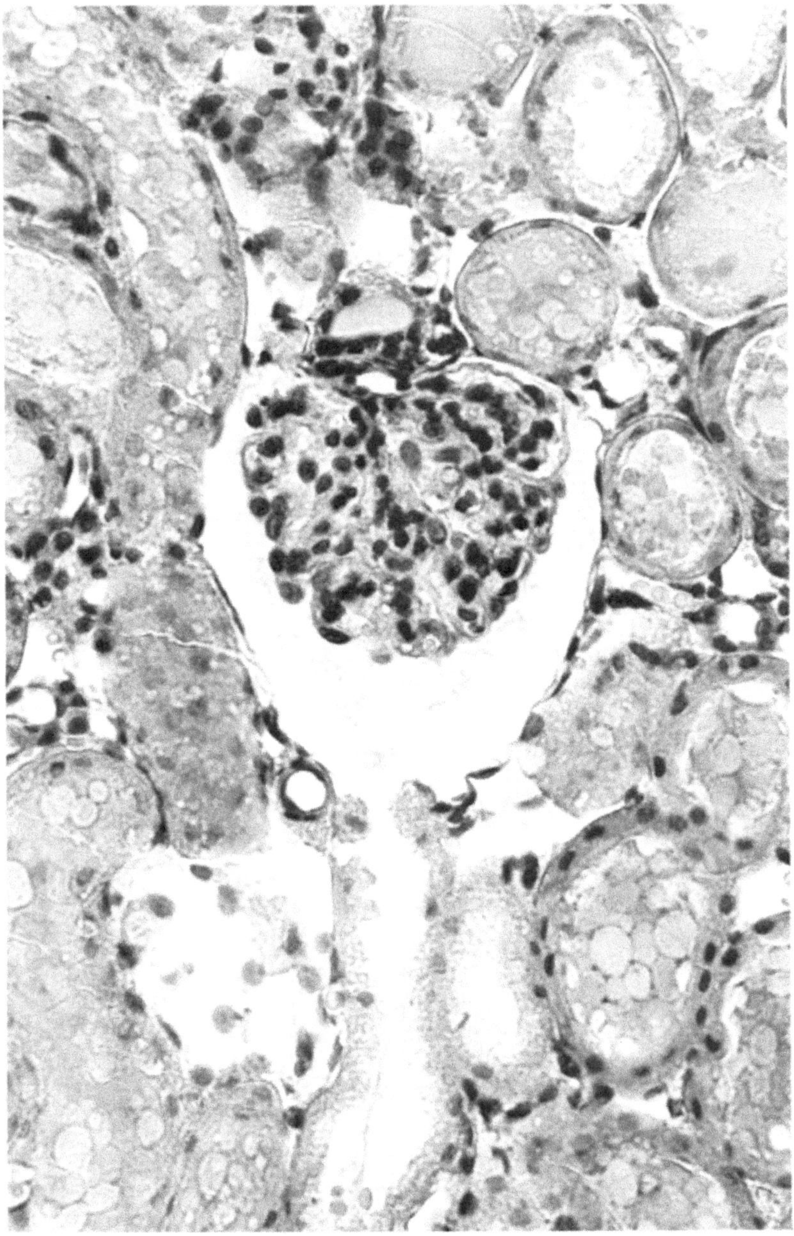

FIGURE 2. Histological section of rat kidney cortex. The kidney was fixed by vascular perfusion with 1% glutaraldehyde in Ringer's solution after 1 hr of blood reflow following 1 hr of renal artery occlusion. Bowman's space surrounding the glomerulus in the middle of the picture is dilated due to an elevated tubular pressure caused by obstruction. The proximal tubule epithelium shows degenerative changes, such as dense nuclei and loss of normal brush border. Many tubule lumens are filled with cell debris, which includes round membrane-bound blebs derived from shed brush border microvilli. Tubule cell necrosis is more prominent 12 or 24 hr after injury. (Hematoxylin and eosin stain.) (Courtesy of Dr. Michael Barrett, Medical College of Georgia, Augusta.)

trol into the orifice of one renal artery. In acute experiments, the catheter can be guided manually. Care must be taken not to obstruct the artery with the catheter. Infusion of norepinephrine is accompanied by prompt blanching of the kidney and a reduction in its size; RBF is reduced to virtually zero for the entire duration of the infusion, due to intense vasoconstriction. The ischemia extends in time well beyond the duration of the infusion. Mauk et al.[222] mentioned that RBF and urine flow rate remained essentially zero for 45 to 60 min after stopping the infusion. In most studies the contralateral kidney has been left intact. This kidney does not really represent a "control" because its function is affected by the presence of its damaged mate (i.e., it will undergo compensatory hypertrophy). Likewise, the function of the norepinephrine infused kidney is almost certainly influenced by the presence of a normal contralateral kidney and the absence of a uremic state.[15]

Direct infusion of norepinephrine into the renal artery of rats has also been studied. Conger et al.[69] infused norepinephrine at a rate of 0.6 μg/min/kg for 90 min, and Steinhausen et al.[314] infused 0.05 to 0.27 μg/min/kg for 45 to 120 min. The infusion was accomplished by means of a 30- to 35-μm tip diameter glass micropipette which was mounted on a micromanipulator and was inserted directly into the renal artery. Conger et al.[69] reported a mortality of 63% in uninephrectomized rats. When kidney function was studied 2 days after the norepinephrine infusion, an inulin clearance less than 10% of control and a near-normal RBF were found.

As in the renal artery occlusion model, tubular obstruction appears to play a major role.[48,69,290,314] RBF is usually somewhat reduced, but this is not the principal explanation for sustained renal insufficiency. Inulin clearance is severely decreased for hours to days after the infusion. A number of investigators [76,92,255] have attempted to prevent the renal impairment by infusing vasodilator drugs or diuretic agents. A solute diuresis appears to have a beneficial effect, possibly by attenuating tubular obstruction.[48,76,92,255,290] Chronic administration of saline and a mineralocorticoid, which supresses renal renin content, did not have a protective effect on renal function.[17] Renin is a proteolytic enzyme produced by the juxtaglomerular cells of the kidney and acts on a plasma protein (angiotensinogen) to produce angiotensin I. Angiotensin I is converted by angiotensin converting enzyme to the octapeptide hormone angiotensin II, a potent vasoconstrictor substance that has been frequently suggested as playing an important role in ATN.[121,242,335]

Glycerol Model

ATN produced by s.c. or i.m. injection of glycerol has been a widely studied model. It has the virtues of ease of induction, reasonable uniformity of results, and similarity in its "clinical" course, reversibility, and morphological findings to human ischemic ATN. In particular, this model has many similarities to human ATN associated with muscle injury (rhabdomyolysis) or massive hemolysis.[50,51] The model was originally developed as a model for studying the hemolytic transfusion reaction.[55] The striking hemoglobinuria produced by glycerol administration led to examination of its renal effects.

ATN is usually induced in rats by i.m. injection of a 50% glycerol solution, 10 mℓ/kg[12,26,36,53,59,63,64,155,160,165,167,175,184,193,244,245,263,267,332,333,343,352,354,358,359] The glycerol is usually injected into the muscles of both thighs. In some studies,[56,111,112,319-321] varying doses of glycerol were given to rats subcutaneously. I.v. or i.p. injection of glycerol is not recommended because it often results in convulsions and death in a few hours.[56] In rabbits, the usual dose is 10 to 17.5 mℓ/kg, usually given subcutaneously.[56,110,159,302,305,337] If rats are deprived of water for 24 hr prior to glycerol injection, severe ATN is more consistently produced.[244,332] Control animals should be similarly water-deprived and injected with isotonic saline in place of the glycerol solution. After the injection, the animals can be allowed free access to water. The glycerol solution is usually made up by mixing equal volumes of pure glycerol (glycerin) and water or isotonic saline; this results in a markedly hyperosmotic solution.

The injection of glycerol produces severe muscle damage at the injection site, with liberation of muscle proteins (myoglobin) and potassium ions and perhaps unknown toxic substances. Hemolysis of red cells also occurs. Fluid accumulates at the injection site and blood volume is decreased.[333,354] Cardiac output and RBF, at least during the initial hours after glycerol injection, are greatly reduced.[12,59,64,167,193,263] Rats show obvious discomfort after glycerol injection, and responses to pain may contribute to altered hemodynamics. The urine is burgundy red within an hour after glycerol injection. The decreased RBF and hemoglobinuria or myoglobinuria appear to be the major factors causing the ATN. This model is often referred to as "hemoglobinuric" or "myohemoglobinuric" ARF.

Rats usually become oliguric or anuric, especially if dehydrated prior to glycerol injection. Nonoliguric ATN may occur in well-hydrated rats. In 1 day after glycerol injection, RBF has been reported to be greatly decreased[12,59,175] or normal.[64,165,167,193] In either case, filtration clearances (creatinine, inulin) are markedly decreased. RBF was improved with inhibitors of the renin-angiotensin system, but no improvement in BUN was produced.[175] Also, infusion of Ringer's solution increased RBF to normal, but inulin clearance remained depressed in established ATN.[263] Renal insufficiency usually lasts for at least 1 week. Ayer et al.[12] reported a mortality of about 70% in water-deprived rats. Thiel et al.[333] found a mortality within 48 hr of 50% in water-deprived rats, but only 14% in nondehydrated rats. ATN induced by glycerol injection is exacerbated by dehydration.

The morphological changes produced by glycerol injection have been studied in rats[56,84,111,267,319,321,343] and rabbits.[110,305] Extensive renal tubular cell necrosis and plugging of tubules with casts and crystals containing heme pigments are found. The tubular necrosis is more widespread and severe and the tubular casts are more prominent than in patients with hemodynamically mediated ATN.[305] The glomeruli appear to be normal.[84,319,343]

Mannitol or saline administration prior to glycerol injection prevents oliguria and reduces mortality.[358] Chronic oral saline loading has been reported to have a protective effect on renal functions by most[95,167,173,223,267,332,354] but not all investigators.[53] The mechanism whereby chronic saline loading protects renal function is not completely understood. It may not be related to renal renin depletion.[35,36] The extracellular volume expansion and solute diuresis produced by chronic saline loading probably contribute to its protective action. Interestingly, Wilkes and Hollenberg[355] found that rats were protected if 1% NaCl was present in the drinking water immediately after i.m. glycerol injection. Chronic drinking of 1% NaCl did not have a protective effect if the drinking fluid was switched to tap water after induction of renal failure.

The pathophysiological mechanisms involved during the maintenance phase of glycerol-induced ATN need further study. The first studies demonstrated a sustained decrease in RBF and suggested preglomerular vasoconstriction as the primary cause of filtration failure.[12,59] Later studies reported that RBF may be normal, even with a markedly decreased GFR.[64,165,167,193] These experiments suggested postglomerular vasodilation or a decrease in glomerular K_f as factors causing the decreased GFR, but experimental evidence is lacking. The possibility that tubular leakiness or obstruction contribute to the suppressed filtration clearances should be reexamined.

Heme Pigments

A number of conditions have been reported in man in which excretion of heme pigments is followed by oliguria and death from ATN. These include crush syndrome (severe muscle trauma), blackwater fever (a complication of malaria), and incompatible blood transfusions. Interest in this type of ARF was greatly stimulated by Bywaters and Beall's report in 1941.[50] They observed, during the London blitz of World War II, that persons who sustained crushing injuries to the limbs often developed a fatal renal failure. Because of the prominence of heme pigment casts in the kidneys, pathologists have called the renal disorder associated with the above conditions "pigment nephropathy, hemoglobinuric nephrosis", or "myoglobinuric nephrosis".

The role of heme pigments (myoglobin, hemoglobin, methemoglobin) in producing renal injury is still not entirely clear. Even large amounts of purified heme pigments are not nephrotoxic and can be infused without danger into normal man or animals.[37,52,131,197,199,211] In the clinical setting, however, there is often shock and trauma in addition to appearance of heme pigment in plasma and urine. When muscle cells or erythrocytes are damaged or destroyed, toxic or vasoconstrictor substances may be released, as well as myoglobin and hemoglobin.

Based on clinical observations and animal studies, certain conditions are now known to favor the development of ATN when heme pigments are liberated or injected. These include dehydration, acidosis, and antecedent tubule damage or ischemia. Dehydration prior to and during i.v. injection of hemoglobin favors the development of ATN.[176,196,199,211,214] Dehydration probably reduces RBF to some extent and also favors the precipitation of administered heme pigment in the urine. Acidosis results in more severe functional disturbances after heme pigment injection.[20,52,74,198,199,258,363] An acidic urine promotes the conversion of hemoglobin to methemoglobin. It also favors the precipitation of heme pigments in the tubular urine.[20] If the kidney is subjected to a mild injury simultaneously with heme pigment administration, then severe damage may result. Renal ischemia or nephrotoxins may cause such injury. The importance of such factors is illustrated by the work of Yuile et al.[363] These investigators found that hemoglobin injected into normal rabbits had no detrimental effects. Clamping of the renal artery for 15 min or administration of a nephrotoxin (sodium tartrate) alone produced no or only modest elevations in blood nonprotein nitrogen. If, however, hemoglobin was injected after moderate renal injury had been produced by temporary ischemia or sodium tartrate, then severe ATN developed. This was especially true if the urine was acidic (pH 5 to 5.5, compared to a normal urine pH in rabbits of 8 to 8.5). Thus, a combination of insults (heme pigment plus ischemic damage or heme pigment plus nephrotoxin) is much more effective in producing renal injury than a single insult.

A number of animal models of heme pigment-induced acute renal failure have been studied (Table 3). This table lists only studies in which ATN followed a single dose of heme pigment; other studies in which multiple doses were given over days to weeks are not included.[20,89,90,196,199] Of the three pigments, myoglobin will produce ATN in the smallest amounts. Methemoglobin is hemoglobin in which the iron atoms have been oxidized to the ferric (Fe^{3+}) form; it produces more severe injury than equal amounts of hemoglobin.[37,154,214] The reasons for these differences in potency are not entirely clear. Myoglobin (mol wt 17,000) is a smaller molecule than the hemoglobins (mol wt 68,000) and is much more filterable by the glomeruli. Hemoglobin bears a net negative charge, whereas methemoglobin bears a net positive charge at pH 7.4. The glomerular capillary wall possesses fixed negative charges that impede the passage of negatively charged macromolecules and enhance the filtration of positively charged macromolecules.[42] This suggests that methemoglobin is more filterable than hemoglobin. Possibly, differences in the degree to which myoglobin, methemoglobin, and hemoglobin are filtered may contribute to their differing potencies in inducing ATN. The amount of administered heme pigment is an important determinant of renal injury. In a transfusion accident, the amount of hemoglobin that would be liberated from hemolyzed

Table 3
HEME PIGMENT MODELS OF ACUTE RENAL FAILURE (ARF)

Animal	Heme pigment	Treatment	With ARF (%)	Died (%)	Ref.
Rabbit	Human myoglobin (impure), 50—240 mg/kg	Acid diet (oats, NH_4Cl)	63	17	52
Rabbit	Horse myoglobin, 43—52 mg/kg	Acid diet (oats, dilute HCl)	100	100	258
Rabbit	Rabbit hemoglobin, 0.45—0.74 g/kg	Acid diet (oats, stale bread); renal artery occlusion for 15—25 min or 0.8—0.95 g/kg of a 20% sodium tartrate solution, s.c.	100	36	363
Dog	Dog hemoglobin, 3—6 g/kg	NH_4Cl or $NaHCO_3$	46	40	131
Rat	Rat hemoglobin, 0.4 g/kg	Dehydrated 24—72 hr, ether anesthesia	—	—	176, 177
Dog	Dog methemoglobin, 3—14 g per dog	NH_4Cl	87	—	37
Rabbit	Rabbit methemoglobin 1 g/kg	Acid diet, 2 days of water and food deprivation	100	62	198
Rat	Human methemoglobin, 0.5 g/kg + 14.3 mg/kg sodium ferrocyanide	Acid diet (oats, horsemeat), 24-hr dehydration	80	13	214
Rat	Human methemoglobin, 2 g/kg	Acid diet (oats, horsemeat), 24-hr dehydration	—	93	253
Rat	Human methemoglobin, 2 g/kg	48-hr dehydration	—	—	277
Rat	Rat methemoglobin, 0.5 g/kg	24—48 hr dehydration	—	—	178
Rat	Methemoglobin, 0.75 g/kg	30—40 hr dehydration	—	—	248
Rabbit	Bovine methemoglobin, 0.5—1.0 g/kg	—	100	—	348

red cells after administering 1 pt of incompatible blood to a 70-kg man is about 1 g/kg. In most animal studies, comparable amounts of heme pigment were administered. As noted before, dehydration and acidosis favor the development of ATN. In most animal studies, the animals have been water-deprived prior to pigment injection for 24 hr or even more. Blood transfusions are also commonly given to volume depleted patients. An acidic urine is usually achieved by placing animals on an acid-forming diet or by administering an acidifying salt (e.g., ammonium chloride). Rabbits (and other herbivores) normally produce an alkaline urine. The urine can be made acidic by starving the rabbits or by supplying a diet of oats and stale bread.

Procedures for preparing hemoglobin and methemoglobin solutions from red cells have been described in detail by Mason et al.[214] Methemoglobin is prepared by adding 4 mol of oxidizing agent (e.g., potassium ferricyanide or sodium nitrite) for each mole of hemoglobin. Hemoglobin solutions should be freshly prepared on the experiment day since hemoglobin is converted to methemoglobin on standing. The heme pigments have characteristic absorption peaks, which allow one to check purity by spectrophotometry. Bacterial contamination of solutions should be avoided. Purified heme pigments are available from biochemical supply houses at nominal cost. The heme pigment is usually dissolved in isotonic saline as an approximately 15 g/dℓ solution and is administered into the tail vein in rats, into an ear vein in rabbits, or into a leg vein in dogs. For i.v. injections into rats, the rat should be restrained and the tail should be warmed with a heat lamp or warm water. The tail vein can be injected with a 26-gauge needle and a 1 mℓ syringe. It is good practice to draw back on the syringe plunger and watch for the appearance of blood in the syringe to confirm the intravascular location of the needle tip. If rats are anesthetized with ether to inject them, this will influence their response to heme pigment injection.[176,177] Control animals should be treated similarly with respect to diet and fluid intake and should be injected with an equal volume of saline. Addition of ferrocyanide to the injected solution increases the damaging effect of small doses of methemoglobin (or hemoglobin), but the mechanism is unknown.[214] The methemoglobin-ferrocyanide model described by Mason et al.[214] has been most carefully characterized.[214,271,272,326,328] Most recent studies have used variations of this model (Table 3), and it deserves further study.

In the heme pigment models, oliguric or nonoliguric ATN may be present. Inulin or endogenous creatinine clearance is reduced soon after injecting the heme pigment and stays below normal for about 1 week. Kidney weight is increased. There is extensive pigment cast formation in the ascending limb of Henle's loop, distal convoluted tubule, and collecting tubule. Part of the cast material may be derived from cell debris. Kidney tubules may be dilated or collapsed. Tubular necrosis is generally milder than in the glycerol model. The histological changes that occur with time have been described in many studies.[179,215,272,277]

ATN associated with heme pigment injection can often be prevented by administering diuretics (e.g., mannitol) just before or a few minutes after the pigment.[253,326] This suggests that a key factor in the early initiation of the disorder is tubular obstruction,[253,283,326,348] which results at a decreased urine flow. According to this view, a prompt and brisk diuresis prevents cast formation and aborts the lesion. Others[65,179] have suggested that renal ischemia associated with pigment injection is the primary mechanism. Goldberg,[140] however, found that RBF was essentially unchanged during the acute oliguric state that followed hemoglobin injection into dogs. He favored tubular obstruction as the explanation for the initiation of renal failure.

The factors responsible for renal insufficiency during the maintenance phase of heme pigment induced ARF are not completely agreed upon. The importance of the pigment casts is controversial. Some authors believe that the casts contribute significantly to the renal failure by obstructing the tubules.[20,177] Others consider that the heme casts are a consequence, not a cause, of the excretory defect.[13] The casts contain not only the administered heme pigments, but also tubule cell debris. It is possible that the heme pigments injure the tubule cells. Breakdown products of the heme pigments, such as hematin or globin, may have a nephrotoxic action.[74,198,226] The chemical composition and toxicity of the pigment casts deserve further study. Tubular obstruction might also secondarily lead to circulatory impairment within the kidney by elevating intrarenal pressure and compressing the blood vessels or by activating a nephron feedback mechanism. This would also contribute to cell damage. Ruiz-Guiñazú et al.[277] reported reduced capillary blood flow and nephron pressures during the maintenance phase of methemoglobin induced ARF and suggested that preglomerular vasoconstriction was of major importance. Jaenike[177] also found that RBF was often reduced. RBF, however, could be restored to normal without restoring GFR to normal, suggesting that decreased RBF alone was not the cause of the impaired renal function. It seems likely that both tubular obstruction and an inadequate filtration pressure contribute to renal insufficiency in the established phase of ATN.[178] The role of tubular leakiness has been examined in one study in which renal function was only moderately impaired. An inconsequential increase in tubular leakiness to inulin was found.[248] In summary, heme pigments may cause renal insufficiency by tubular obstruction, toxic effects, or by disturbing RBF. The relative importance of these actions needs further study.

Nephrotoxins

Mercuric Chloride

All forms of mercury (elemental mercury, inorganic and organic mercury compounds) are poisonous if absorbed. In general, except for elemental mercury and alkyl mercury compounds, the main target organ for mercury toxicity is the kidney. The inorganic mercury salt mercuric chloride ($HgCl_2$, mercuric bichloride, corrosive sublimate) is especially toxic because of its solubility and formation of the mercuric ion (Hg^{2+}). Most of the poisonous actions of mercury are associated with Hg^{2+}. In man, accidental or intentional ingestion of $HgCl_2$ has led to death after doses of 1 to 2 g. Death, after 5 to 12 days, is usually the result of renal failure. The toxic effects of $HgCl_2$ in man are vividly described by Harmon[153] and Oliver et al.[251] The kidney has a very high affinity for Hg^{2+}. Soon after administration of $HgCl_2$, kidney tissue Hg^{2+} levels greatly exceed blood levels. The kidney, compared to other organs, also retains Hg^{2+} for much longer times. This explains why renal damage is so prominent after mercury intoxication.[183,274] The main mechanism of mercury's action is its formation of covalent bonds with sulfur. In very low concentrations, mercury inactivates sulfhydryl enzymes and thus interferes with cellular metabolism and function. Mercury also binds to carboxyl, amino, imidazole, and phosphoryl groups and thereby disrupts cell functions.

$HgCl_2$ has often been administered to experimental animals to produce and study nephrotoxic ATN. The weight of the administered salt is most often indicated in the literature references; sometimes doses are expressed in terms of weight of mercuric ion. Since the formula weight of $HgCl_2$ is 271.5 and atomic weight of mercury is 200.6, mercury comprises 74% of the weight of the salt. In different studies, the dose and route of administration have varied widely. In rats, the majority of studies have used a "low" s.c. dose of 4 to 5 mg $HgCl_2$ per kilogram.[25,28,29,33,34,54,64,94,125-127,142,166,169,183,192,219,224,281,298,313,329,364,365] Some investigators have given this dose intramuscularly.[270,339] This dose produces an oliguric or nonoliguric ATN, with a 25 to 50% mortality after 7 to 9 days.[25,34,224] Some studies have employed a s.c. dose of 1.5 to 2.5 mg/kg; this dose produces less renal damage and no mortality.[94,106,144,145,313,365] A higher s.c. or i.m. dose of $HgCl_2$, 6 to 16 mg/kg, has been administered to rats[27,33,54,64,91,130,142,219,298] and produces much more severe renal damage. $HgCl_2$ has also been given intravenously to rats in doses from 0.4 to 5.0 mg/kg.[8,81,82,137,152,302] An i.v. dose of 1.5 mg/kg was a sublethal dose (5% mortality),[81,82] but the same dose given intraperitoneally produced a 100% mortality in another laboratory.[201] An i.v. dose of 2.5 to 4.0 mg/kg was uniformly fatal.[81,82] The i.v. and i.p. routes are more lethal than s.c. or i.m. routes because higher blood levels are reached more rapidly and presumably more mercury accumulates in the kidneys. Male rats appear to be more sensitive to the nephrotoxic effect of mercury than female rats.[152] In rabbits, ATN was induced by i.v. injection of 1.5 to 4.1 mg of $HgCl_2$ per kilogram.[302,347] In dogs, $HgCl_2$ has been given intravenously in doses of 0.5 to 5 mg/kg.[16,21,23,24,280,297,340] A 2-mg/kg dose produced a 68% mortality after 3 to 10 days; a 4-mg/kg dose was uniformly fatal.[21,280] Conn et al.[72] gave $HgCl_2$ intramuscularly in a dose of 15 to 30 mg/kg to dogs, but most of these animals died from extrarenal effects before studies could be done.

To administer $HgCl_2$, it is usually dissolved as a 1 to 5 mg/m*l* solution in isotonic saline. To give a dose of 4 mg/kg to a 250-g rat, for example, one would administer either 1 m*l* of a 1 mg/m*l* solution or 0.2 m*l* of a 5 mg/m*l* solution. Control animals should be injected with an equal volume of isotonic saline. It is sometimes a good idea to stain the injection solution with a nontoxic dye (e.g., lissamine green) so as to detect possible loss from the injection site. $HgCl_2$ is a violent poison, much more dangerous than elemental mercury, and care should be used in handling it (do not breathe dust; keep away from food; wash hands before eating or smoking).

The morphological changes produced in the kidneys by mercury intoxication have been described by many investigators.[28,29,33,46,79 82,94,102,106,137,142,144,145,201,224,250,251,270,281,298,308,347,364,365] Low doses of mercury produce selective necrosis of the pars recta of the proximal tubule. Larger doses also damage the convoluted portion of the proximal tubule. Necrosis of the distal tubule is only rarely observed, even when large amounts of mercury are administered.[33,102,106,137,144,152,224,251,270,308] Damage to the kidney proximal tubule cells is related to their accumulation of mercury.[308] Distal nephrons are filled with casts of cell debris. Glomerular changes, such as swelling of the podocytes and glomerular endothelium and a loss of endothelial fenestrae, have been described by some authors.[16,54,365]

The response to $HgCl_2$ injection varies, of course, with the dose. Intoxicated animals may show bloody diarrhea, lowered body temperature, anorexia, dehydration, loss of weight, weakness, and lethargy. In the initial phase of ATN, animals will often show a diuresis, due to impaired reabsorption of filtered salt and water. (Organic mercurials were at one time widely used as effective diuretics.) GFR decreases progressively soon after injection of $HgCl_2$.[27,106,340] RBF usually is described as falling only modestly, *if* at all, in the initial phase.[16,21,24,64,166,192,297,302,340] The fall in RBF, when it occurs, may be either related to dehydration, to a cardiotoxic action of mercury, or to an effect on vascular smooth muscle. In the maintenance phase of ATN (days 1 to 3, or longer, depending on the dose), animals may be oliguric or nonoliguric. In general, higher mercury doses are associated with oliguria. The inulin clearance is markedly decreased. The reasons for this decrease are complex and controversial. Evidence has been obtained for several mechanisms: back-leak of filtered substances,[25,33,266,313,347] tubular obstruction,[33,127,219] a decrease in glomerular K_f,[54] and a hemodynamic disturbance. The significance of tubule obstruction in this model is not certain. RBF is often normal or greater than normal in the established phase of this disorder,[23,24,64,72,166,192] but sometimes it is somewhat reduced.[16,21,33,72] If RBF is increased to supernormal values by saline infusion, GFR is not improved.[16] This result suggests that renal vasoconstriction is not essential to the renal failure. If, however, there is a combination of preglomerular vasoconstriction and postglomerular vasodilation, then RBF may be normal, but GFR will be suppressed because of a low glomerular capillary pressure.[24,166] The disagreement in the literature concerning the factors responsible for a low GFR reflects the fact that probably all of the proposed mechanisms may be involved, but to different extents, depending on the dose, time of study, and other experimental conditions. Even in the same kidney, different proximal tubules may be damaged to differing extents,[169] suggesting that the pathophysiological mechanisms which cause the apparent filtration failure may vary from one nephron to another.

The time for recovery varies with the dose of $HgCl_2$ (extent of injury), but it is usually heralded by polyuria. With a low $HgCl_2$ dose in the rat, functional recovery usually begins on the 3rd to 4th day after injection.[25] Regeneration of kidney tubule cells begins between 1 to 3 days after injury. The surviving cells rapidly reline the tubule, and the damaged nephron will recover completely, provided that the tubular basement membrane is intact. The regenerating cells come mainly from the zone of necrosis. Although a normal GFR may be observed 9 days after $HgCl_2$ administration, restoration of normal cell morphology and enzyme patterns may take several weeks.[79 82] During the recovery process, there is an increase in anaerobic glycolysis and hexose monophosphate shunt activity.[8]

Several studies have demonstrated that chronic saline ingestion (1% sodium chloride in drinking water for several weeks) reduces the severity of ATN after $HgCl_2$ administration.[16,29,34,94,125,126,166,329,330] Recent studies suggest that this protective effect is independent of renal renin suppression.[34,91] Continuous i.v. saline loading also has a protective effect.[91] The beneficial effect of saline could be related to reduced mercury levels in the kidney[183] or relief or prevention of tubular obstruction. At the moment, the mechanism of this beneficial effect is not clear. The diuretic drug furosemide, in high or low doses, did not improve renal function when administered after $HgCl_2$ to rats.[339] Dithiothreitol, a dithiol sugar capable of chelating heavy metals and regenerating reduced sulfhydryl groups, had a protective effect when administered intraperitoneally 30 min after $HgCl_2$[28,187] Clonidine, a centrally active sympathomimetic agent, afforded partial protection of renal function and structure in mercury poisoned rats.[106]

Uranium

Uranium is a well-known nephrotoxin that has often been used to produce ATN. In experimental studies of ATN, the soluble uranium salts uranyl nitrate (UN) or uranyl acetate (UA) are usually administered. UN is a hexahydrate, $UO_2(NO_3)_2 \cdot 6H_2O$, and has a formula weight of 502; it contains 0.47 mg of uranium per milligram of salt. UA, $UO_2(C_2H_3O_2)_2 \cdot 2H_2O$ is a dihydrate and has a formula weight of 424; it contains 0.56 mg of uranium per milligram of salt. Doses in the literature are most often given as weight of the hydrated salt, as will be done here.

Extensive studies on uranium toxicity have been performed in experimental animals.[345] The kidney is the site of the most severe injury; low doses of uranium result in death from renal failure. The toxicity of UN varies with the dose and route of administration, species, strain, and age. The LD_{50} (dose which will cause 50% mortality) for i.v. administration is approximately 0.2 mg/kg in rabbits, 2 mg/kg in rats, and 20 to 40 mg/kg in mice.[345] In rats, a s.c. dose of 10 mg/kg resulted in a 60% mortality.[147] In dogs, a s.c. dose of 5 mg/kg resulted in a 92% mortality,[98] and an i.v. dose of 6.4 mg/kg led to 100% mortality.[143] Many of the recent studies of UN-induced ATN have employed i.v., i.p., or s.c. doses of 5 to 25 mg/kg in rats[11,34,38,54,122-124,146,147,187,219,220,225,248] and 5 to 10 mg/kg intravenously in dogs.[105,128,186,222,309,311] In rabbits, doses have varied between about 0.02 and 2 mg/kg of UA[236,316,317] or UN[32,132,194] usually given intravenously. In many of these studies the doses are quite high and a lethal (irreversible) ATN was probably induced. The mortality data cited above suggest that the order of susceptibility to ATN from UN varies 100-fold from rabbit (most sensitive) to mouse (least sensitive). Even within the same species, significant strain differences in susceptibility may be present.[345] The age of the animal affects the toxicity of uranium; uranium is less toxic in younger than in older dogs and rats.[206,209,345] This may occur in part because more uranium is deposited in the skeleton in younger animals, thereby leaving less toxin to damage the kidneys.[234]

UN or UA is usually dissolved in isotonic saline so that 1 m*l* contains the dose to be given per kilogram of body weight. For example, for a 10 mg/kg dose, a 10 mg/m*l* solution would be prepared, and 1 m*l* would be injected per kilogram. Control animals should be injected with an equal volume of isotonic saline. As with any potent nephrotoxin, uranium should be handled with care (see mercuric chloride). Uranium salts are also hazardous because they are radioactive. After i.v. injection, uranium complexes with bicarbonate in the blood and is filtered by the glomeruli. About two thirds of the injected uranium is excreted in the urine within 24 hr in rats, rabbits, and cats. About one fifth is taken up by bone, from which it is slowly released.[234,254]

The morphological changes produced by uranium poisoning have been described in many studies.[11,31,146,147,249,278,309,345] Morphologic changes are detectable as early as 1 hr after UN injection and become progressively more severe over several days.[146] The most obvious histological change is necrosis of proximal tubules, particularly the pars recta segment.[207,249]With large doses of uranium, all nephron segments become involved.[146] Tubular casts become prominent at the height of the renal failure.[31,146] Glomerular changes include abnormalities in the epithelial[146,309] and endothelial cells.[11,54] The severity of renal damage clearly depends upon the uranium dose.[278]

The initial phase of uranium-induced ATN usually involves polyuria for the first day or two.[31,38,124,186,207,219,220,225,248,278,316] With large doses, polyuria is usually followed by oliguria or anuria and death.[38,309,316] Inulin clearance decreases soon after administering the toxin.[11,38,122,123,186,219] RBF during the initial stage (3 to 7 hr post-injection) has been reported to be decreased by some investigators,[124,128,186,204,309,311] but normal by others.[38,317] Mauk et al.[222] produced a marked increase in RBF in uranium-poisoned dogs by infusing the vasodilator prostaglandin E. This did not, however, prevent the initial fall in inulin clearance or the marked suppression in renal function 48 hr later. Hence, the pathophysiological significance of an early fall in RBF is uncertain. RBF is often normal during the established phase of ATN,[40,156,222,309,316,317] but is decreased in some studies.[124,128] Proteinuria, glucosuria, and aminoaciduria are observed at various stages of the disorder; these findings suggest glomerular and proximal tubular damage.[32,38,194,236] The reasons for the reduced filtration clearances are controversial, probably owing to variations in the species studied, the dose of uranium, and the time of study. Many studies suggest an abnormal tubular permeability to the chemical substances used to measure filtration rate as the explanation for reduced clearances,[38,40,105,156,236,309,311,316] but the significance of tubular leakiness is disputed by other studies.[122,124,132,248,317] Morphological studies suggest that tubular obstruction may become important in the late maintenance phase of ATN,[31,146] but functional evidence is lacking. A decrease in glomerular K_f may contribute to the reduced GFR.[38,54,309] Avasthi et al.[11] found that there was a decrease in the number and size of glomerular endothelial fenestrae, which may be the morphological basis for the reduction in K_f. In uranium-induced ATN, both tubular and glomerular injury appear to contribute to the renal insufficiency.

Recovery of renal function in surviving animals is relatively slow. Kidney regeneration after a single i.v. injection of uranium salt begins about 1 day after the first degenerative changes. In the rat, recovery of renal function was quite incomplete 1 week after s.c. injection of 5 mg of UN per kilogram[34] or 2 weeks after 10 mg of UN per kilogram.[147] Recovery takes about 3 weeks in the rabbit.[345] Many animals never completely recover normal renal function and structure, and a chronic nephritis develops.[207,249] For example, in dogs receiving 4 mg of UN per kilogram subcutaneously, 71% of the surviving animals did not recover normal kidney function.[208] The recovery pattern for the kidneys after uranium poisoning remains to be studied with micropuncture techniques.

Continuous saline loading (1% NaCl in drinking water) reduced the increase in BUN after UN injection.[34,278] It also resulted in protection of GFR and preservation of normal endothelial cell ultrastructure.[11] Alkali administration (sodium bicarbonate or sodium citrate) reduced the nephrotoxicity of uranium.[98,143] This effect may be related to increased urinary excretion of uranium and reduced deposition in the kidneys.[234] Administration of dithiothreitol prevented the induction of ATN by UN.[187]

The effects of mercury and uranium on the kidney show many similarities, but also many differences. Both metals rapidly accumulate in the kidneys after their administration. Uranium is excreted rapidly, but mercury is excreted slowly, by the kidneys.[183,187,234] Both metals selectively damage the last third of the proximal tubule; high doses produce more widespread tubule damage. Proximal tubular necrosis is most extensive 5 days after administering uranium,[146] whereas peak necrosis with mercury is observed after 1 day.[298] Tubular obstruction is more evident early in the mercury model. Uranium produces glomerular damage; this is seen only with high doses of mercury. Uranium poisoned animals typically show polyuria for several days before the onset of oliguria; mercury poisoned animals are often oliguric or anuric 1 day after mercury injection. BUN climbs more slowly in the uranium model,[34] and repair occurs more slowly in the uranium model than in the mercury model.[81,145,147] After uranium intoxication a chronic nephritis frequently is present.[207,250]

Other Models

Table 4 lists several additional models of ARF. The list is not exhaustive. The diversity of models parallels the variety of conditions that cause the disorder in human beings. In many instances, the same agents are known or suspected of causing ARF in man. The experimental analogues are informative in defining the renal changes which occur and in devising approaches to prevent or treat ATN. In the table, "type of study" indicates the main approach used. In most studies, both morphological and functional approaches were employed to define the nature and degree of injury. There are many additional chemical substances which will produce ARF, e.g., certain analgesics, antibiotics, and cancer chemotherapy drugs, but space does not permit their consideration here.

Constant i.v. infusion of epinephrine in the dog for 6 hr produces ATN. The course of the disorder after termination of the infusion has not been reported. Renal injury is diminished by prior splenectomy or i.v. infusion of a vasodilator prostaglandin.[30,212,213]

Dextran 40 (low-molecular-weight dextran) is a glucose polymer, with a weight average molecular weight of about 40,000 daltons, which is used as a plasma volume expander. It is by itself not demonstrably nephrotoxic. If, however, dextran 40 infusion is combined with a decreased renal perfusion pressure (due to hemorrhage or aortic constriction), then ATN may develop. Under these conditions, precipitation of dextran 40 in the urine is believed to play a key role in the development of renal insufficiency.[61]

Globin, the protein constituent of hemoglobin, produces ATN when injected intravenously in rats.[158,226] Although tubular obstruction is an obvious contributor to renal dysfunction, globin also appears to exert a toxic effect on cells of the thick ascending limb and distal convoluted tubule.[226] Results with globin may be relevant to the heme pigment models of ATN.

High plasma urate levels in man, such as may occur during treatment for leukemias, may result in ARF. Unlike man, the common laboratory animals possess uricase, an enzyme that rapidly breaks down uric acid. For this reason, experimental models have often used a uricase inhibitor (e.g., oxonic acid) in combination with urate administration. Obstruction of both tubules and the renal vasculature with precipitated uric acid and urate crystals is a primary and early pathogenetic event in urate-induced ARF.[67,68,73,307] A small dose of uric acid in the rabbit, harmless in itself, may induce ARF if it is combined with tourniquet shock.[41]

Folic (pteroylglutamic) acid, a vitamin, when injected in a large i.v. dose into rats, produces reversible ARF characterized by only mild tubular damage. Intratubular precipitation of the folic acid and consequent tubular obstruction appears to be the primary cause of renal failure.[170] A nephrotoxic effect on tubular cells may also be important.[287,289,291]

Glaphenine, an analgesic compound, produces ARF in man or the rat, if administered in large doses. Tubular obstruction by glaphenine derivatives is a prominent factor in the pathogenesis of renal insufficiency.[136]

As discussed before, many of the heavy metals produce ATN. Chromium salts produce selective damage to the pars convoluta of the proximal tubule.[19,33,82,109] ATN induced by potassium dichromate has been examined in a number of studies in rats (Table 4). Abnormal tubular leakiness, tubular obstruction, and a moderately reduced RBF contribute to the renal insufficiency in the maintenance phase of ATN.[33,158,300] *Cis*-diamminedichloroplatinum II (cisplatin; *cis*-DDP; NSC-119875) is a chemothera-

Table 4
OTHER MODELS OF ACUTE RENAL FAILURE[a]

Insult	Dose	Time of study	Animal	Type of study	Ref.
Epinephrine	4 μg/kg/min i.v. for 6 hr	0—6 hr	Dog	Clearance, blood flow	30
Dextran 40	500 mℓ 10% dextran 40 i.v. plus decreased renal perfusion	4—48 hr	Dog	Clearance	61
Human globin	0.75 g/kg i.v.	1 min—6 hr	Rat	Electron microscopy	226
	0.25 g/kg i.v.	3 hr	Rat	Micropuncture	158
Uric acid	0.08 g/kg ethanolamine urate i.v. + tourniquet on a single hind leg	1—6 days	Rabbit	Blood and urine chemistry	41
	Oxonic acid 0.25 g/kg i.p. + 2 mg/kg/min i.v. + 300 mg/kg lithium urate i.v.	1 hr	Rat	Micropuncture, clearance	67, 68
	5% oxonic acid plus 2.5% uric acid in food for 7 days	8th day	Rat	Micropuncture, clearance	307
	3% oxonic acid + 5% uric acid in food for 2 days	3rd day	Rat	Micropuncture	73
Folic acid	0.22 g/kg i.v.	6 hr—7 days	Rat	Clearance	168
	0.25 g/kg/i.v.	10 min—19 days	Rat	Enzymology, electron microscopy	289
	0.25 g/kg i.v.	1 hr, 24 hr	Rat	Enzymology	287
	0.25 g/kg i.v.	2 hr—9 days	Rat	Histology	291
	0.25 g/kg i.v.	0—25.5 hr	Rat	Micropuncture	170
	0.20 g/kg i.p.	6 hr—14 days	Mouse	Biochemistry	229
Glaphenine	0.80 g/kg i.g.	1.5 hr—8 days	Rat	Micropuncture, microspheres	136
Chromium salts	10 mg/kg $K_2Cr_2O_7$ s.c.	1—42 days	Rat	Histochemistry, histology	19
	7—16 mg/kg $K_2Cr_2O_7$ s.c.	3, 7 days	Rat	Micropuncture, microdissection	33
	15 mg/kg $K_2Cr_2O_7$ s.c.	1 day	Rat	Micropuncture	158
	15 mg/kg $K_2Cr_2O_7$ i.v.	1—28 days	Rat	Histology	82
	15 mg/kg $K_2Cr_2O_7$ s.c.	3 hr—16 days	Rat	Immunohistochemistry	292
	10—20 mg/kg Na_2CrO_4 i.p.	1—48 hr	Rat	Electron microscopy	109
	15 mg/kg $K_2Cr_2O_7$ s.c.	1—14 days	Rat	Clearance, microspheres	4, 300
Cisplatin	5—10 mg/kg or 1.5—3 mg/kg/day for 5 days i.v.	1—8 days	Dog	Blood and urine chemistry, histology	284
	7.6 mg/kg i.p.	1—15 days	Rat	Histology, blood chemistry	349
	3 mg/kg i.v.	1—21 days	Dog	Blood and urine chemistry	83
	6 mg/kg i.v.	1—10 days	Rat	Clearance, histology	257
	6 mg/kg i.p.	1—7 days	Rat	Electron microscopy	96
	10 mg/kg i.p.	1—10 days	Rat	Micropuncture, electron microscopy	62, 182

Table 4 (continued)
OTHER MODELS OF ACUTE RENAL FAILURE[a]

Insult	Dose	Time of study	Animal	Type of study	Ref.
	5 mg/kg i.p.	1—30 days	Rat	Micropuncture	279
	5—6 mg/kg i.p.	1—8 days	Rat	Clearance, balance	141

[a] Abbreviations: i.g., intragastric; i.p., intraperitoneal; i.v., intravenous; s.c., subcutaneous.

peutic drug found to be effective against certain solid tumors, including testicular cancer. It has considerable antitumor activity because of its property of inhibiting or interfering with DNA synthesis. The nephrotoxicity of this compound limits its clinical use.[210,276,351] The drug is a coordination complex of platinum surrounded by chlorine atoms and ammonia groups in the *cis*-position in the horizontal plane. The nephrotoxic effect of cisplatin is exerted primarily on the S3 segment (pars recta) of the proximal tubule.[62,96,349] Experimental ATN after a single dose of cisplatin has been examined in a number of studies (Table 4). Abnormal tubular leakiness and a primary decrease in GFR play major roles in the impaired renal function.[62,182,279]

OUTSTANDING ISSUES

Despite extensive study of ATN during the past decade, many issues remain unresolved. Some of these will be briefly mentioned below.

How or why ATN develops in man in many clinical settings is not well understood. Although many factors which contribute to the initiation of ATN have been identified and studied in experimental models (e.g., renal ischemia, heme pigments, and nephrotoxins), the precise manner in which renal injury is produced is often incompletely understood.

The role of endotoxin, a component of the cell wall of Gram-negative bacteria, in the etiology of ATN deserves further study. Elevated endotoxin levels will reduce RBF and may cause or accentuate renal ischemia and reduced GFR in human ATN. The role of endotoxin should be studied further in animal models.[350]

Tubular cell damage appears to play a key role in the initiation and maintenance of ATN. More studies on the biochemical events that underlie cell injury need to be done. In particular, the effects of ischemic and nephrotoxic types of injury should be examined. Such studies might include measurements of cell adenine nucleotide levels, mitochondrial calcium content, oxygen consumption, and cell electrolyte concentrations in different experimental models. These parameters may provide indexes of the effectiveness of various agents in preventing tubular cell injury.

How tubular damage leads to a reduced GFR is often poorly understood. The idea that a tubuloglomerular feedback mechanism plays a role in restraining GFR is an attractive one.[335] It has received some experimental support,[216,220,361,362] but needs further experimental verification. Of course, tubular leakiness and tubular obstruction can explain reduced GFR and clearances in many instances.

Ways in which ATN can be interrupted or prevented require further study. A variety of manipulations prior to or during the initiation phase have been found to be effective in protecting the kidneys against development of severe ATN in experimental animals. These include chronic saline loading, anesthetics,[296] beta-adrenergic antagonists,[60,108,135,171,185,303] clonidine,[106,174,304,306] dopamine,[172] diuretic drugs,[18,114,191,218] mannitol,[48,76,114,151,263,326] and ATP-$MgCl_2$.[299] The effectiveness of these interventions depends on the experimental model and the time of application. In many cases, the mode of action is not completely understood.

In the established phase of ATN, many factors are now recognized as contributing to filtration failure, depending on the model and time of study. It is clear that no single pathophysiologic mechanism can explain the low GFR in all types of ATN. A challenge to the future will be to define better the relative quantitative importance of tubular leakiness, tubular obstruction, reduced glomerular K_f, and renal hemodynamic disturbances in explaining the low GFR in various models. These factors may be interrelated in ways which are not completely understood (Figure 1). The relative weight of these different factors probably varies with the course of the disease in a single model. It may also vary from nephron to nephron in models where damage is not uniform.

The role of altered renal hemodynamics in ATN needs further study. It is now clear that in many models total RBF is normal, or nearly so, during established ATN. Some investigators have suggested that despite a normal RBF, glomerular capillary pressure is reduced due to combined preglomerular vasoconstriction and postglomerular vasodilation.[24,164,264] No clear-cut evidence in support of this idea has, however, been obtained in micropuncture studies. Further studies on nephron pressures and flows and determinations of pre- and postglomerular vascular resistances should be done, using micropuncture techniques, in ATN models with a normal RBF. The intrarenal distribution of blood flow may be disturbed in ATN.[71,331] Study of this problem will, however, require the development and use of more reliable methods for measuring regional blood flow.[10]

In those models in which RBF is clearly reduced, the reasons for the increased renal vascular resistance are not well understood. A popular idea has been that the renin-angiotensin system causes renal vasoconstriction in ATN,[121,242,335] but the bulk of recent evidence does not support an important role for this system either in the initiation or maintenance of ATN.[26,34-36,91,203,310,329] It is also possible that a lack of vasodilator prostaglandins[113,157,240,338] or enhanced formation of vasoconstrictor substances, such as thromboxane,[228] might contribute to a reduced RBF.

The role of changes in glomerular K_f in filtration failure needs to be clarified. Some evidence has been obtained that glomerular K_f decreases with high doses of uranium or mercury[38,54] or after temporary renal artery occlusion.[282,356] The relation between changes in glomerular morphology and K_f[11,306] deserves further study.

Much information is still lacking as to how many nephrotoxic drugs produce renal injury. Further study of how various nephrotoxins are distributed, metabolized, and excreted and of how they interact with cells and cellular components to produce injury are needed. The reasons for the distinctive patterns of damage produced by the heavy metals remain to be elucidated. Interestingly, renal ischemia also produces necrosis of a specific tubule segment (proximal tubule S_3).[97,342] Why this is so is not precisely known. It is possible that the selective effects of heavy metals are related not only to uptake by certain nephron segments, but are also related to alterations in intrarenal hemodynamics.

An important area for future investigation concerns means for enhancing recovery from ATN. Once ATN is established, it seems that little can be done to alter its course. Two new experimental maneuvers hold some promise, however, and there may be others. I.v. infusion of amino acids was found to be effective in improving renal function in rats with ATN induced by a low dose of $HgCl_2$.[336] ATP-$MgCl_2$ accelerated renal recovery even when infused 1 day after ischemia-induced ATN in rats.[138] The reasons for these beneficial effects are not known for certain. Biochemical studies on the factors that promote renal cell regeneration in the injured nephron would be interesting.

If tubular obstruction is important in the maintenance phase of ATN, then its relief might result in an improvement of GFR. At present, we have no means to remove the obstruction and no real understanding of how casts are removed during the course of recovery from ATN.

Histological studies have detected cell necrosis very late in time after an initiating insult.[88,180,305] This suggests that during prolonged ATN there is ongoing tubular necrosis, perhaps due to recurrent ischemia. The basis for this is not understood, although it is possible that impaired renal blood flow autoregulation[1] and episodes of hypotension are involved.[269] A better understanding of the factors that lead to continuing cell death, and how this can be prevented, should result in improved survival after this serious form of ARF.

REFERENCES

1. **Adams, P. L., Adams, F. F., Bell, P. D., and Navar, L. G.,** Impaired renal blood flow autoregulation in ischemic acute renal failure, *Kidney Int.*, 18, 68, 1980.
2. **Anderson, R. J., Gordon, J. A., Kim, J., Peterson, L. M., and Gross, P. A.,** Renal concentration defect following nonoliguric acute renal failure in the rat, *Kidney Int.*, 21, 583, 1982.
3. **Anderson, R. J., Linas, S. L., Berns, A. S., Henrich, W. L., Miller, T. R., Gabow, P. A., and Schrier, R. W.,** Nonoliguric acute renal failure, *N. Engl. J. Med.*, 296, 1134, 1977.
4. **Appel, G. B., Siegel, N. J., Appel, A. S., and Hayslett, J. P.,** Studies on the mechanism of nonoliguric experimental acute renal failure, *Yale J. Biol. Med.*, 54, 273, 1981.
5. **Arendshorst, W. J., Finn, W. F., and Gottschalk, C. W.,** Nephron stop-flow pressure response to obstruction for 24 hours in the rat kidney, *J. Clin. Invest.*, 53, 1497, 1974.
6. **Arendshorst, W. J., Finn, W. F., and Gottschalk, C. W.,** Pathogenesis of acute renal failure following temporary renal ischemia in the rat, *Circ. Res.*, 37, 558, 1975.
7. **Arendshorst, W. J., Finn, W. F., and Gottshchalk, C. W.,** Micropuncture study of acute renal failure following temporary renal ischemia in the rat, *Kidney Int.*, 10(Suppl. 6), S-100, 1976.
8. **Ash, S. R. and Cuppage, F. E.,** Shift toward anaerobic glycolysis in the regenerating rat kidney, *Am. J. Pathol.*, 60, 385, 1970.
9. **Ash, S. R., Cuppage, F. E., Hodes, M. E., and Selkurt, E. E.,** Culture of isolated renal tubules: a method of assessing viability of normal and damaged cells, *Kidney Int.*, 7, 55, 1975.
10. **Aukland, K.,** Methods for measuring renal blood flow: total flow and regional distribution, *Annu. Rev. Physiol.*, 42, 543, 1980.
11. **Avasthi, P. S., Evan, A. P., and Hay, D.,** Glomerular endothelial cells in uranyl nitrate-induced acute renal failure in rats, *J. Clin. Invest.*, 65, 121, 1980.
12. **Ayer, G., Grandchamp, A., Wyler, T., and Truniger, B.,** Intrarenal hemodynamics in glycerol-induced myohemoglobinuric acute renal failure in the rat, *Circ. Res.*, 29, 128, 1971.
13. **Ayer, G. D. and Gauld, A. G.,** Uremia following blood transfusion. The nature and the significance of the renal changes, *Arch. Pathol.*, 33, 513, 1942.
14. **Badenoch, A. W. and Darmady, E. M.,** The effects of temporary occlusion of the renal artery in rabbits and its relationship to traumatic uremia, *J. Pathol. Bacteriol.*, 59, 79, 1947.
15. **Baehler, R. W., Williams, R. H., Work, J., Gottschall, J., and Chuang, V.,** Studies on the natural history of the norepinephrine model of acute renal failure in the dog, *Nephron*, 26, 266, 1980.
16. **Baehler, R. W., Kotchen, T. A., Burke, J. A., Galla, J. H., and Bhathena, D.,** Considerations on the pathophysiology of mercuric chloride-induced acute renal failure, *J. Lab. Clin. Med.*, 90, 330, 1977.
17. **Baehler, R. W., Kotchen, T. A., and Ott, C. E.,** Failure of chronic sodium chloride loading to protect against norepinephrine-induced acute renal failure in dogs, *Circ. Res.*, 42, 23, 1978.
18. **Bailey, R. R., Natale, R., Turnbull, D. I., and Linton, A. L.,** Protective effect of frusemide in acute tubular necrosis and acute renal failure, *Clin. Sci. Mol. Med.*, 45, 1, 1973.
19. **Baines, A. D.,** Cell renewal following dichromate induced renal tubular necrosis. An enzyme histochemical study, *Am. J. Pathol.*, 47, 851, 1965.
20. **Baker, S. L. and Dodds, E. C.,** Obstruction of the renal tubules during the excretion of hemoglobin, *Br. J. Exp. Pathol.*, 6, 247, 1925.
21. **Bálint, P.,** Pathogenesis of mercuric chloride-induced renal failure in the dog, *Acta Med. Acad. Sci. Hung.*, 25, 287, 1968.
22. **Bálint, P., Châtel, R., Fekete, A., and Forgács, I.,** Haemodynamics and oxygen consumption of the kidney in post-ischaemic renal failure, *Clin. Sci.*, 26, 471, 1964.

23. Bálint, P., Fekete, Á., and Harza, T., Mercuric chloride-induced renal failure and intrarenal distribution of blood flow, *Acta Physiol. Acad. Sci. Hung.*, 36, 193, 1969.
24. Bálint, P., Szöcs, E., and László, K., Effect of mercuric chloride intoxication on intrarenal pressure gradients and vascular resistances, *Acta Physiol. Acad. Sci. Hung.*, 43, 155, 1973.
25. Bank, N., Mutz, B. F., and Aynedjian, H. S., The role of "leakage" of tubular fluid in anuria due to mercury poisoning, *J. Clin. Invest.*, 46, 695, 1967.
26. Baranowski, R. L., Westenfelder, C., and Kurtzman, N. A., Intrarenal renin and angiotensins in glycerol-induced acute renal failure, *Kidney Int.*, 14, 576, 1978.
27. Barenberg, R. L., Solomon, S., Papper, S., and Anderson, R., Clearance and micropuncture study of renal function in mercuric chloride treated rats, *J. Lab. Clin. Med.*, 72, 473, 1968.
28. Barnes, J. L., McDowell, E. M., McNeil, J. S., Flamenbaum, W., and Trump, B. F., Studies on the pathophysiology of acute renal failure. IV. Protective effect of dithiothreitol following administration of mercuric chloride in the rat, *Virchows Arch. (Cell Pathol.)*, 32, 201, 1980.
29. Barnes, J. L., McDowell, E. M., McNeil, J. S., Flamenbaum, W., and Trump, B. F., Studies on the pathophysiology of acute renal failure. V. Effect of chronic saline loading on the progression of proximal tubular injury and functional impairment following administration of mercuric chloride in the rat, *Virchows Arch. (Cell Pathol.)*, 32, 233, 1980.
30. Bell, R. D., Mandal, A. K., and Parker, D. E., The effect of splenectomy on renal function in epinephrine-induced renal failure, *Proc. Soc. Exp. Biol. Med.*, 167, 12, 1981.
31. Bencosme, S. A., Stone, R. S., Latta, H., and Madden, S. C., Acute tubular and glomerular lesions in rat kidneys after uranium injury, *Arch. Pathol.*, 69, 470, 1960.
32. Berke, H. and Rothstein, A., Amino aciduria in uranium poisoning. II. The response to different amounts of uranium given intravenously and by inhalation, *J. Pharmacol. Exp. Ther.*, 96, 198, 1949.
33. Biber, T. U. L., Mylle, M., Baines, A. D., Gottschalk, C. W., Oliver, J. R., and MacDowell, M. C., A study by micropuncture and microdissection of acute renal damage in rats, *Am. J. Med.*, 44, 664, 1968.
34. Bidani, A., Churchill, P., and Fleischmann, L., Sodium-chloride-induced protection in nephrotoxic acute renal failure: independence from renin, *Kidney Int.*, 16, 481, 1979.
35. Bidani, A. K., Fleischmann, L. E., Churchill, P., and Becker-McKenna, B., Natriuresis-induced protection in acute myohemoglobinuric renal failure without renal cortical renin content depletion in the rat, *Nephron*, 22, 529, 1978.
36. Bidani, A. K. and Churchill, P. C., Na intake, renal renin, and the severity of myohemoglobinuric renal failure in rats, *Am. J. Physiol.*, 241, F34, 1981.
37. Bing, R. J., The effect of hemoglobin and related pigments on renal functions of the normal and acidotic dog, *Bull. Johns Hopkins Hosp.*, 74, 161, 1944.
38. Blantz, R. C., The mechanism of acute renal failure after uranyl nitrate, *J. Clin. Invest.*, 55, 621, 1975.
39. Blantz, R. C. and Tucker, B. J., Measurements of glomerular dynamics, in *Methods in Pharmacology*, Vol. 4B, Martinez-Maldonado, M., Ed., Plenum Press, New York, 1978, 141.
40. Bobey, M. E., Longley, L. P., Dickes, R., Price, J. W., and Hayman, J. M., Jr., The effect of uranium poisoning on plasma diodrast clearance and renal plasma flow in the dog, *Am. J. Physiol.*, 139, 155, 1943.
41. Boda, D., Háry, J., and Szinay, G., Acute renal failure induced by urate infusion in the rabbit. An experimental study of the part played by urates in the induction of shock-kidney, *Acta Med. Acad. Sci. Hung.*, 23, 69, 1966.
42. Brenner, B. M., Hostetter, T. H., and Humes, H. D., Molecular basis of proteinuria of glomerular origin, *N. Engl. J. Med.*, 298, 826, 1978.
43. Brenner, B. M. and Stein, J. H., Eds., *Acute Renal Failure*, Churchill Livingstone, New York, 1980.
44. Brun, C., *Acute Anuria*, Munksgaard, Copenhagen, 1954.
45. Bulger, R. E. and Dobyan, D., Morphological techniques for study of the kidney, in *Methods in Pharmacology, Vol. 4B, Martinez-Maldonado, M., Ed.*, Plenum Press, New York, 1978, 1.
46. Bulger, R. E. and Siegel, F. L., Alterations of the renal papilla during mercuric chloride-induced acute tubular necrosis, *Lab. Invest.*, 33, 712, 1975.
47. Bull, G. M., Joekes, A. M., and Lowe, K. G., Renal function studies in acute tubular necrosis, *Clin. Sci.*, 9, 379, 1950.
48. Burke, T. J., Cronin, R. E., Duchin K. L., Peterson, L. N., and Schrier, R. W., Ischemia and tubule obstruction during acute renal failure in dogs: mannitol in protection, *Am J. Physiol.*, 238, F305, 1980.
49. Burwell, R. G., Changes in the proximal tubule of the rabbit kidney after temporary complete renal ischemia, *J. Pathol. Bacteriol.*, 70, 387, 1955.
50. Bywaters, E. G. L. and Beall, D., Crush injuries with impairment of renal function, *Br. Med. J.*, 1, 427, 1941.

51. Bywaters, E. G. L. and Dible, J. H., The renal lesion in traumatic anuria, *J. Pathol. Bacteriol.*, 54, 111, 1942.
52. Bywaters, E. G. L. and Stead, J. K., The production of renal failure following injection of solutions containing myohemoglobin, *Q. J. Exp. Physiol.*, 33, 53, 1944.
53. Cabili, S. and Charney, A. N., Lack of an effect of saline loading on glycerol-induced acute renal failure, *Nephron*, 30, 73, 1982.
54. Cachia, R., Savin, V. J., Patak, R. V., and Ridge, S. M., Effect of mercuric chloride ($HgCl_2$) and uranyl nitrate (UN) on ultrafiltration coefficient (K_f) in isolated glomeruli, *Kidney Int.*, 19, 236, 1981.
55. Cameron, G. R. and Finckh, E. S., The production of an acute hemolytic crisis by the subcutaneous injection of glycerol, *J. Pathol. Bacteriol.*, 71, 165, 1956.
56. Carroll, R., Kovăcs, K., and Tapp, E., The pathogenesis of glycerol-induced renal tubular necrosis, *J. Pathol. Bacteriol.*, 89, 573, 1965.
57. Cerilli, G. J., Geever, E. F., and Gagnon, J., The effects of suprarenal aortic occlusion and hemorrhagic hypotension on the renal function of primates, *J. Surg. Res.*, 2, 233, 1962.
58. Chan, L., Ledingham, J. G. G., Dixon, J. A., Thulborn, K. R., Waterton, J. C., Radda, G. K., and Ross, B. D., Acute renal failure: a proposed mechanism based upon ^{31}P nuclear magnetic resonance studies in the rat, in *Acute Renal Failure*, Eliahou, H. E., Ed., John Libbey, London, 1982, 35.
59. Chedru, M. F., Baethke, R., and Oken, D. E., Renal cortical blood flow and glomerular filtration in myohemoglobinuric acute renal failure, *Kidney Int.*, 1, 232, 1972.
60. Chevalier, R. L. and Finn, W. F., Effects of propranolol on post-ischemic acute renal failure, *Nephron*, 25, 77, 1980.
61. Chinitz, J. L., Kin, K. E., Onesti, G., and Swartz, C., Pathophysiology and prevention of dextran-40-induced anuria, *J. Lab. Clin. Med.*, 77, 76, 1971.
62. Chopra, S., Kaufman, J. S., Jones, T. W., Hong, W. K., Gehr, M. K., Hamburger, R. J., Flamenbaum, W., and Trump, B. F., Cis-diamminedichlorplatinum-induced acute renal failure in the rat, *Kidney Int.*, 21, 54, 1982.
63. Churchill, P., Bidani, A., Fleischmann, L., and Becker-McKenna, B., Glycerol-induced acute renal failure in the two kidney Goldblatt rat, *Am. J. Physiol.*, 233, F247, 1977.
64. Churchill, S., Zarlengo, M. D., Carvalho, J. S., Gottlieb, M. N., and Oken, D. E., Normal renocortical blood flow in experimental acute renal failure, *Kidney Int.*, 11, 246, 1977.
65. Cirksena, W. J., Keller, H. I., Bernier, G., and Teschan, P. E., Pathogenetic studies in a model of pigment nephropathy in the rat, in *Pathogenesis and Clinical Findings with Renal Failure*, Gessler, U., Schröder, K., and Weidinger, H., Eds., Georg Thieme Verlag, Stuttgart, 1971, 105.
66. Clarkson, A. R., MacDonald, M. K., Fuster, V., Cash, J. D., and Robson, J. S., Glomerular coagulation in acute ischemic renal failure, *Q. J. Med.*, 39, 585, 1970.
67. Conger, J. D. and Falk, S. A., Intrarenal dynamics in the pathogenesis and prevention of acute urate nephropathy, *J. Clin. Invest.*, 59, 786, 1977.
68. Conger, J. D., Falk, S. A., Guggenheim, S. J., and Burke, T. J., A micropuncture study of the early phase of acute urate nephropathy, *J. Clin. Invest.*, 58, 681, 1976.
69. Conger, J. D., Robinette, J. B., and Guggenheim, S. J., Effect of acetylcholine on the early phase of reversible norepinephrine-induced acute renal failure, *Kidney Int.*, 19, 399, 1981.
70. Conger, J. D. and Schrier, R. W., Pathogenesis and diagnosis of acute renal failure, in *Pediatric Nephrology, Vol. 4, Renal Failure. Current Concepts in Diagnosis and Management*, Strauss, J., Ed., Garland STPM Press, New York, 1978, 3.
71. Conger, J. D. and Schrier, R. W., Renal hemodynamics in acute renal failure, *Annu. Rev. Physiol.*, 42, 603, 1980.
72. Conn, H. L., Jr., Wilds, L., and Helwig, J., A study of the renal circulation, tubular function and morphology, and urinary volume and composition in dogs following mercury poisoning and transfusion of human blood, *J. Clin. Invest.*, 33, 732, 1954.
73. Cook, M. A. and Adkinson, J. T., A micropuncture study of a dietary induced, hyperuricemic model of acute renal failure in the rat, *Proc. Soc. Exp. Biol. Med.*, 163, 187, 1980.
74. Corcoran, A. C. and Page, I. H., Renal damage from ferroheme pigments myoglobin, hemoglobin, hematin, *Tex. Rep. Biol. Med.*, 3, 528, 1945.
75. Cox, J. W., Baehler, R. W., Sharma, H., O'Dorisio, T., Osgood, R. W., Stein, J. H., and Ferris, T. F., Studies on the mechanism of oliguria in a model of unilateral acute renal failure, *J. Clin. Invest.*, 53, 1546, 1974.
76. Cronin, R. E., de Torrente, A., Miller, P. D., Bulger, R. E., Burke, T. J., and Schrier, R. W., Pathogenic mechanisms in early norepinephrine-induced acute renal failure: functional and histological correlates of protection, *Kidney Int.*, 14, 115, 1978.
77. Cronin, R. E., Erickson, A. M., de Torrente, A., McDonald, K. M., and Schrier, R. W., Norepinephrine-induced acute renal failure: a reversible ischemic model of acute renal failure, *Kidney Int.*, 14, 187, 1978.

78. Cronin, R. E. and Schrier, R. W., Vascular aspects of the pathogenesis of acute renal failure, in *Kidney Disease. Hematologic and Vascular Problems*, McIntosh, R. M., Guggenheim, S. J., and Schrier, R. W., Eds., John Wiley & Sons, New York, 1977, 149.
79. Cuppage, F. E., Chiga, M., and Tate, A., Cell cycle studies in the regenerating rat nephron following injury with mercuric chloride, *Lab. Invest.*, 26, 122, 1972.
80. Cuppage, F. E., Cunningham, N., and Tate, A., Nucleic acid synthesis in the regenerating nephron following injury with mercuric chloride, *Lab. Invest.*, 21, 449, 1969.
81. Cuppage, F. E. and Tate, A., Repair of the nephron following injury with mercuric chloride, *Am. J. Pathol.*, 51, 405, 1967.
82. Cuppage, F. E. and Tate, A., Repair of the nephron in acute renal failure: comparative regeneration following various forms of acute tubular injury, *Pathol. Microbiol.*, 32, 327, 1968.
83. Cvitkovic, E., Spaulding, J., Bethune, V., Martin, J., and Whitmore, W. F., Improvement of cis-dichlordiammineplatinum (NSC 119875): therapeutic index in an animal model, *Cancer*, 39, 1357, 1977.
84. Dach, J. L. and Kurtzman, N. A., A scanning electron microscopic study of the glycerol model of acute renal failure, *Lab. Invest.*, 34, 406, 1976.
85. Dahlager, J. I., Sequential [^{125}I]-*o*-iodohippurate renograms from rabbit kidneys after temporary renal ischemia, *J. Nucl. Med.*, 19, 1324, 1978.
86. Dalgaard, O. Z. and Pedersen, K. J., Renal tubular degeneration. Electron microscopy in ischaemic anuria, *Lancet*, 2, 484, 1959.
87. Darmady, E. M., The renal changes in anuria following medical conditions, *Proc. R. Soc. Med.*, 50, 498, 1957.
88. Darmady, E. M. and Stranack, F., Microdissection of the nephron in disease, *Br. Med. Bull.*, 13, 21, 1957.
89. DeGowin, E. L., Warner, E. D., and Randall, W. L., Renal insufficiency from blood transfusion. II. Anatomic changes in man compared with those in dogs with experimental hemoglobinuria, *Arch. Intern. Med.*, 61, 609, 1938.
90. DeNavasquez, S., The excretion of haemoglobin with special reference to the "transfusion" kidney, *J. Pathol. Bacteriol.*, 51, 413, 1940.
91. De Rougemont, D., Wunderlich, P. F., Torhorst, J., Keller, M., Peters-Häfeli, L., Thiel, G., and Brunner, F. P., $HgCl_2$-induced acute renal failure in the rat. Effects of water diuresis, saline loading and diuretic drugs, *J. Lab. Clin. Med.*, 99, 646, 1982.
92. de Torrente, A., Miller, P. D., Cronin, R. E., Paulsen, P. E., Erickson, A. L., and Schrier, R. W., Effects of furosemide and acetylcholine in norepinephrine-induced acute renal failure, *Am. J. Physiol.*, 235, F131, 1978.
93. De Wardener, H. E., *The Kidney*, 4th ed., Churchill Livingstone, London, 1973.
94. DiBona, G. F., McDonald, F. D., Flamenbaum, W., Dammin, G. J., and Oken, D. E., Maintenance of renal function in salt loaded rats despite severe tubular necrosis induced by $HgCl_2$, *Nephron*, 8, 205, 1971.
95. DiBona, G. F. and Sawin, L. L., The renin-angiotensin system in acute renal failure in the rat, *Lab. Invest.*, 25, 528, 1971.
96. Dobyan, D. C., Levi, J., Jacobs, C., Kosek, J., and Weiner, M. W., Mechanism of *cis*-platinum nephrotoxicity. II. Morphologic observations, *J. Pharmacol. Exp. Ther.*, 213, 551, 1980.
97. Dobyan, D. C., Nagle, R. B., and Bulger, R. E., Hypovolemic models of acute tubular necrosis in the rat kidney, *Virchows Arch. (Cell Pathol.)*, 25, 271, 1977.
98. Donnelly, G. L. and Holman, R. L., The stimulating influence of sodium citrate on cellular regeneration and repair in the kidney injured by uranium nitrate, *J. Pharmacol. Exp. Ther.*, 75, 11, 1942.
99. Donohoe, J. F., Venkatachalam, M. A., Bernard, D. B., and Levinsky, N. G., Tubular leakage and obstruction after renal ischemia: structural-functional correlations, *Kidney Int.*, 13, 208, 1978.
100. Dunnill, M. S., A review of the pathology and pathogenesis of acute renal failure due to acute tubular necrosis, *J. Clin. Pathol.*, 27, 2, 1974.
101. Dunnill, M. S., *Pathological Basis of Renal Disease*, W. B. Saunders, London, 1976, 313.
102. Edwards, J. G., The renal tubule (nephron) as affected by mercury, *Am. J. Pathol.*, 18, 1011, 1942.
103. Eisenbach, G. M., Kitzlinger, B., and Steinhausen, M., Renal blood flow after temporary ischemia of rat kidneys. Renal venous outflow and clearance techniques, *Pflügers Arch.*, 347, 223, 1974.
104. Eisenbach, G. M. and Steinhausen, M., Micropuncture studies after temporary ischemia of rat kidneys, *Pflügers Arch.*, 343,11,1973.
105. Eisner, G. M., Slotkoff, L. M., and Lilienfeld, L. S., Distribution volumes in the dog kidney during anuria produced by uranium nitrate, *Am. J. Physiol.*, 214, 929, 1968.
106. Eknoyan, G., Bulger, R. E., and Dobyan, D. C., Mercuric chloride-induced acute renal failure in the rat. I. Correlation of functional and morphologic changes and their modification by clonidine, *Lab. Invest.*, 46, 613, 1982.

107. Eliahou, H. E., Ed., *Acute Renal Failure*, John Libbey, London, 1982.
108. Eliahou, H. E., Iaina, A., Solomon, S., and Gavendo, S., Alleviation of anoxic experimental acute renal failure in rats by β-adrenergic blockade, *Nephron*, 19, 158, 1977.
109. Evan, A. P. and Dail, W. G., Jr., The effects of sodium chromate on the proximal tubules of the rat kidney. Fine structural damage and lysozymuria, *Lab. Invest.*, 30, 704, 1974.
110. Fajers, C. M., Experimental studies in hemoglobinuric nephrosis. II. The enhanced effect on the kidneys of glycerol-induced hemoglobinemia combined with unilateral nephrectomy, *Acta Pathol. Microbiol. Scand.*, 46, 17, 1959.
111. Finckh, E. S., Experimental acute tubular nephrosis following subcutaneous injection of glycerol, *J. Pathol. Bacteriol.*, 73, 69, 1957.
112. Finckh, E. S., The failure of experimental renal tubulonecrosis to produce oliguria in the rat, *Australas. Ann. Med.*, 9, 283, 1960.
113. Fine, L. G., Acquired prostaglandin E_2 (medullin) deficiency as the cause of oliguria in acute tubular necrosis. A hypothesis, *Isr. J. Med. Sci.*, 6, 346, 1970.
114. Fink, M., Are diuretics useful in the treatment or prevention of acute renal failure?, *South. Med. J.*, 75, 329, 1982.
115. Finn, W. F., Acute renal failure, in *Strauss and Welt's Diseases of the Kidney*, Earley, L. E. and Gottschalk, C. W., Eds., 3rd ed., Little, Brown, Boston, 1979.
116. Finn, W. F., Enhanced recovery from postischemic acute renal failure. Micropuncture studies in the rat, *Circ. Res.*, 46, 440, 1980.
117. Finn, W. F., Postischemic acute renal failure. Initiation, maintenance, and recovery, *Invest. Urol.*, 17, 427, 1980.
118. Finn, W. F., Nephron heterogeneity in polyuric acute renal failure, *J. Lab. Clin. Med.*, 98, 21, 1981.
119. Finn, W. F., Arendshorst, W. J., and Gottschalk, C. W., Pathogenesis of oliguria in acute renal failure, *Circ. Res.*, 36, 675, 1975.
120. Finn, W. F. and Chevalier, R. L., Recovery from post-ischemic acute renal failure in the rat, *Kidney Int.*, 16, 113, 1979.
121. Flamenbaum, W., Pathophysiology of acute renal failure, *Arch. Intern. Med.*, 131, 911, 1973.
122. Flamenbaum, W., Hamburger, R. J., Huddleston, M. L., Kaufman, J., McNeil, J. S., Schwartz, J. H., and Nagle, R., The initiation phase of experimental acute renal failure: an evaluation of uranyl nitrate-induced acute renal failure in the rat, *Kidney Int.*, 10(Suppl. 6), S-115, 1976.
123. Flamenbaum, W., Hamburger, R., and Kaufman, J., Distal tubule (Na^+) and juxtaglomerular apparatus renin activity in uranyl nitrate induced acute renal failure in the rat, *Pflügers Arch.*, 364, 209, 1976.
124. Flamenbaum, W., Huddleston, M. L., McNeil, J. S., and Hamburger, R. J., Uranyl nitrate-induced acute renal faiure in the rat: micropuncture and renal hemodynamic studies, *Kidney Int.*, 6, 408, 1974.
125. Flamenbaum, W., Kotchen, T. A., Nagle, R., and McNeil, J. S., Effect of potassium on the renin-angiotensin system and $HgCl_2$-induced acute renal failure, *Am. J. Physiol.*, 224, 305, 1973.
126. Flamenbaum, W., Kotchen, T. A., and Oken, D. E., Effect of renin immunization on mercuric chloride and glycerol-induced acute renal failure, *Kidney Int.*, 1, 406, 1972.
127. Flamenbaum, W., McDonald, F. D., DiBona, G. F., and Oken, D. E., Micropuncture study of renal tubular factors in low dose mercury poisoning, *Nephron*, 8, 221, 1971.
128. Flamenbaum, W., McNeil, J. S., Kotchen, T. A., and Saladino, A. J., Experimental acute renal failure induced by uranyl nitrate in the dog, *Circ. Res.*, 31, 682, 1972.
129. Flamenbaum, W. and Wilson, D., Models of acute renal failure, in *Proc. 7th Int. Congr. Nephrology*, 1978, 687.
130. Flanigan, W. J. and Oken, D. E., Renal micropuncture study of the development of anuria in the rat with mercury-induced acute renal failure, *J. Clin. Invest.*, 44, 449, 1965.
131. Flink, E. B., Blood transfusion studies. III. The relationship of hemoglobinemia and of the pH of the urine to renal damage produced by injection of hemoglobin solutions into dogs, *J. Lab. Clin. Med.*, 32, 223, 1947.
132. Foulkes, E. C., Glomerular filtration and renal plasma flow in uranium poisoned rabbits, *Toxicol. Appl. Pharmacol.*, 20, 380, 1971.
133. Friedman, E. A. and Eliahou, H. E., Eds., Proc. Conf. on Acute Renal Failure, Publ. No. 74-608, Department of Health, Education and Welfare, National Institutes of Health, Washington, D.C., 1973.
134. Friedman, S. M., Johnson, R. L., and Friedman, C. L., The pattern of recovery of renal function following renal artery occlusion in the dog, *Circ. Res.*, 2, 231, 1954.
135. Gaál, K. and Siklós, J., Effect of beta-receptor antagonists on $HgCl_2$-induced acute renal failure in rats, *Renal Physiol.*, 5, 245, 1982.

136. Ganeval, D., Grünfeld, J. P., Eloy, L., Lacour, B., Russo-Marie, F., Noël, L. H., and Anagnostopoulos, T., Glaphenine-induced acute renal failure in the rat: a new experimental model, *Am. J. Physiol.*, 243, F416, 1982.
137. Ganote, C. E., Reimer, K. A., and Jennings, R. B., Acute mercuric chloride nephrotoxicity. An electron microscopic and metabolic study, *Lab. Invest.*, 31, 633, 1974.
138. Gaudio, K. M., Taylor, M. R., Chaudry, I. H., Kashgarian, M., and Siegel, N. J., Accelerated recovery of single nephron function by the postischemic infusion of ATP-$MgCl_2$, *Kidney Int.*, 22, 13, 1982.
139. Gerlach, E., Deuticke, B., and Dreisbach, R. H., Zum Verhalten von Nucleotiden und ihren dephosphorylierten Abbauprodukten in der Niere bei Ischämie und kurzzeitiger post-ischämischer Wiederdurchblutung, *Pflügers Arch.*, 278, 296, 1963.
140. Goldberg, M., Studies of the acute renal effects of hemolyzed red blood cells in dogs including estimations of renal blood flow with krypton^{85}, *J. Clin. Invest.*, 41, 2112, 1962.
141. Gordon, J. A., Peterson, L. N., and Anderson, R. J., Water metabolism after cisplatin in the rat, *Am. J. Physiol.*, 243, F36, 1982.
142. Gritzka, T. L. and Trump, B. F., Renal tubular lesions caused by mercuric chloride. Electron microscopic observations: degeneration of the pars recta, *Am. J. Pathol.*, 52, 1225, 1968.
143. Gustafson, G. E., Koletsky, S., and Free, A. H., Effect of sodium citrate on uranium poisoning in dogs, *Arch. Intern. Med.*, 74, 416, 1944.
144. Haagsma, B. H. and Pound, A. W., Mercuric chloride-induced renal tubular necrosis in the rat, *Br. J. Exp. Pathol.*, 60, 341, 1979.
145. Haagsma, B. H. and Pound, A. W., Mercuric chloride-induced tubulonecrosis in the rat kidney: the recovery phase, *Br. J. Exp. Pathol.*, 61, 229, 1980.
146. Haley, D. P., Morphologic changes in uranyl nitrate-induced acute renal failure in saline- and water-drinking rats, *Lab. Invest.*, 46, 196, 1982.
147. Haley, D. P., Bulger, R. E., and Dobyan, D. C., The long-term effects of uranyl nitrate on the structure and function of the rat kidney, *Virchows Arch. (Cell Pathol.)*, 41, 181, 1982.
148. Hamilton, P. B., Phillips, R. A., and Hiller, A., Duration of renal ischemia required to produce uremia, *Am. J. Physiol.*, 152, 517, 1948.
149. Hanley, M. J., Isolated nephron segments in a rabbit model of ischemic acute renal failure, *Am. J. Physiol.*, 239, F17, 1980.
150. Hanley, M. J., Studies on acute disease models, *Kidney Int.*, 22, 536, 1982.
151. Hanley, M. J. and Davidson, K., Prior mannitol and furosemide infusion in a model of ischemic acute renal failure, *Am. J. Physiol.*, 241, F556, 1981.
152. Harber, M. H. and Jennings, R. B., Renal response of the rat to mercury. The effect of sex and sex hormones, *Arch. Pathol.*, 79, 218, 1965.
153. Harmon, E. L., Human mercuric chloride poisoning by intravenous injection, *Am. J. Pathol.*, 4, 321, 1928.
154. Harrison, H. E., Bunting, H., Ordway, N. K., and Albrink, W. S., The pathogenesis of the renal injury produced in the dog by hemoglobin or methemoglobin, *J. Exp. Med.*, 86, 339, 1947.
155. Hayes, J. M., Boonshaft, B., Maher, J. F., O'Connell, J. M. B., and Schreiner, G. E., Resistance to glycerol induced hemoglobinuric acute renal failure, *Nephron*, 7, 155, 1970.
156. Hayman, J. M., Jr., Shumway, N. P., Dumke, P., and Miller, M., Experimental hyposthenuria, *J. Clin. Invest.*, 18, 195, 1939.
157. Held, E., Protective effects of renomedullary autotransplants upon the course of postischemic acute renal failure in rabbits, *Kidney Int.*, 10(Suppl. 6), S-201, 1976.
158. Henry, L. N., Lane, C. E., and Kashgarian, M., Micropuncture studies of the pathophysiology of acute renal failure in the rat, *Lab. Invest.*, 19, 309, 1968.
159. Hobbs, J. B., Chusilp, S., Kincaid-Smith, P., and McIver, M. A., The mechanism of glycerol-induced acute renal failure, *J. Lab. Clin. Med.*, 88, 44, 1976.
160. Hofbauer, K. G., Konrads, A., Bauereiss, K., Möhring, B., Möhring, J., and Gross, F., Vasopressin and renin in glycerol-induced acute renal failure in the rat, *Circ. Res.*, 41, 424, 1977.
161. Holck, H. G. O., Kanân, M. A., Mills, L. M., and Smith, E. L., Studies upon the sex-difference in rats in tolerance to certain barbiturates and to nicotine, *J. Pharmacol. Exp. Ther.*, 60, 323, 1937.
162. Hollenberg, N. K., Adams, D. F., Oken, D. E., Abrams, H. L., and Merrill, J. P., Acute renal failure due to nephrotoxins. Renal hemodynamic and angiographic studies in man, *N. Engl. J. Med.*, 282, 1329, 1970.
163. Hollenberg, N. K., Epstein, M., Rosen, S. M., Basch, R. I., Oken, D. E., and Merrill, J. P., Acute oliguric renal failure in man: evidence for preferential renal cortical ischemia, *Medicine*, 47, 455, 1968.
164. Hsu, C. H. and Kurtz, T. W., Renal hemodynamics in experimental acute renal failure, *Nephron*, 27, 204, 1981.

165. Hsu, C. H., Kurtz, T. W., Goldstein, J. R., Keinath, R. D., and Weller, J. M., Intrarenal hemodynamics in acute myohemoglobinuric renal failure, *Nephron*, 17, 65, 1976.
166. Hsu, C. H., Kurtz, T. W., Rosenzweig, J., and Weller, J. M., Renal hemodynamics in $HgCl_2$-induced acute renal failure, *Nephron*, 18, 326, 1977.
167. Hsu, C. H., Kurtz, T. W., and Waldinger, T. P., Cardiac output and renal blood flow in glycerol-induced acute renal failure in the rat, *Circ. Res.*, 40, 178, 1977.
168. Hsueh, W. and Rostorfer, H. H., Chemically induced renal hypertrophy in the rat, *Lab. Invest.*, 29, 547, 1973.
169. Huguenin, M., Thiel, G., and Brunner, F. P., $HgCl_2$-induced acute renal failure studied by split drop micropuncture technique in the rat, *Nephron*, 20, 147, 1978.
170. Huguenin, M. E., Birbaumer, A., Brunner, F. P., Thorhorst, J., Schmidt, U., Dubach, U. C., and Thiel, G., An evaluation of the role of tubular obstruction in folic acid-induced acute renal failure in the rat. A micropuncture study, *Nephron*, 22, 41, 1978.
171. Iaina, A. Solomon, S., and Eliahou, H. E., Reduction in severity of acute renal failure in rats by beta-adrenergic blockade, *Lancet*, 2, 157, 1975.
172. Iaina, A., Solomon, S., Gavendo, S., and Eliahou, H. E., Reduction in severity of acute renal failure (ARF) in rats by dopamine, *Biomedicine*, 27, 137, 1977.
173. Iaina, A., Solomon, S., Serban, I., and Eliahou, H. E., Chronic saline loading in anoxic renal failure in rats, *Isr. J. Med. Sci.*, 12, 1457, 1976.
174. Ideura, T., Solez, K., and Heptinstall, R. H., The effect of clonidine on tubular obstruction in postischemic acute renal failure in the rabbit demonstrated by microradiography and microdissection, *Am. J. Pathol.*, 98, 123, 1980.
175. Ishikawa, I. and Hollenberg, N. K., Pharmacologic interruption of the renin-angiotensin system in myohemoglobinuric acute renal failure, *Kidney Int.*, 10(Suppl 6), S-183, 1976.
176. Jaenike, J. R., The renal lesion associated with hemoglobinemia. I. Its production and functional evolution in the rat, *J. Exp. Med.*, 123, 523, 1966.
177. Jaenike, J. R., The renal lesion associated with hemoglobinemia: a study of the pathogenesis of the excretory defect in the rat, *J. Clin. Invest.*, 46, 378, 1967.
178. Jaenike, J. R., Micropuncture study of methemoglobin-induced acute renal failure in the rat, *J. Lab. Clin. Med.*, 73, 459, 1969.
179. Jaenike, J. R. and Schneeberger, E. E., The renal lesion associated with hemoglobinemia. II. Its structural characteristics in the rat, *J. Exp. Med.*, 123, 537, 1966.
180. Jones, D. B., Ultrastructure of human acute renal failure, *Lab. Invest.*, 46, 254, 1982.
181. Karlberg, L., Källskog, Ö., Norlén, B. J., and Wolgast, M., Nephron function in postischemic acute renal failure, *Scand. J. Urol. Nephrol.*, 16, 167, 1982.
182. Kaufman, J., Chopra, S., and Flamenbaum, W., CIS-diamminedichlorplatinum-induced acute renal failure, in *Acute Renal Failure*, Eliahou, H. E., Ed., John Libbey, London, 1982, 69.
183. Kirschbaum, B. B., Sprinkle, F. M., and Oken, D. E., Renal function and mercury level in rats with mercuric chloride nephrotoxicity, *Nephron*, 26, 28, 1980.
184. Klein, H. and Greven, J., Renal effects of mannitol in the early stage of glycerol-induced acute renal failure in the rat, *Nephron*, 23, 255, 1979.
185. Klein, L. A., Propranolol protection in acute renal failure, *Invest. Urol.*, 15, 401, 1978.
186. Kleinman, J. G., McNeil, J. S., and Flamenbaum, W., Uranyl nitrate acute renal failure in the dog: early changes in renal function and hemodynamics, *Clin. Sci. Mol. Med.*, 48, 9, 1975.
187. Kleinman, J. G., McNeil, J. S., Schwartz, J. H., Hamburger, R. J., and Flamenbaum, W., Effect of dithiothreitol on mercuric chloride- and uranyl nitrate-induced acute renal failure in the rat, *Kidney Int.*, 12, 115, 1977.
188. Knapp, R., Hollenberg, N. K., Busch, G. J., and Abrams, H. L., Prolonged unilateral acute renal failure induced by intra-arterial norepinephrine infusion in the dog, *Invest. Radiol.*, 7, 164, 1972.
189. Koletsky, S. and Gustafson, G. E., The effects of temporary cessation of renal blood flow in rats, *J. Clin. Invest.*, 26, 1072, 1947.
190. Koretsky, A. P., Strauss, W., Basus, V., Murphy, J., Bendel, P., James, T., and Weiner, M. W., Feasibility of ^{31}P NMR to study metabolic effects of renal ischemia in vivo, in *Acute Renal Failure*, Eliahou, H. E., Ed., John Libbey, London, 1982, 42.
191. Kramer, H. J., Schüürmann, J., Wassermann, C., and Düsing, R., Prostaglandin-independent protection by furosemide from oliguric ischemic renal failure in conscious rats, *Kidney Int.*, 17, 455, 1980.
192. Kurtz, T. W. and Hsu, C. H., Systemic hemodynamics in nephrotoxic acute renal failure, *Nephron*, 21, 100, 1978.
193. Kurtz, T. W., Maletz, R. M., and Hsu, C. H., Renal cortical blood flow in glycerol-induced acute renal failure in the rat, *Circ. Res.*, 38, 30, 1976.
194. Laake, H., Experimental investigations of the excretory and reabsorptive function of the renal tubules in normal and nephrotic rabbits, *Acta Med. Scand. Suppl.*, 168, 1945.

195. Ladefoged, J. and Winkler, K., Hemodynamics in acute renal failure, *Scand. J. Clin. Lab. Invest.*, 26, 83, 1970.
196. Lalich, J. J., Urine volume, non-protein nitrogen and pigment cast studies in rabbits with hemoglobinuric nephrosis, *Am. J. Med. Sci.*, 219, 65, 1950.
197. Lalich, J. J., The influence of in vitro hemoglobin modification on hemoglobinuric nephrosis in rabbits, *J. Lab. Clin. Med.*, 40, 102, 1952.
198. Lalich, J. J., The role of oxyhemoglobin and its derivatives in the pathogenesis of experimental hemoglobinuric nephrosis, *Am. J. Pathol.*, 31, 153, 1955.
199. Lalich, J. J. and Schwartz, S. I., The role of aciduria in the development of hemoglobinuric nephrosis in dehydrated rabbits, *J. Exp. Med.*, 92, 11, 1950.
200. Lang, F., Greger, R., Lechene, C., and Knox, F. G., Micropuncture techniques, in *Methods in Pharmacology*, Vol. 4B, Martinez-Maldonado, M., Ed., Plenum Press, New York, 1978, 75.
201. Lapp, H. and Schafé, K., Morphologische, histochemische und Speicherungs-Untersuchungen über den Verlauf der Sublimatnephrose bei der Ratte, *Beitr. Pathol. Anat. Allg. Pathol.*, 123, 77, 1960.
202. Levinsky, N. G., Pathophysiology of acute renal failure, *N. Engl. J. Med.*, 296, 1453, 1977.
203. Levinsky, N. G., Alexander, E. A., and Venkatachalam, M. A., Acute renal failure, in *The Kidney, Vol. 1*, 2nd ed., Brenner, B. M. and Rector, F. C., Jr., Eds., W. B. Saunders, Philadelphia, 1981, 1181.
204. Lindner, A., Cutler, R. E., and Bell, A. J., Attenuation of nephrotoxic acute renal failure in the dog with angiotensin-converting enzyme inhibitor (SQ-20, 881), *Circ. Res.*, 51, 216, 1982.
205. Lucké, B., Lower nephron nephrosis (the renal lesions of the crush syndrome, of burns, transfusions, and other conditions affecting the lower segments of the nephrons), *Mil. Surg.*, 99, 371, 1946.
206. MacNider, W. de B., A consideration of the relative toxicity of uranium nitrate for animals of different ages, *J. Exp. Med.*, 26, 1, 1917.
207. MacNider, W. de B., A review of acute experimental nephritis, *Physio. Rev.*, 4, 595, 1924.
208. MacNider, W. de B., The development of the chronic nephritis induced in the dog by uranium nitrate. A functional and pathological study with observations on the formation of urine by the altered kidneys, *J. Exp. Med.*, 49, 387, 1929.
209. MacNider, W. de B., The factor of age in determining the toxicity of certain poisons, *J. Gerontol.*, 1, 189, 1946.
210. Madias, N. E. and Harrington, J. T., Platinum nephrotoxicity, *Am. J. Med.*, 65, 307, 1978.
211. Maluf, N. S. R., Factors inducing renal shut-down from lysed erythrocytes: an experimental study, *Ann. Surg.*, 130, 49, 1949.
212. Mandal, A. K., The spleen and acute renal failure: mechanisms of renal protection by splenectomy. Involvement of prostaglandins, *Prostaglandins Leukotrienes Med.*, 9, 85, 1982.
213. Mandal, A. K. and Miller, J., Protection against ischemic acute renal failure by prostaglandin infusion, *Prostaglandins Leukotrienes Med.*, 8, 361, 1982.
214. Mason, A. D., Jr., Alexander, J. W., and Teschan, P. E., Studies in acute renal failure. I. Development of a reproducible lesion in experimental animals, *J. Surg. Res.*, 3, 430, 1963.
215. Mason, A. D., Jr., Teschan, P. E., and Muirhead, E. E., Studies in acute renal failure. III. Renal histologic alterations in acute renal failure in the rat, *J. Surg. Res.*, 3, 450, 1963.
216. Mason, J., Tubulo-glomerular feedback in the early stages of experimental acute renal failure, *Kidney Int.*, 10(Suppl 6), S-106, 1976.
217. Mason, J., Beck, F., Dörge, A., Rick, R., and Thurau, K., Intracellular electrolyte composition following renal ischemia, *Kidney Int.*, 20, 61, 1981.
218. Mason, J., Kain, H., Welsch, J., and Schnermann, J., The early phase of experimental acute renal failure. VI. The influence of furosemide, *Pflügers Arch.*, 392, 125, 1981.
219. Mason, J., Olbricht, C., Takabatake, T., and Thurau, K., The early phase of experimental acute renal failure. I. Intratubular pressure and obstruction, *Pflügers Arch.*, 370, 155, 1977.
220. Mason, J., Takabatake, T., Olbricht, C., and Thurau, K., The early phase of experimental acute renal failure. III. Tubuloglomerular feedback *Pflügers Arch.*, 373, 69, 1978.
221. Mattern, W. D. and Finn, W. F., Changing perceptions of acute renal failure, *Kidney*, 11, 25, 1978.
222. Mauk, R. H., Patak, R. V., Fadem, S. Z., Lifschitz, M. D., and Stein, J. H., Effect of prostaglandin E administration in a nephrotoxic and a vasoconstrictor model of acute renal failure, *Kidney Int.*, 12, 122, 1977.
223. McDonald, F. D., Thiel, G., Wilson, D. R., DiBona, G. F., and Oken, D. E., The prevention of acute renal failure in the rat by long-term saline loading: a possible role of the renin-angiotensin axis, *Proc. Soc. Exp. Biol. Med.*, 131, 610, 1969.
224. McDowell, E. M., Nagle, R. B., Zalme, R. C., McNeil, J. S., Flamenbaum, W., and Trump, B. F., Studies on the pathophysiology of acute renal failure. I. Correlation of ultrastructure and function in the proximal tubule of the rat following administration of mercuric chloride, *Virchows Arch. (Cell Pathol.)*, 22, 173, 1976.

225. Mendelsohn, F. A. O. and Smith, E. A., Intrarenal renin, angiotensin II, and plasma renin in rats with uranyl nitrate-induced and glycerol-induced acute renal failure, *Kidney Int.*, 17, 465, 1980.
226. Menefee, M. G., Mueller, C. B., Miller, T. B., Myers, J. K., and Bell, A. L., Experimental studies in acute renal failure. II. Fine structure changes in tubules associated with renal failure induced by globin, *J. Exp. Med.*, 120, 1139, 1964.
227. Morel, F., Chabardès, D., and Imbert-Teboul, M., Methodology for enzymatic studies of isolated tubular segments: adenylate cyclase in *Methods in Pharmacology*, Vol. 4B, Martinez-Maldonado, M., Ed., Plenum Press, New York, 1978, 297.
228. Morrison, A. R. Benabe, J. E., and Hoffman, M. K., Identification of thromboxane A_2 in glycerol induced acute renal failure in the rabbit, *Adv. Prostaglandin Thromboxane Res.*, 7, 1131, 1980.
229. Mullin, E. M., Bonar, R. A., and Paulson, D. F., Acute tubular necrosis. An experimental model detailing the biochemical events accompanying renal injury and recovery, *Invest. Urol.*, 13, 289, 1976.
230. Munck, O., Renal circulation in acute renal failure measured with radioactive krypton^{85} and xenon^{133}, in *Acute Renal Failure*, Shaldon, S. and Cook, G. C., Eds., F. A. Davis, Philadelphia, 1964, 15.
231. Myers, B. D., Carrie, B. J., Yee, R. R., Hilberman, M., and Michaels, A. S., Pathophysiology of hemodynamically mediated acute renal failure in man, *Kidney Int.*, 18, 495, 1980.
232. Myers, B. D., Chui, F., Hilberman, M., and Michaels, A. S., Transtubular leakage of glomerular filtrate in human acute renal failure, *Am. J. Physiol.*, 237, F319, 1979.
233. Neely, W. A. and Turner, M. D., The effect of arterial, venous, and arteriovenous occlusion on renal blood flow, *Surg. Gynecol. Obstet.*, 108, 669, 1959.
234. Neuman, W. F., Fleming, R. W., Dounce, A. L., Carlson, A. B., O'Leary, J., and Mulryan, B., The distribution and excretion of injected uranium, *J. Biol. Chem.*, 173, 737, 1948.
235. Nihei, H., Honda, N., Suzuki, K., Nagase, M., and Yoshitoshi, Y., Renal hemodynamics and medullary osmolal gradient in ischemic acute renal failure in rabbits, *Jpn. Heart J.*, 16, 44, 1975.
236. Nomiyama, K. and Foulkes, E. C., Some effects of uranyl acetate on proximal tubular function in rabbit kidney, *Toxicol. Appl. Pharmacol.*, 13, 89, 1968.
237. Oken, D. E., Modern concepts of the role of nephrotoxic agents in the pathogenesis of acute renal failure, *Prog. Biochem. Pharmacol.*, 7, 219, 1972.
238. Oken, D. E., Acute renal failure (vasomotor nephropathy): micropuncture studies of the pathogenetic mechanisms, *Annu. Rev. Med.*, 26, 307, 1975.
239. Oken, D. E., On the passive back flow theory of acute renal failure, *Am. J. Med.*, 58, 77, 1975.
240. Oken, D. E., Role of prostaglandins in the pathogenesis of acute renal failure, *Lancet*, 1, 1319, 1975.
241. Oken, D. E., Acute renal failure caused by nephrotoxins, *Environ. Health Perspect.*, 15, 101, 1976.
242. Oken, D. E., Local mechanisms in the pathogenesis of acute renal failure, *Kidney Int.*, 10(Suppl. 6), S-94, 1976.
243. Oken, D. E., Pathogenetic mechanisms in acute renal failure, in *Toxicology of the Kidney*, Hook, J. B., Ed., Raven Press, New York, 1981, 117.
244. Oken, D. E., Arce, M. L., and Wilson, D. R., Glycerol-induced hemoglobinuric acute renal failure in the rat. I. Micropuncture study of the development of oliguria, *J. Clin. Invest.*, 45, 724, 1966.
245. Oken, D. E., DiBona, G. F., and McDonald, F. D., Micropuncture studies of the recovery phase of myohemoglobinuric acute renal failure in the rat, *J. Clin. Invest.*, 49, 730, 1970.
246. Oken, D. E., Mende, C. W., Taraba, I., and Flamenbaum, W., Resistance of acute renal failure afforded by prior renal failure: examination of the role of renal renin content, *Nephron*, 15, 131, 1975.
247. Olbricht, C. H. J., Experimental models of acute renal failure, *Contrib. Nephrol.*, 19, 110, 1980.
248. Olbricht, C., Mason, J., Takabatake, T., Hohlbrugger, G., and Thurau, K., The early phase of experimental acute renal failure. II. Tubular leakage and the reliability of glomerular markers, *Pflügers Arch.*, 372, 251, 1977.
249. Oliver, J., The histogenesis of chronic uranium nephritis with especial reference to epithelial regeneration, *J. Exp. Med.*, 21, 425, 1915.
250. Oliver, J., Correlations of structure and function and mechanisms of recovery in acute tubular necrosis, *Am. J. Med.*, 15, 535, 1953.
251. Oliver, J., MacDowell, M., and Tracy, A., The pathogenesis of acute renal failure associated with traumatic and toxic injury. Renal ischemia, nephrotoxic damage and the ischemuric episode, *J. Clin. Invest.*, 30, 1307, 1951.
252. Parekh, N. and Veith, U., Renal hemodynamics and oxygen consumption during postischemic acute renal failure in the rat, *Kidney Int.*, 19, 306, 1981.
253. Parry, W. L., Schaefer, J. A., and Mueller, C. B., Experimental studies of acute renal failure. I. The protective effect of mannitol, *J. Urol.*, 89, 1, 1963.
254. Passow, H., Rothstein, A., and Clarkson, T. W., The general pharmacology of the heavy metals, *Pharmacol. Rev.*, 13, 185, 1961.

255. Patak, R. V., Fadem, S. Z., Lifschitz, M. D., and Stein, J. H., Study of factors which modify the development of norepinephrine-induced acute renal failure in the dog, *Kidney Int.*, 15, 227, 1979.
256. Pedersen, F. and Ladefoged, J., Renal hemodynamics in acute renal failure in man measured by intra-arterial injection, *Scand. J. Urol. Nephrol.*, 7, 187, 1973.
257. Pera, M. F., Jr., Zook, B. C., and Harder, H. C., Effects of mannitol or furosemide on the nephrotoxicity and physiological disposition of *cis*-dichlorodiammineplatinum-(II) in rats, *Cancer Res.*, 39, 1269, 1979.
258. Perri, G. C. and Gorini, P., Uraemia in the rabbit after injection of crystalline myoglobin, *Br. J. Exp. Pathol.*, 33, 440, 1952.
259. Preuss, H. G., Tourkantonis, A., Hsu, C. H., Shim, P. S., Barzyk, P., Tio, F., and Schreiner, G. E., Early events in various forms of experimental acute tubular necrosis in rats, *Lab. Invest.*, 32, 286, 1975.
260. Randall, H. M., Jr., Metabolic and functional effects of acute renal ischemia in dog kidney slices, *Am. J. Physiol.*, 223, 756, 1972.
261. Reimer, K. A., Ganote, C. E., and Jennings, R. B., Alterations in renal cortex following ischemic injury. III. Ultrastructure of proximal tubules after ischemia or autolysis, *Lab. Invest.*, 26, 347, 1972.
262. Reimer, K. A. and Jennings, R. B., Alterations in renal cortex following ischemic injury. II. PAH uptake, O_2 consumption, and water content in slices of cortex after ischemia or autolysis, *Lab. Invest.*, 25, 185, 1971.
263. Reineck, H. J., O'Connor, G. J., Lifschitz, M. D., and Stein, J. H., Sequential studies on the pathophysiology of glycerol induced acute renal failure, *J. Lab. Clin. Med.*, 96, 356, 1980.
264. Reubi, F. C., The pathogenesis of anuria following shock, *Kidney Int.*, 5, 106, 1974.
265. Reubi, F. C., Vorburger, C., and Tuckman, J., Renal distribution volumes of indocyanine green, (^{51}Cr) EDTA, and ^{24}Na in man during acute renal failure after shock. Implications for the pathogenesis of anuria, *J. Clin. Invest.*, 52, 223, 1973.
266. Richards, A. N., Direct observations of change in function of the renal tubule caused by certain poisons, *Trans. Assoc. Am. Physicians*, 44, 64, 1929.
267. Richards, C. J. and DiBona, G. F., Acute renal failure: structural-functional correlation, *Proc. Soc. Exp. Biol. Med.*, 146, 880, 1974.
268. Riley, A. L., Effect of ischemia on renal blood flow in the rat, *Nephron*, 21, 107, 1978.
269. Robinette, J. B. and Conger, J. D., Effects of minimal hypotension on the course of acute renal failure in rats with innervated and denervated kidneys, *Kidney Int.*, 23, 207, 1983.
270. Rodin, A. E. and Crowson, C. N., Mercury nephrotoxicity in the rat. I. Factors influencing the localization of the tubular lesions, *Am. J. Pathol.*, 41, 297, 1962.
271. Rosen, S., Mailloux, L. U., Lawson, N. L., and Teschan, P. E., Acute renal failure in the rat. I. Gross characteristics, *Lab. Invest.*, 18, 438, 1968.
272. Rosen, S., Mailloux, L. U., Lawson, N. L., and Teschan, P. E., Acute renal failure in the rat. II. Light and electron microscopic observations, *Lab. Invest.*, 18, 444, 1968.
273. Roth, R. A., Finlayson, B., and Hackett, R. L., Analysis of the rate of ischemic injury in rat kidney utilizing survival statistics, *Invest. Urol.*, 12, 125, 1974.
274. Rothstein, A. and Hayes, A. D., The metabolism of mercury in the rat studied by isotope techniques, *J. Pharmacol. Exp. Ther.*, 130, 166, 1960.
275. Rotter, W., Lapp, H., and Zimmermann, H., Pathogenese und morphologisches Substrat des akuten Nierenversagens und seine Erholungszeit, *Dtsch. Med. Wochenschr.*, 87, 669, 1962.
276. Rozencweig, M., Von Hoff, D. D., Slavik, M., and Muggia, F. M., *Cis*-diamminedichloroplatinum (II), *Ann. Intern. Med.*, 86, 803, 1977.
277. Ruiz-Guiñazú, A., Coelho, J. B., and Paz, R. A., Methemoglobin-induced acute renal failure in the rat. *In vivo* observation, histology, and micropuncture measurements of intratubular and postglomerular vascular pressures, *Nephron*, 4, 257, 1967.
278. Ryan, R., McNeil, J. S., Flamenbaum, W., and Nagle, R., Uranyl nitrate-induced acute renal failure in the rat: effect of varying doses and saline loading, *Proc. Soc. Exp. Biol. Med.*, 143, 289, 1973.
279. Safirstein, R., Miller, P., and Dikman, S., Cisplatin nephrotoxicity in rats, in *Acute Renal Failure*, Eliahou, H. E., Ed., John Libbey, London, 1982, 91.
280. Sansum, W. D., The principles of treatment in mercuric chlorid poisoning. An experimental study, *JAMA*, 70, 824, 1918.
281. Sato, T., McDowell, E. M., McNeil J. S., Flamenbaum, W., and Trump, B. F., Studies on the pathophysiology of acute renal failure. III. A study of the juxtaglomerular apparatus of the rat nephron following administration of mercuric chloride, *Virchows Arch. (Cell Pathol.)*, 24, 279, 1977.
282. Savin, V. J., Patak, R. V., and Marr, G., Glomerular filtration in ischemic renal failure, *Kidney Int.*, 16, 776, 1979.
283. Schaefer, J. A., Parry, W. L., and Mueller, C. B., Experimental studies of acute renal failure. III. Histopathology of the renal lesion produced by methemoglobin, *J. Surg. Res.*, 6, 247, 1966.

284. Schaeppi, U., Heyman, I. A., Fleischman, R. W., Rosenkrantz, H., Ilievski, V., Phelan, R., Cooney, D. A., and Davis, R. D., *Cis*-dichlorodiammineplatinum (II) (NSC-119875): preclinical toxicologic evaluation of intravenous injection in dogs, monkeys and mice, *Toxicol. Appl. Pharmacol.*, 25, 230, 1973.
285. Schimmelpfennig, W. and Hagemann, I., Über zwei experimentelle Methode zur Erzeugung eines "Akutes Nierenversagens" bei der Ratte, *Acta Biol. Med. Ger.*, 15, 781, 1965.
286. Schirmer, H. K. A., Marshall, R. E., and Ballinger, W. F., II, Energy provision by glycolysis in stressed renal tissue, *Surg. Gynecol. Obstet.*, 125, 1013, 1967.
287. Schmidt, U. and Dubach, U. C., Acute renal failure in the folate-treated rat: early metabolic changes in various structures of the nephron, *Kidney Int.*, 10(Suppl. 6), S-39, 1976.
288. Schmidt, U. and Horster, M., Sodium-potassium-activated adenosine triphosphatase: methodology for quantification in microdissected renal tubule segments from freeze-dried and fresh tissue, in *Methods in Pharmacology*, Vol. 4B, Martinez-Maldonado, M., Ed., Plenum Press, New York, 1978, 259.
289. Schmidt, U., Torhorst, J., Huguenin, M., and Dubach, U. C., Acute renal failure after folate: Na K ATPase in isolated rat renal tubule. Ultramicrochemical and clinical studies, *Eur. J. Clin. Invest.*, 3, 169, 1973.
290. Schrier, R. W., Cronin, R. E., Miller, P., de Torrenté, A., Burke, T., and Bulger, R., Role of solute excretion in prevention of norepinephrine-induced acute renal failure, *Yale J. Biol. Med.*, 51, 355, 1978.
291. Schubert, G. E., Folic acid-induced acute renal failure in the rat: morphological studies, *Kidney Int.*, 10(Suppl. 6), S-46, 1976.
292. Schwartz, R. H., Lewis, R. A., and Schenk, E. A., Tamm-Horsfall mucoprotein. III. Potassium dichromate-induced renal tubular damage, *Lab. Invest.*, 27, 214, 1972.
293. Seybold, D. and Gebler, U., Eds., *Acute Renal Failure*, S. Karger, Basel, 1982.
294. Sheehan, H. L. and Davis, J. C., Complete permanent renal ischemia, *J. Pathol. Bacteriol.*, 76, 569, 1958.
295. Sheehan, H. L. and Davis, J. C., Intermittent complete renal ischemia, *J. Pathol. Bacteriol.*, 79, 77, 1960.
296. Sheehan, H. L. and Davis, J. C., The protective effect of anesthesia on experimental renal ischemia, *J. Pathol. Bacteriol.*, 79, 337, 1960.
297. Sherwood, T., Lavender, J. P., and Russell, S. B., Mercury-induced renal vascular shut-down: observations in experimental acute renal failure, *Eur. J. Clin. Invest.*, 4, 1, 1974.
298. Siegel, F. L. and Bulger, R. E., Scanning and transmission electron microscopy of mercuric chloride-induced acute tubular necrosis in rat kidney, *Virchows Arch. (Cell Pathol.)*, 18, 243, 1975.
299. Siegel, N. J., Glazier, W. B., Chaudry, I. H., Gaudio, K. M., Lytton, B., Baue, A. E., and Kashgarian, M., Enhanced recovery from acute renal failure by the postischemic infusion of adenine nucleotides and magnesium chloride in rats, *Kidney Int.*, 17, 338, 1980.
300. Siegel, N. J., Gunstream, S. K., Handler, R. I., and Kashgarian, M., Renal function and cortical blood flow during the recovery phase of acute renal failure, *Kidney Int.*, 12, 199, 1977.
301. Smith, H. W., *Principles of Renal Physiology*, Oxford University Press, New York, 1956.
302. Solez, K., Altman, J., Rienhoff, H. Y., Riela, A. R., Finer, P. M., and Heptinstall, R. H., Early angiographic and renal blood flow changes after $HgCl_2$ or glycerol administration, *Kidney Int.*, 10(Suppl. 6), S-153, 1976.
303. Solez, K., D'Agostini, R. J., Stawowy, L., Freedman, M. T., Scott, W. W., Jr., Siegelman, S. S., and Heptinstall, R. H., Beneficial effect of propranolol in a histologically appropriate model of post-ischemic acute renal failure, *Am. J. Pathol.*, 88, 163, 1977.
304. Solez, K., Ideura, T., Silvia, C. B., Hamilton, B., and Saito, H., Clonidine after renal ischemia to lessen acute renal failure and microvascular damage, *Kidney Int.*, 18, 309, 1980.
305. Solez, K., Morel-Maroger, L., and Sraer, J. D., The morphology of "acute tubular necrosis" in man: analysis of 57 renal biopsies and a comparison with the glycerol model, *Medicine*, 58, 362, 1979.
306. Solez, K., Racusen, L. C., and Whelton, A., Glomerular epithelial cell changes in early postischemic acute renal failure in rabbits and man, *Am J. Pathol.*, 103, 163, 1981.
307. Spencer, H. W., Yarger, W. E., and Robinson, R. R., Alterations of renal function during dietary-induced hyperuricemia in the rat, *Kidney Int.*, 9, 489, 1976.
308. Staemmler, M., Die akuten Nephrosen. I. Die Sublimatnephrose, *Virchows Arch. (Pathol. Anat.)*, 328, 1, 1956.
309. Stein J. H., Gottschall, J., Osgood, R. W., and Ferris, T. F., Pathophysiology of a nephrotoxic model of acute renal failure, *Kidney Int.*, 8, 27, 1975.
310. Stein, J. H., Lifschitz, M. D., and Barnes, L. D., Current concepts on the pathophysiology of acute renal failure, *Am. J. Physiol.*, 234, F171, 1978.
311. Stein, J. H. and Sorkin, M. I., Pathophysiology of a vasomotor and nephrotoxic model of acute renal failure in the dog, *Kidney Int.*, 10(Suppl. 6), S-86, 1976.

312. Steinhausen, M., Eisenbach, G. M., and Böttcher, W., High-frequency microcinematographic measurements on peritubular blood flow under control conditions and after temporary ischemia of rat kidneys, *Pflügers Arch.*, 339, 273, 1973.
313. Steinhausen, M., Eisenbach, G. M., and Helmstädter, V., Concentration of lissamine green in proximal tubules of antidiuretic and mercury poisoned rats and the permeability of these tubules, *Pflügers Arch.*, 311, 1, 1969.
314. Steinhausen, M., Thederan, H., Nolinski, D., Dallenbach, F. D., and Schwaier, A., Further evidence of tubular blockage after acute ischemic renal failure in *Tupaia belangeri* and rats, *Virchows Arch. (Pathol. Anat.)*, 381, 13, 1978.
315. Stitzer, S. O. and Martinez-Maldonado, M., Clearance methods in the rat, in *Methods in Pharmacology*, Vol. 4B, Martinez-Maldonado, M., Ed., Plenum Press, New York, 1978, 23.
316. Sudo, M., Honda, N., Hishida, A., and Nagase, M., Renal hemodynamics in uranyl acetate-induced acute renal failure of rabbits, *Kidney Int.*, 11, 35, 1977.
317. Sudo, M., Honda, N., Hishida, A., and Nagase, M., Renal hemodynamics in oliguric and nonoliguric acute renal failure of rabbits, *Nephron*, 25, 144, 1980.
318. Summers, W. K. and Jamison, R. L., The no reflow phenomenon in renal ischemia, *Lab. Invest.*, 25, 635, 1971.
319. Suzuki, T. and Mostofi, F. K., Electron microscopic studies of acute tubular necrosis. Early changes in the glomeruli of rat kidney after subcutaneous injection of glycerin, *Lab. Invest.*, 23, 8, 1970.
320. Suzuki, T. and Mostofi, F. K., Electron microscopic studies of acute tubular necrosis. Early changes in lower tubules of rat kidney after subcutaneous injection of glycerin, *Lab. Invest.*, 23, 15, 1970.
321. Suzuki, T. and Mostofi, F. K., Electron microscopic studies of acute tubular necrosis. Vascular changes in the rat kidney after subcutaneous injection of glycerin, *Lab. Invest.*, 23, 29, 1970.
322. Tanner, G. A., Nephron obstruction and tubuloglomerular feedback, *Kidney Int.*, 22(Suppl. 12), S-213, 1982.
323. Tanner, G. A., Sloan, K. L., and Sophasan, S., Effects of renal artery occlusion on kidney function in the rat, *Kidney Int.*, 4, 377, 1973.
324. Tanner, G. A. and Sophasan, S., Kidney pressures after temporary renal artery occlusion in the rat, *Am. J. Physiol.*, 230, 1173, 1976.
325. Tanner, G. A. and Steinhausen, M., Tubular obstruction in ischemia-induced acute renal failure in the rat, *Kidney Int.*, 10(Suppl. 6), S-65, 1976.
326. Teschan, P. E. and Lawson, N. L., Studies in acute renal failure. Prevention by osmotic diuresis, and observations on the effect of plasma and extracellular volume expansion, *Nephron*, 3, 1, 1966.
327. Teschan, P. E. and Mason, A. D., Mechanism of renal lesions induced by shock: hemodynamic data, in *Proc. 1st Int. Congr. Nephrology*, 1960, 303.
328. Teschan, P. E. and Mason, A. D., Jr., Studies in acute renal failure. II. Incidence, mortality, urinary and plasma chemical alterations and clinical characteristics of reversible acute renal failure in the rat, *J. Surg. Res.*, 3, 442, 1963.
329. Thiel, G., Brunner, F., Wunderlich, P., Huguenin, M., Bienko, B., Torhorst, J., Peters-Haefeli, L., Kirchertz, E. J., and Peters, G., Protection of rat kidneys against $HgCl_2$-induced acute renal failure by induction of high urine flow without renin suppression, *Kidney Int.*, 10(Suppl. 6), S-191, 1976.
330. Thiel, G., Huguenin, M., Brunner, F., Peters, L., Peters, G., Eckert, H., Torhorst, J., and Rohr, H. P., Etude du mechanisme de l'insuffisance renale aigue a $HgCl_2$ chez le rat, *J. Urol. Nephrol.*, 79, 967, 1973.
331. Thiel, G. and Mason, J., Symposium on medullary circulation in acute renal failure, *Nephron*, 31, 289, 1982.
332. Thiel, G., McDonald, F. D., and Oken, D. E., Micropuncture studies of the basis for protection of renin depleted rats from glycerol induced acute renal failure, *Nephron*, 7, 67, 1970.
333. Thiel, G., Wilson, D. R., Arce, M. L., and Oken, D. E., Glycerol induced hemoglobinuric acute renal failure in the rat. II. The experimental model, predisposing factors, and pathophysiologic features, *Nephron*, 4, 276, 1967.
334. Thurau, K., Ed., Proceedings of an international symposium on experimental models and pathophysiology of acute renal failure, *Kidney Int.*, 10(Suppl. 6), 1976.
335. Thurau, K. and Boylan, J. W., Acute renal success. The unexpected logic of oliguria in acute renal failure, *Am. J. Med.*, 61, 308, 1976.
336. Toback, F. G., Amino acid enhancement of renal regeneration after acute tubular necrosis, *Kidney Int.*, 12, 193, 1977.
337. Torres, V. E., Strong, C. G., Romero, J. C., and Wilson, D. M., Changes in plasma renin substrate, plasma and renal renin, and plasma osmolarity during glycerol-induced acute renal failure in rabbits, *Mayo Clin. Proc.*, 50, 111, 1975.
338. Torres, V. E., Strong, C. G., Romero, J. C., and Wilson, D. M., Indomethacin enhancement of glycerol-induced acute renal failure in rabbits, *Kidney Int.*, 7, 170, 1975.

339. Ufferman, R. C., Jaenike, J. R., Freeman, R. B., and Pabico, R. C., Effects of furosemide on low-dose mercuric chloride acute renal failure in the rat, *Kidney Int.*, 8, 362, 1975.
340. Vanholder, R. C., Praet, M. M., Pattyn, P. A., Leusen, I. R., and Lameire, N. H., Dissociation of glomerular filtration and renal blood flow in $HgCl_2$-induced acute renal failure, *Kidney Int.*, 22, 162, 1982.
341. Van Slyke, D. D., The effects of shock on the kidney, *Ann. Intern. Med.*, 28, 701, 1948.
342. Venkatachalam, M. A., Bernard, D. B., Donohoe, J. F., and Levinsky, N. G., Ischemic damage and repair in the rat proximal tubule: differences among the S_1, S_2, and S_3 segments, *Kidney Int.*, 14, 31, 1978.
343. Venkatachalam, M. A., Rennke, H. G., and Sandstrom, D. J., The vascular basis for acute renal failure in the rat. Preglomerular and postglomerular vasoconstriction, *Circ. Res.*, 38, 267, 1976.
344. Vertel, R. M. and Knochel, J. P., Nonoliguric acute renal failure, *JAMA*, 200, 598, 1967.
345. Voegtlin, C. and Hodge, H. C., *Pharmacology and Toxicology of Uranium Compounds*, McGraw-Hill, New York, 1949.
346. Vogt, M. T. and Farber, E., On the molecular pathology of ischemic renal cell death, *Am. J. Pathol.*, 53, 1, 1968.
347. Wada, T., Aizawa, K., Kan, K., Kitamoto, K., Kuroda, S., Ogawa, M., and Kato, E., Morphologic evidence to support the role of tubular leakage as a cause of anuria induced by mercury poisoning, *Am. J. Pathol.*, 77, 175, 1974.
348. Wada, T., Kan, K., Aizawa, K., Kuroda, S., Ino, Y., Inamoto, H., Kitamoto, K., Ogawa, M., and Yamayoshi, W., Morphologic demonstration of tubular obstruction in acute renal failure, *Am. J. Pathol.*, 87, 323, 1977.
349. Ward, J. M. and Fauvie, K. A., The nephrotoxic effects of *cis*-diamminedichloroplatinum (II) (NSC-119875) in male F344 rats, *Toxicol. Appl. Pharmacol.*, 38, 535, 1976.
350. Wardle, N., Acute renal failure in the 1980's: the importance of septic shock of endotoxemia, *Nephron*, 30, 193, 1982.
351. Weiss, R. B. and Poster, D. S., The renal toxicity of cancer chemotherapeutic agents, *Cancer Treat. Rev.*, 9, 37, 1982.
352. Westenfelder, C., Arevalo, G. J., Crawford, P. W., Zerwer, P., Baranowski, R. L., Birch, F. M., Earnest, W. E., Hamburger, R. K., Coleman, R. D., and Kurtzman, N. A., Renal tubular function in glycerol induced acute renal failure, *Kidney Int.*, 18, 432, 1980.
353. Westenfelder, C. and Baranowski, R. L., Intrarenal oxygen (O_2) profiles during the initiation phase of $HgCl_2$-induced acute renal failure, *Clin. Res.*, 30, 755, 1982.
354. Westenfelder, C., Crawford, P. A., Hamburger, R. K., Baranowski, R. L., and Kurtzman, N. A., Tubular function in glycerol-induced acute renal failure in rats: effect of saline loading and prior acute renal failure, *Clin. Sci.*, 62, 667, 1982.
355. Wilkes, B. M. and Hollenberg, N. K., Saline- and glycerol-induced acute renal failure: "protection" occurs after insult, *Nephron*, 30, 352, 1982.
356. Williams, R. H., Thomas, C. E., Navar, L. G., and Evan, A. P., Hemodynamic and single nephron function during the maintenance phase of ischemic acute renal failure in the dog, *Kidney Int.*, 19, 503, 1981.
357. Wilson, D. R., Renal function during and following obstruction, *Annu. Rev. Med.*, 28, 329, 1977.
358. Wilson, D. R., Thiel, G., Arce, M. L., and Oken, D. E., Glycerol induced hemoglobinuric acute renal failure in the rat. III. Micropuncture study of the effects of mannitol and isotonic saline on individual nephron function, *Nephron*, 4, 337, 1967.
359. Wilson, D. R., Thiel, G., Arce, M. L., and Oken, D. E., The role of the concentration mechanism in the development of acute renal failure: micropuncture studies using diabetes insipidus rats, *Nephron*, 6, 128, 1969.
360. Windhager, E. E., *Micropuncture Techniques and Nephron Function*, Butterworths, London, 1968.
361. Wunderlich, P., Tubuloglomerular feedback in the rat initiated by serum activity in patients with acute renal failure, *Kidney Int.*, 22(Suppl. 12), S-219, 1982.
362. Wunderlich, P. F., Brunner, F. P., Davis, J. M., Häberle, D. A., Thölen, H., and Thiel, G., Feedback activation in rat nephrons by sera from patients with acute renal failure, *Kidney Int.*, 17, 497, 1980.
363. Yuile, C. L., Gold, M. A., and Hines, E. G., Hemoglobin precipitation in renal tubules, *J. Exp. Med.*, 82, 361, 1945.
364. Zalme, R. C., McDowell, E. M., Nagle, R. B., McNeil, J. S., Flamenbaum, W., and Trump, B. F., Studies on the pathophysiology of acute renal failure. II. A histochemical study of the proximal tubule of the rat following administration of mercuric chloride, *Virchows Arch. (Cell Pathol.)*, 22, 197, 1976.
365. Zimmermann, H. D., Schmidt, E., Weller, E., Becker, C., and Dieker, P., Intra- and extrarenal vascular changes in the acute renal failure of the rat caused by mercury chloride, *Virchows Arch.* (Pathol. Anat.), 372, 259, 1977.

AMINOGLYCOSIDE-INDUCED TOXIC NEPHROPATHY

J. Edmond Riviere

INTRODUCTION

Aminoglycosides are a therapeutically essential class of antibiotics whose usefulness is often restricted by their nephrotoxic potential. Recent clinical trials in man have estimated the incidence of nephrotoxicity to be approximately 10 to 20% of treated patients. Aminoglycosides are thus of interest to nephrologists who must either administer them in such a fashion as to avoid toxicosis or manage and treat the patient who has developed a drug-induced toxic nephropathy. Because of their clinical importance, extensive applied and basic research efforts have been directed toward studying this toxicosis in experimental animals. As a result, aminoglycoside-induced toxic nephropathy (AgTN) has become an accepted model of nonoliguric acute renal failure in man.

There are two primary uses for animal models of AgTN. The first is to model the physiologic determinants and clinical expression of the syndrome in man. If extrapolations are to be meaningful, such studies should utilize appropriately scaled drug doses in experimental animals which approximate those used clinically in man. The second use of animal models is to investigate the mechanisms of nonoliguric acute renal failure (ARF). For studies of this nature, supratherapeutic doses may be employed to produce a dose-related toxic nephropathy. Before an appropriate animal is selected, it is important that the objective of the experiment be clearly defined and the proper dose of drug selected. Care must be exercised in extrapolating the findings from one type of study to the other.

This chapter will first review the chemistry and the comparative pharmacology and pharmacokinetics of the aminoglycosides as a prelude to discussing their toxicology. The toxic syndrome seen clinically in man will then be presented followed by a brief discussion of the postulated mechanisms of AgTN at a nephron and cellular level. With this background, the syndromes in the various animal species will be critically presented. Finally, nonaminoglycoside drug-induced nephrotoxic renal failure will be briefly discussed. This review is written from a comparative pharmacology perspective and will stress how the disease syndrome can be experimentally produced in different animals. For other aspects of AgTN, several recently published reviews should be consulted.[58,118,127,178,286]

CHEMISTRY

Aminoglycoside antibiotics are a class of antimicrobial compounds produced from strains of *Streptomyces*, *Micromonospora*, and *Bacillus*. Chemically, they are aminocyclitols: hydroxyl and amino or guanidine substituted cyclohexane with amino sugars joined by glycosidic linkages to one or more of the hydroxyl groups. This molecule has excellent water but poor lipid solubility. The pK_a of these weak bases is approximately 8. All are stable over a wide range of pHs and are also relatively resistant to thermodegradation.[139,146,181,193]

Table 1 is a list of the common aminoglycosides. Of these, gentamicin, tobramycin, and amikacin are currently in common use. Kanamycin and streptomycin are less frequently encountered. Netilmicin, sisomicin, and dibekacin are newer compounds which may be clinically introduced. The majority of toxicologic studies have been conducted on gentamicin and tobramycin, and therefore this chapter will largely be focussed on these drugs.

Table 1
AMINOGLYCOSIDE ANTIBIOTICS

1.	Amikacin
2.	Butikacin
3.	Butirosin
4.	Dibekacin
5.	Fortimicin
6.	Gentamicin
7.	Kanamycin
8.	Lividomycin
9.	Neomycin
10.	Netilmicin
11.	Paromomycin
12.	Propikacin
13.	Ribostamycin
14.	Sagamycin (Gentamicin C_{2b})
15.	Seldomycin
16.	Sisomicin
17.	Sorbistin
18.	Streptomycin
19.	Tobramycin

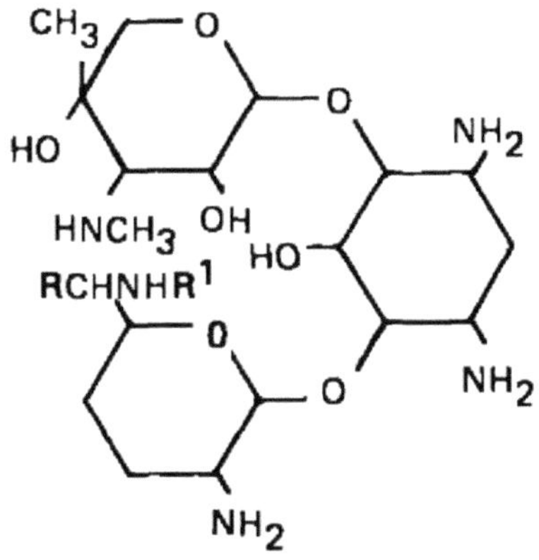

C_1 : $R=R^1=CH_3$

C_2 : $R=CH_3, R^1=H$

C_{1a}: $R=R^1=H$

FIGURE 1. Structure of the gentamicin complex.

Gentamicin is a complex of three components produced by *Micromonospora purpurea:* C_1, C_{1a}, and C_2. As can be seen in Figure 1, they differ only in their degree of methylation of carbon 4 of the sugar moiety. Component C_1 has two methyl substituents, C_2 has one, and C_{1a} is not methylated at this position. The average molecular weight is 463 g/mol. Tobramycin is produced by *Streptomyces tenebrarius* and is structurally similar to kanamycin, being 3′-deoxykanamycin B. Both gentamicin and tobramycin are chemically derivatives of deoxystreptamine, as are all of the other clinically used aminoglycosides except streptomycin.

A knowledge of chemical structure is important in determining antimicrobial activity, resistance patterns, and inherent propensity to cause toxicity. Nephrotoxicity, in vitro binding to proximal tubule brush border vesicles, binding to acidic membrane phospholipids (postulated aminoglycoside receptor), and inhibition of mitochondrial function, may correlate to an increased number of free amino groups on the aminoglycoside molecule. The most ionized drugs (i.e., neomycin with six groups) are more toxic and show greater binding affinity than the least ionized members of the class (i.e., streptomycin with three groups).[16,63,86,118,121,137,150,161,277] Other structural characteristics account for differences in toxicity within groups of drugs with similar total ionization potentials (i.e., netilmicin, tobramycin, and gentamicin all with five ionizable groups). Within the gentamicin complex itself, component C_1 appears relatively less nephrotoxic and C_2 appears more nephrotoxic.[90,134] More specific information on aminoglycoside structure-toxicity relationships is not presently available.

Before an animal model can be selected to study AgTN, the comparative pharmacology and pharmacokinetics of these drugs must be reviewed in the relevant species. Extrapolation of data between animals and man is fraught with difficulties when doses are not properly scaled to account for differences in underlying physiology. Second, the parameters of the proper pharmacokinetic model describing drug disposition in the species in question must be known so that the optimal dose regimen can be selected. This is especially important when modeling the clinical syndrome in man which results from the use of therapeutic doses.

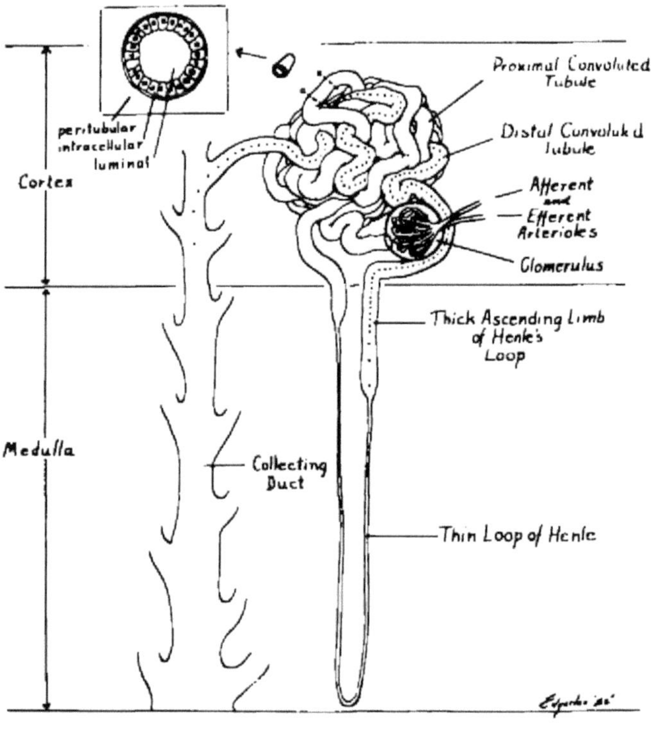

FIGURE 2. Schematic of a cortical-medullary nephron.

Whole animal and human renal clearance,[31,49,52,103,147,200,233,244] isolated perfused rat kidney,[59a,174a] and micropuncture studies[190,238,239] have demonstrated that aminoglycosides are eliminated nonmetabolized from the body, in all animal species thus far studied, primarily by renal glomerular filtration. Some degree of proximal tubular reabsorption occurs and results in intracellular sequestration or storage in the tubule cells without a significant transepithelial flux from the intraluminal to peritubular space (Figure 2). Net aminoglycoside secretion along more distal nephron segments may also occur. The factor confounding interpretation of such renal clearance studies is the degree of aminoglycoside serum protein binding, generally considered negligible,[101a,179a,183,190] and the recently observed restricted ultrafiltration of these cationic molecules across the glomerular filtration barrier.[191,209] Proximal tubule luminal absorption of aminoglycoside appears quantitatively to be the primary mechanism of intracellular uptake; however, selective peritubular or basolateral reabsorption, evident in isolated tissue slice studies, does occur and may be of toxicologic significance in specific situations. Reabsorption requires metabolic energy and occurs along the mid convoluted and straight portions of the proximal tubule.[14,22,114a,127,131,136,190,238,241,243,244,266,269,283,287]

The initial event in luminal reabsorption is the binding of drug to the proximal tubule plasma membrane at so-called "aminoglycoside receptors". These have been recently identified as a group of acidic membrane phospholipids, the polyphosphoinositides. The interaction between the cationic aminoglycosides and anionic phospholipids appears to be electrostatic and proportional to the cationic charge of the drug. This interaction is saturable and is competitively inhibited by the divalent cation calcium, by spermine, poly-L-lysine, and other aminoglycosides. Once bound to these receptors, the drug is absorbed intracellularly by an energy dependent process, presumably pinocytosis.[22,76,86,118,149,150,190,229] Autoradiography has shown drug to appear in apical vesi-

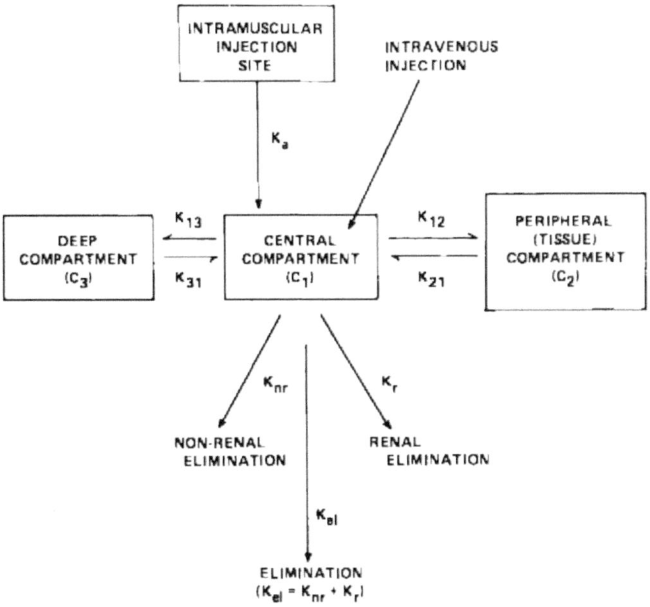

FIGURE 3. Open three-compartment pharmacokinetic model.

cles within 10 min after injection and ultimately is incorporated into secondary lysosomes (cytosegresomes) within 1 hr.[136,241,243,269] Cytosegresomes are the primary locus of intracellular aminoglycoside storage. However, a recent study employing combined immunofluorescence and section freeze-dry autoradiography techniques has demonstrated that after parenteral administration of from 4 to 100 mg/kg gentamicin, freely diffusible (nonlysosome associated) gentamicin is present in S_1 and S_2 proximal tubule sections for 1 to 6 hr.[274] After this time, gentamicin appears to be primarily sequestered in lysosomes. This intralysosomal binding within the proximal tubules results in drug sequestration and is primarily responsible for the prolonged half-life of aminoglycosides seen in the renal cortical tissue.

It appears clear that the renal handling of aminoglycosides is similar across species lines. In order to quantitatively describe aminoglycoside disposition, a three compartment pharmacokinetic model such as depicted in Figure 3 is useful.[44,89,123,124,160,183,193,209,233-235,285] The assumptions inherent to its derivation have been adequately discussed elsewhere.[98,184,271] In this scheme, the body is represented as a system of theoretical compartments, each containing a fraction of the administered dose. The "size" of the various compartments is reflected by the kinetic parameters governing the drug's entry into and exit from each area and they are not strictly defined in physiologic or anatomic terms. A drug absorbed from an extravascular site (s.c., i.m.) into the blood (central compartment) distributes between the two other compartments and is then eliminated from the central compartment by renal and nonrenal processes. For aminoglycosides, the sole route of excretion is renal. The rates of transfer of a drug between individual compartments (k_{21}, k_{12}, k_{31}, k_{13}) is directly proportional, i.e., first-order kinetics, to the drug's concentration in each compartment. If the aminoglycoside were given intravenously, the extravascular absorption rate constant, ka, would be zero. By definition, the magnitude of k_{13} and k_{31} are less than k_{12} and k_{21}, implying that the rate of transfer into and out of compartment three is slower than the corresponding rates of compartment two. Compartment three is thus termed a "deep" compartment. In relation to aminoglycosides, it has been postulated by

Schentag and co-workers[230,232,235-237] that k_{13} is related to renal uptake of drug and may be increased in prenephrotoxic patients. The relatively large value for k_{31} is reflected in the prolonged terminal elimination half-life previously discussed. Studies have not yet been conducted which clearly define the physiologic nature or determinants of this slow terminal phase. Dose independency of this phase implying a first-order, linear pharmacokinetic process, also has not been demonstrated. In fact, the duration of this phase may be longer after multiple dose administration in man[89,123] (n = 4) and sheep[36] (n = 6). This was not detected in a single nephrotoxic patient given single and multiple doses.[1]

Values for the above constants are obtained by analyzing repeated blood samples for aminoglycoside concentration and then subjecting the data to a nonlinear regression analysis. A triphasic curve results with intercepts A, B, and C and slopes α, β, and γ. These slopes may be used to obtain the values of intercompartmental rate constants according to the differential equations which define the model selected. The half-lives for each of these phases can be obtained from the relation:

$$T^1/_2 = \frac{0.693}{\text{slope}} \tag{1}$$

For aminoglycosides, $T1/2(\alpha)$ is the distribution half-life, the $T1/2(\beta)$ reflects the classic elimination phase governed largely by renal elimination (k_{el}), and $T1/2(\gamma)$ reflects the slow release of drug from the "deep" tissue compartment (k_{31}). Caution must be exercised when using these half-life values because they are hybrid constants and their interpretation is dependent upon the pharmacokinetic model selected. The three-compartment scheme presented is only valid if sufficient blood samples are collected over a long enough period of time to discern all three phases of the curve. If blood samples were only collected for 8 hr, a two-compartment model consisting of α and β phases would be used. If the α phase were missed, i.e., sample collection started after 1 hr, a one-compartment model would suffice. Likewise, if samples were collected from 1 hr to 10 days, a two-compartment model consisting of only β and γ phases would result. The experimental design of the study must be known before parameters can be interpreted. In the latter case, k_{12} and k_{21} would be equivalent to the three-compartment k_{13} and k_{31}. Table 2 summarizes these relationships. When a three-compartment analysis is employed, Vd(ss) and to a much greater extent Vd(area) will be larger than corresponding two compartment values. Total body clearance will remain relatively stable.[1,36,89,115,218] Finally, in one study in calves,[301] a very rapid initial distribution phase has been observed resulting in a three-compartment model consisting of π, α, and β phases. Such a rapid distribution phase may represent a rapid equilibration of drug between plasma and red blood cells,[145] a process which could significantly affect the value of plasma gentamicin concentrations used in pharmacokinetic studies at very early time points.

Model independent pharmacokinetic parameters may be utilized to analyze the data: volume of distribution (Vd) and total body clearance (Cl_B). These are useful in designing dosage regimens in different species. Second, they have physiological significance in that they can be correlated to specific processes in the animal. Methods of calculating these parameters have been presented elsewhere.[18,94,98,120,184,271,272] Cl_B is the sum of all processes clearing drug from the body. For aminoglycoside, Cl_B has been assumed to equal the renal clearance of drug (Cl_R) which approximates the glomerular filtration rate (GFR) of the animal. However, this is not true because of limitations inherent to the pharmacokinetic derivation of Cl_B which would overestimate the time Cl_R even

Table 2
COMPARISON OF AMINOGLYCOSIDE MULTICOMPARTMENT MODELS

Number of compartments	Study duration	Parameters: Slopes	Parameters: Microconstants
3	2—3 weeks	α,β,γ $\lambda_1,\lambda_2,\lambda_3$	k_{12},k_{13} k_{21},k_{31},k_{el}
2	1 day	$\alpha,\beta,-$	k_{12} $k_{21},-,k_{el}$
2	2—3 weeks	$-,\alpha,\beta$	k_{12} $k_{21},-,k_{el}$
1	1 day	k or k_{el}	$-,-,k_{el}$

when renal excretion is the sole route of elimination.[209] The second parameter, Vd, is actually a proportionality constant which enables one to relate dose to expected plasma drug concentration.

$$Vd = \frac{\text{dose}}{\text{concentration of drug in blood}} \tag{2}$$

When comparing aminoglycoside disposition between species, these two parameters are often not available and only "elimination half-life" is reported. Half-life is related to Cl_B and Vd by the following relationship:

$$T^1/_2 = \frac{(0.693)(Vd)}{(Cl_B)} \tag{3}$$

When extrapolating T1/2 between species, one must remember that two physiologically independent parameters determine its value.

A second problem which arises when comparing aminoglycoside disposition and toxicity between species involves the study of size and its physiologic consequences, the field of allometry. Although not previously studied with respect to gentamicin, the application of allometry to the field of pharmacokinetics has been investigated for other drugs.[34,70,71,278] In general, drug half-life is related to body weight (B) raised to the 0.25 power for compounds primarily eliminated by the kidney ($B^{0.25}$). This relationship allows for comparisons to be made between species differing greatly in weight. When scaled in these terms, half-life becomes invariant and, thus, independent of species. The reader is referred to the excellent review by Boxenbaum[34] for the physiologic basis of this relationship. Table 3 is a compilation of gentamicin pharmacokinetic data in 15 species. A log-log plot of the mean T1/2(β) vs. B of mouse, rat, rabbit, cat, dog, sheep, man, and horse was constructed. Regression of this data yielded the following allometric equation describing gentamicin disposition ($p = 0.0001$; $R = 0.977$).

$$T^1/_2(\beta)\ \text{min} = 42.3\ B^{0.249} \tag{4}$$

In selecting data from Table 3 to be used in this analysis, outliers were excluded and data from this author's laboratory were used when available (rat, rabbit, dog, sheep, and horse). Body weights varied by a factor of 2×10^4. This equation is very similar to that of Dedrick's classical study with methotrexate, a compound also primarily eliminated from the body by renal mechanisms ($T1/2 = 54.6B^{0.228}$, 5 species).[34,70,71]

This analysis strongly suggests that a scaling factor of $B^{0.25}$ be used when comparing aminoglycoside disposition parameters and toxic thresholds between species. The hybrid parameter T1/2 was used because insufficient data exists to directly compare Vd and Cl_B. However, as suggested by Boxenbaum,[34] this may be advantageous in allometry because T1/2 is dependent upon a variety of biological energy sources. Finally, it can be concluded from an examination of this data, that T1/2(β) is dose independent in all animal species thus far studied.

Although the variation in T1/2 between species is fairly well explained on an allometric basis, a closer examination of the data in Table 3 illustrates significant variability. The route of drug administration is important as can be seen by comparing the prolonged T1/2 in rats given gentamicin i.p. compared to i.m. or i.v. Note that in many toxicologic investigations, drug is administered i.p. Volume of distribution varies widely depending on the method of calculation employed. However, even when the same volume parameter is compared (Vd_{ss}) with data from the same laboratory, species specific differences are evident. As an example, when sheep and horses are compared, Cl_B is similar, yet sheep have a relatively smaller Vd_{ss} compared to horses. This results in the shorter T1/2(β) for sheep. This highlights the importance of Equation 3 in interpreting the physiological basis of T1/2. Note that in ruminants such as sheep, body weight could also be adjusted for rumen mass since aminoglycosides would not partition into this body space.[36]

Allometric relationships appear sensitive to predicting T1/2 when body mass varies several orders of magnitude. However, when comparing species of similar mass, other physiologic parameters can refine the analysis. When available, pharmacokinetic studies conducted in the species of interest should be employed to design toxicologic studies.

Age is an important consideration in selecting an animal model. Young animals appear to have a prolonged T1/2(β)[62,169] compared to adults which is probably the result of a relatively decreased renal clearance and expanded volume of distribution.[62,214,297,301]

Another major source of variability in aminoglycoside pharmacokinetic studies is the assay method employed to measure drug concentrations. These can be divided into five classes: radiotracer studies, microbiological (plate diffusion, microdilution), immunological (radioimmunoassay, fluoroimmunoassay, enzyme linked immunoassay), enzymatic (transferases), and chromatographic (high performance liquid chromatography). The advantages and disadvantages of each have recently been reviewed and are beyond the scope of this chapter.[142,170,174,186,203,224,230,289] Except for radiotracer methods, radioimmunoassay is the most sensitive analytical technique available. Only chromatographic techniques have the specificity to separate closely related aminoglycosides or components (i.e., gentamicin). When comparing single drug concentrations between species, the analytical method used must be taken into consideration. Rate parameters in pharmacokinetic studies are not as sensitive to absolute errors in concentrations as are volume parameters. Intralaboratory variation is also great even if the same analytical technique is employed. In the end, selection of a specific assay method is often dependent on availability of equipment and experience of the operators. However, caution must be exercised when comparing results to other laboratories.

AMINOGLYCOSIDE PHARMACOLOGY IN HUMANS

Numerous workers have investigated clinical aminoglycoside pharmacokinetics in

Table 3
COMPARATIVE PHARMACOKINETICS OF GENTAMICIN

Species	$T^1/_2(\beta)$ (hr)	Vd (ℓ/100 kg)	Cl_β (mℓ/min/kg)	Dose (mg/kg)	Route	Number of animals	Assay method	Model[a]	Comments	Ref.
Mouse	0.2	26[b]	20	4	i.v.	—	Radio	3	$T^1/_2(\gamma)$ = 173 hr (Vd and Cl_β based on α and B phase only)	126
Rat[c]	0.8	26[b]	4.8	3	i.v.	8	Radio	2		144
	0.9	—	—	100	s.c.	5	Micro	1		252
	0.5	—	—	10	s.c.	4[d]	Micro	1		160
Rat[e]	0.5	—	—	5	i.v.	3[d]	Micro	1		169
	0.7	—	—	5	i.v.	3[d]	Micro	1	Young rats (90 g)	169
	1.5	—	—	4	i.p.	6	Micro	1		84
	0.6	—	—	6	i.m.	10	Micro	1		107
	0.8	21	3.3	10/16/25/40	s.c.	33	RIA	1		45
Guinea pig	0.9	30[f]	3.4	40	i.v.	5	RIA	3	$T^1/_2(\gamma)$ = 7.9 hr	53
Rabbit	0.8	31[f]	4.6	3	i.v.	5	Micro	2		104
	1.7	—	—	1.5	i.m.	5	Micro	1		41
	1.1	67[b]	3.0	20	i.v.	4	HPLC	3	$T^1/_2(\gamma)$ = 14 hr	115
	0.8	—	—	1.5	i.m.	5	Micro	1		42
	0.8	—	—	2.5/7.5	s.c.	18	Micro	1		91
Cat	1.2	—	—	40	i.a.	2	Micro	1		273
Dog	1.0	34[b]	4.1	10	i.v.	11	RIA	2	$T^1/_2(\gamma)$ = 31 hr (218)	210, 212
	1.0	—	—	45	i.v.	93	ENZ	2	Gentamicin, netilmicin, tobramycin	200
	1.3	26[b]	2.9	10	i.v.	7	Micro	2		11
	0.7	—	3.4[g]	3.8	i.m.	4	RIA	1	30 days old	62
	0.8	—	2.4[g]	3.8	i.m.	4	RIA	1	20 days old	62
	1.4	—	2.6[g]	3.8	i.m.	3	RIA	1	10 days old	62
	1.9	—	—	1.5	i.m.	10	Micro	1		195
Sheep	2.4[h]	24[b]	1.0	10	i.v.	6	RIA	3	$T^1/_2(\gamma)$ = 30 hr	36
	1.4[i]	19[f]	1.6	2.2	i.v.	7	Micro	2		293

	1.8[f]	16	—	20	i.v.	4	Micro	1		300
Calves	3.9	75[b]	0.7	3	i.v.	7	Micro	3[k]	4—5 weeks old	301
Horses	3.2	31[b]	1.6	4.5	i.v.	7	RIA	2	Foals	214
	3.4	26[b]	1.1	4.5	i.v.	2	RIA	2	Suckling foals	214
	2.5	24[b]	1.2	5	i.v.	6	RIA	2		194
Chickens	4.4	—	—	0.2	i.c.	10[d]	Micro	2	1 day old	255
Quail	0.7	—	—	5/10/20	i.m.	3/3/3	Micro	1		68
Pheasants	1.2	—	—	5/10/20	i.m.	5/7/3	Micro	1		68
Cranes	2.8	—	—	5/10/20	i.m.	8/1/1	Micro	1		68
Leopard frogs	26	—	—	10	i.c.	2	RIA	1	Hydrated	219
Gopher snakes	82	—	—	5/2.5	s.c.	5/5	Micro	1		39

Note: All animals are adults unless indicated otherwise; i.v., intravenous; i.m., intramuscular; s.c., subcutaneous; i.a., intra-arterial; i.c., intracardiac; i.p., intraperitoneal; Radio, radiolabeled tracer study; Micro, microbiological assay; HPLC, high performance liquid chromatography; RIA, radioimmunoassay; ENZ, radiolabeled adenylase assay.

[a] Number of compartments in pharmacokinetic model.
[b] Vd(ss).
[c] Sprague-Dawley strain.
[d] Number of animals sacrificed per time point.
[e] Wistar strain.
[f] Vd(area).
[g] Renal clearance using urine excretion data.
[h] Suffolk breed.
[i] Cross-bred.
[j] Awassi breed.
[k] Compartments 1 and 2 are distribution; compartment 3 is β phase elimination.

human patients utilizing one-, two-, or three-compartment models.[5,31,37,52,54,61,102,103,117,125,147,152,157,168,183,193,196,205,206,230,232,246,264,280,294,295,298] There is a general consensus that all of the aminoglycosides studied (amikacin, gentamicin, kanamycin, netilmicin, sisomicin, streptomycin, and tobramycin) have a similar pharmacokinetic profile. The serum β phase half-life has a value of 1.7 to 2.5 hr independent of dose, the apparent volume of distribution is 20 to 25 ℓ/100 kg, and the Cl_B is 70 to 95 mℓ/min/1.73 m^2, a value slightly less than the corresponding glomerular filtration rate. These drugs are not appreciably bound to serum proteins and 80 to 90% of a single dose is excreted unchanged in the urine in a 24-hr period. Amikacin and kanamycin peak and trough blood concentrations are always three to four times greater than the other aminoglycosides because a greater dose is required for therapeutic equivalency. Therefore, in a multiple dose regimen where drug is given every 8 hr (2 mg/kg i.m. or 30-min i.v. infusion) peak concentrations are approximately 6 to 8 μg/mℓ and trough concentrations are 0.5 to 1.5 μg/mℓ in subjects with normal renal function. Corresponding values for 7.5 mg/kg dose of amikacin or kanamycin are 20 to 25 μg/mℓ at peak and 6 to 10 μg/mℓ at trough. These concentrations are important to consider because toxicologic studies in animals designed to model the human clinical AgTN syndrome must utilize dosages which, when adjusted for body weight, produce peak and trough serum concentrations equivalent to those seen therapeutically in man. Aminoglycosides are well absorbed after i.m. or s.c. administration, but not after oral dosing. Tissue concentrations of drug are generally lower than corresponding serum concentrations in all organs except the well-perfused renal cortex, lungs, and liver. Interstitial fluid drug concentrations are 17 to 30% of simultaneous serum concentrations. Aminoglycoside penetration into intraocular and cerebrospinal fluid is minimal. In contrast, therapeutic concentrations of drug up to 80% of serum values can be achieved in peritoneal, pericardial, pleural, perilymph, or synovial fluids. Penetration into bile and bronchial secretions is intermediate. As would be expected, aminoglycosides can be effectively removed from the body by peritoneal dialysis or hemodialysis. Concentrations of drug in urine are always greater than corresponding serum drug concentrations unless severe renal parenchymal damage is present.

The gamma phase half-life has been reported to vary from 27 to greater than 600 hr, depending on the aminoglycoside studied, the number of doses given, and on the route of administration. The value of this half-life appears to be independent of renal function. The percent of the area under the serum drug concentration vs. time curve (AUC) attributed to this phase is only 5 to 10%. If an aminoglycoside is dosed on the basis of T1/2(β) alone, trough concentrations will increase over time due to accumulation governed by dosing at intervals shorter than T1/2(γ).[1,8,59,89,125,168,193,196,230-235,237]

There is a great amount of intraindividual variation in pharmacokinetic parameters.[12,13,128,295,297-299] Therefore, although population pharmacokinetic estimates are useful, they do not necessarily predict drug disposition accurately in an individual patient. An example is volume of distribution, which may range from 4 to 74 ℓ/100 kg depending on the state of hydration, obesity, and concurrent disease processes.

T1/2(β) reported for man with an average body weight of 70 kg is similar to the value predicted from allometric data. Sheep would appear to be a useful animal model for studying aminoglycoside disposition in man because of similar body weights and similarity in pharmacokinetic parameters without allometric adjustment. Comparisons with other species are valid if dosage is properly scaled to account for differences in body weight.

AMINOGLYCOSIDE NEPHROTOXICITY IN HUMANS

Estimates of the incidence of AgTN in man range from 2 to 26% of treated patients, in spite of the fact that patients receiving a drug are carefully monitored to avoid toxicity. Viewing the problem from a different perspective, approximately 10% of all nephrologic consultations may result from a nephrotoxic etiology. Aminoglycoside toxic nephropathy in man is characterized clinically initially by enzymuria arising from proximal tubule brush border membranes and lysosomes. This may be accompanied by β_2 microglobulinuria and possibly glycosuria and aminoaciduria. All of these events are signs of proximal tubular injury. Increased fractional sodium excretion, hypokalemia and hypomagnesemia, have also been observed in man, the latter suggestive of renal potassium and magnesium wasting. Polyuria and cylindruria are the next observable signs as AgTN develops. These mild changes, not directly correlated to dose, are followed by a significant decrease in glomerular filtration rate, reflected in elevated serum creatinine or urea nitrogen concentrations or decreased creatinine clearance. These latter changes are not recognized until 7 to 10 days after drug therapy has been initiated. This nonoliguric acute renal failure is generally reversible if the patient is managed properly. In a smaller percentage of cases, an acute oliguric renal failure marked by a uremic crisis may occur and cause fatality. These abrupt episodes are often dose related or occur in debilitated patients who have complicating medical or surgical conditions, such as intravascular volume depletion, severe metabolic acidosis, endotoxemia, or a decompensated coexisting renal disease with or without dialysis. These patients require aggressive therapy for survival. It is often very difficult to ascertain the etiology of renal insufficiency in severely debilitated septicemic patients receiving aminoglycosides.[5,6,93,96,108,123,140,156,167,199,230-232,236,237,247-251,262,265,297,298]

The histopathologic appearance of AgTN is one of a focal and patchy tubular necrosis unless very high doses of drug are used, in which case massive tubular necrosis is present. In the typical case, most tubules appear normal and show no evidence of necrosis. In tubules showing changes, affected cells may be heavily vacuolated, flattened, or sloughed, while other cells in the same tubule cross-section are normal. Some tubules may appear dilated with a patchy mild interstitial infiltration of mononuclear cells which may occur in the area of the cortical medullary junction. Ultrastructurally, proximal tubule cells of gentamicin-treated patients contain lysosomes containing membranous whorls, termed myeloid bodies or cytesegresomes. These are considered a marker of aminoglycoside exposure rather than of overt AgTN. Renal biopsy is not recommended because of the focal nature of the lesions.[112,187]

Differences in reported incidences of drug-induced toxicity are often dependent on the demographics of the patient population studied with increased incidences of toxicity noted in groups composed of older, more severely ill patients.[140,237,251,298] A problem in assessing the incidence of drug-induced toxicity in man is the lack of a uniform definition of renal dysfunction. Many authors define a rise in serum creatinine concentration of 0.5 mg/100 mℓ or greater over pretreatment values as toxicologically significant. However, this criteria for toxicity may be too sensitive in patients with preexisting renal disease. The relationship between serum creatinine and creatinine clearance is nonlinear and thus a fixed increment cannot be used at all levels of renal function. An alternative technique is to use a graded serum creatinine nephrotoxicity scale where an increase of $\geq$0.4 to 0.5 mg/100 mℓ (initial serum creatinine $<$3) or an increase of $\geq$0.9 to 1.0 mg/100 mℓ (initial serum creatinine $\geq$3) is indicative of toxicosis.[140,247,249,250] Measurement of creatinine clearance itself is the most reliable indicator.[129,265] Other workers have employed fractional or percentage increases over pretreatment values. A difficulty with using either serum creatinine or creatinine clearances as a marker for GFR in man is the dependence on factors other than glomerular filtration. The rate of creatinine production is sex dependent and decreases with age. Additionally, the human kidney secretes and possibly reabsorbs creatinine.

Table 4
POSSIBLE RISK FACTORS ASSOCIATED WITH AMINOGLYCOSIDE NEPHROTOXICITY

Age
Volume contraction, shock
Acidosis
Sodium, potassium depletion
Sepsis
Renal transplantation
Prior renal insufficiency
Prior aminoglycoside exposure
Hepatic disease
Total dose of drug administered
Duration of treatment
Concurrent administration of loop diuretics, methoxyflurane, cephalosporin antibiotics, or other nephrotoxic drugs

In the presence of renal dysfunction, the secretory component may mask the decline in filtration resulting in an overestimation of true GFR.[30,43,47,55,97,282] There is a lag in the time that it takes for a decrease in GFR to be reflected in a rise in serum creatinine because of creatinine pharmacokinetics. This factor would mask drug-induced nephrotoxicity in patients with preexisting renal disease whose creatinine half-life is already relatively prolonged.[48] Finally, concurrent liver disease can result in an overprediction of GFR from serum creatinine.[116] These limitations could have significant effects on the dose of aminoglycoside administered, if creatinine-based nomograms or algorithms were used, and on the reported incidence of drug-induced toxicity in clinical trials. Other renal function tests should be concurrently employed if an accurate estimate of toxicity is to be made. Tests which are more specific for tubular damage include urine enzymes, β_2 microglobulin urine excretion, and urine cast counts.

A third variable affecting the observed incidence of toxicity in humans is the dose regimen of aminoglycosides employed and how this dose is altered in the presence of preexisting renal disease. When an intermittent dose schedule is employed, trough aminoglycoside concentrations of >2 μg/mℓ (>20 μg/mℓ amikacin) have been associated with nephrotoxicity.[101b,108,123,250,251,260a] An elevated trough concentration may be the result of, rather than the cause of, a decrease in drug elimination. Human clinical studies have not been able to adequately explain this relationship. Patient numbers are often too small to permit valid statistical inferences to be made to the general population. Finally, the relationship of frequency of drug administration to the incidence of nephrotoxicity has not been rigorously tested in human clinical trials.

Several risk factors have been identified which may potentiate AgTN (Table 4).[37,63,82,83,108,118,140,198,199,202,230,251,287,297] The significance of each is difficult to interpret because of study design weaknesses in many of the trials which defined them. This list is not definitive. For example, some recent evidence indicates that prior aminoglycoside exposure may protect against rather than potentiate AgTN.[232,251] The timing of the second aminoglycoside exposure is critical in this respect. Renal insufficiency is not a risk factor in some studies;[251] however, detection of toxicosis in these patients is often confounded by the underlying renal disease and the inadequacies of creatinine as a glomerular filtration marker. Additionally, different dosage nomograms used to decrease drug doses in the presence of renal insufficiency have different toxic potentials, further confounding results. All nomograms rely on the use of creatinine as a glomerular filtration marker and these dosages may not be optimal in disease states where creatinine disposition is altered (see discussion above). It is also probable that different renal diseases have varying sensitivities to AgTN, a stratification which has not been conducted in clinical studies. Prerenal disease with glomerular shutdown might protect against toxicity by decreasing the filtered load and subsequent tubular reabsorption of drug,[230] while chronic renal disease states with compensatory hypertrophy may have increased tubular reabsorption on an individual nephron basis.[208] Post-mortem studies uniformly show decreased renal parenchymal concentrations of aminoglycosides in patients with chronic renal diseases,[20,166,283-285,288] an event predicted by the decreased total functional nephron mass available for binding. These reduced drug concentrations are in whole cortical homogenates and do not directly reflect individual remnant nephron concentrations.

Another important consideration in assessing the nephrotoxicity of aminoglycosides in clinical trials is the specific drug employed. Numerous investigations have addressed this question, an assessment of which is beyond this chapter's scope. There is a consensus that the following order is acceptable, with the most toxic of the clinically employed aminoglycosides listed first: neomycin > gentamicin, kanamycin, sisomicin, (amikacin) > tobramycin (netilmicin). The precise relative toxicities of amikacin and netilmicin must await further trials.[37,69,118,196,230,236,247-249,297]

Schentag and co-workers[231] have postulated that multicompartment pharmacokinetic analysis can be employed to identify potentially nephrotoxic patients because of their propensity to accumulate aminoglycosides in the "deep" tissue compartment. In a series of studies utilizing gentamicin,[59,93,232,236,237] tobramycin,[232,236,237] or amikacin,[93] patients who developed AgTN (increase in serum creatinine ≥ 0.5 mg/dℓ; elevation in β_2 microglobulin in urine) also had a tendency, based on pharmacokinetic parameters, to accumulate a significantly greater amount of drug than did nontoxic patients. They further concluded that aminoglycoside serum concentrations alone, once within the "therapeutic" range, were of limited value in preventing AgTN. Rate of tissue accumulation was the important factor. It is not known if actual renal tissue concentrations of aminoglycosides are elevated in prenephrotoxic human patients before nephrotoxicity ensues. However, the apparent relationship of elevated initial trough serum concentrations with the onset of nephrotoxicity supports this hypothesis, if the higher trough concentration in serum reflects a distribution equillibrium with accumulated drug in tissue.

In summary, clinical studies in man indicate a significant incidence of AgTN in the face of diligent clinical monitoring and sophisticated pharmacokinetic dosing techniques. The toxicosis of clinical significance today cannot be predicted by any specific criteria, although certain risk factors and pharmacokinetic parameters show promise. Animal models designed to study this condition must take into consideration the clinical expression of the disease, should employ a human therapeutic equivalent (low) drug dose, and should monitor the onset of toxicity with techniques applicable in the clinical setting. High-dose animal studies have their place in determining mechanisms of renal toxicosis and in identifying risk factors associated with aminoglycoside use. However, low-dose studies must be used to model the expression of the nephropathy which occurs in the human clinical setting.

PATHOGENESIS AT THE CELLULAR AND NEPHRON LEVEL

The final topic to consider before discussing specific animal models of AgTN is the pathogenesis of this toxicosis at the level of the individual tubule cell and nephron.

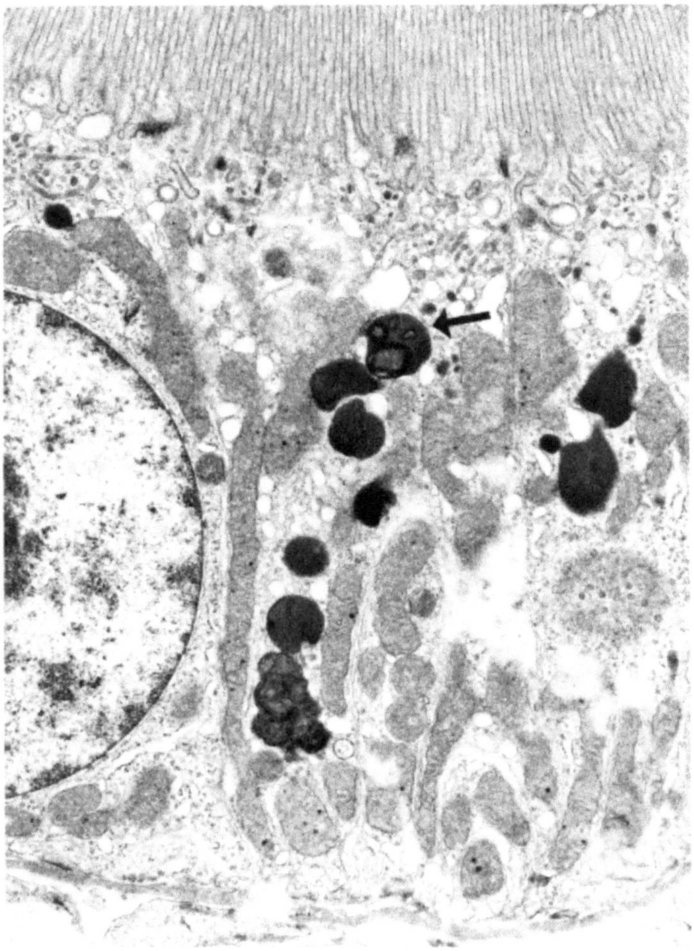

FIGURE 4. Electron micrograph of a proximal tubule from a gentamicin-treated dog showing cystosomes with myeloid figures. (Magnification × 8200.)

These events appear to be similar in all species studied and are pertinent to the use of AgTN as a model of nonoliguric acute renal failure.

Controversy exists as to the precise mechanism by which aminoglycosides initially damage proximal renal tubule cells. Abundant evidence suggests that lysosomal dysfunction is a component of this early phase of cellular injury.[9,42a,87,118,127,143,150,171,173,178,179,266,267] This view is consistent with the concept that lysosomes are the primary locus of aminoglycoside sequestration in proximal tubule cells. Lysosomes are also the first organelle to demonstrate morphologic changes (myeloid body or cytesegresome formation) after exposure to the drugs (Figure 4). These studies demonstrate that lysosomal enzyme activities (i.e., sphingomyelinase, cathepsin B, α-D-galactosidose) are decreased and that the structural latency of lysosomes, reflected by leakage of *N*-acetyl-β-D-glucosaminadase into cytosol, is increased. Inhibition of lysosomal enzymes may cause an intralysosomal accumulation of membrane-associated lipids which would be reflected morphologically as myeloid body formation. However, this process by itself should not be acutely lethal to the cell. Decreased lysosomal function may also result in a decreased ability to degrade endogenous intracellular proteins and exogenous, low molecular weight proteins reabsorbed from the tubular filtrate; events which would perturb nephron function.[57,58] The increase in lysosomal permeability could result in proximal tubule cell dysfunction, although this event is probably a late change in AgTN occurring after cell necrosis has been initiated by another factor.[118] Myeloid body formation is most likely a marker of aminoglycoside exposure rather than toxicosis. The appearance of lysosomal enzymes in the urine of AgTN patients is secondary to proximal tubule cell necrosis, apical plasma membrane damage, or lysosome exocytosis.

A second possible target of aminoglycosides are the mitochondria because, in vitro and in vivo, aminoglycosides decrease mitochondrial respiration, thereby impairing the tubule cell's bioenergetic profile.[16,67,118,127,132,227,245,275-277] This could selectively produce tubule dysfunction which would initially be detectable biochemically, but not morphologically. In acute studies, morphological changes in mitochondria occur after changes are seen in the Golgi apparatus and ribosomes.[135,270] The mechanism of this toxicity may be secondary to a direct aminoglycoside interaction with mitochondrial membrane phospholipids, to a competitive interaction with the divalent cations magnesium or calcium, or to an alteration in the intracellular millieu which would indirectly affect mitochondrial function. The magnitude of aminoglycoside effects on mitochondrial respiration is roughly correlated to the net positive charge of the specific drug. In order for this subcellular lesion to be considered as significantly contributing to AgTN, it must be demonstrated that aminoglycosides can accumulate to a great enough intracellular concentration, after relatively low dose administration, to adversely affect mitochondrial respiration.

The third possible site of initial intracellular aminoglycoside interaction is with the proximal tubule cell plasma membrane's phospholipids and enzymes.[87,118,133a,149,150,166a,227a,228,229,244,292] Binding of aminoglycosides to membrane polyphosphoinositides could perturb the regulation of membrane permeability thereby promoting cellular dysfunction. Aminoglycosides induce a phospholipidosis which may be secondary to inhibition of cytoplasmic phospholipase activity. This event affecting multiple membrane systems may affect other cellular metabolic processes. Aminoglycoside binding to apical tubular membranes may also decrease binding and subsequent uptake of filtered proteins, further perturbing nephron function[57] which could severely affect proximal tubule function. Aminoglycosides also inhibit basolateral membrane Na/K ATPase activity in vitro and in vivo when used at high doses or incubating concentrations.[4,46,65,151,291,292] This could be due to competition with divalent cations necessary for ATPase activity, or secondary to interaction with membrane phospholipids which would alter enzyme activity by altering the membrane environment. Aminoglycosides have also been found to inhibit adenylate cyclase activity in proximal tubule basolateral membranes and in toad bladder epithelium in vitro.[119,222,253] The enzyme interactions at the basolateral membrane could result in significant cellular dysfunction by altering intracellular electrolyte balance or osmolality.

A final possible site of aminoglycoside interaction with the nephron is at the level of the glomerulus where gentamicin has been demonstrated to reduce the glomerular ultrafiltration coefficient,[15] and to reduce the number and size of glomerular endothelial fenestrae.[10,155,158,159] These effects may be mediated by a charge interaction between the cationic aminoglycosides and the anionic endothelial cell surfaces, or alternatively, could be a feedback response to a primary tubular injury. The mediator of this mechanism is not known. The renin-angiotensin system has been suggested;[32,262] however, a recent study in rats indicates that this system is not the initiating factor in decreasing GFR although it may participate in the expression of the syndrome.[154]

The relative contribution of lysosomal, mitochondrial, or membrane tubular mechanisms and glomerular injury to clinical AgTN is not known. The relative importance of each as a primary insult is largely a function of the pattern of intracellular distribution of toxicologically active aminoglycosides. The recent finding of free gentamicin in proximal tubule cytosol for up to 6 hr after administration[274] makes nonlysosomal mechanisms of AgTN possible. Although lysosomes contain the greatest quantities of intracellular aminoglycosides, this depot may initially function as a protective mechanism. The significance of luminal vs. peritubular or basolateral absorption must be further assessed and the toxicologic potential of each determined in relatively low-dose studies. The effect of these two processes on the magnitude of calculated pharmacokinetic parameters must also be further evaluated so that the pathophysiologic basis of the "prenephrotoxic" patient can be determined. In all probability, cellular dysfunction is a result of a combination of the above processes. Whatever the mechanism, dysfunction of the proximal tubule cell ultimately results in a decrease in nephron function, the sum of which determines whole kidney function.

ANIMAL MODELS

Animal models used to study AgTN have primarily been laboratory animal species, i.e., rats, rabbits, and dogs. In general, the syndrome of AgTN is similar across species lines, with any peculiarities being a result of pharmacokinetic factors or inherent differences in underlying renal morphology or physiology. Until recently, most studies in laboratory animals have been largely descriptive in nature, concentrating on the histological and functional changes seen after drug administration.

Rats

Without question, the rat has been the primary laboratory animal species used to study AgTN. Most of the mechanistic studies cited in the previous section utilized this species. The clinical and pathologic expression of AgTN is identical to that previously described for humans except for the shorter time course of the syndrome in rats.

In the rat, histopathologic lesions are dependent on dose and duration of drug administration. Ultrastructural alterations, primarily restricted to intralysosomal myeloid body formation (cytesegresomes), have been noticed 80 min after administration of a single 40 mg/kg dose of gentamicin to Sprague-Dawley rats.[270] As dose was increased to 160 mg/kg, ultrastructural changes consisted of dilation of endoplasmic reticulum, dilation or swelling of mitochondria, and increased numbers of cytesegresomes. Similarly, ultrastructural cytesegresome formation has been observed 48 hr after a single 10 mg/kg dose of gentamicin administered to Fischer 344 rats.[135] As previously discussed, early lysosomal changes should be interpreted with caution since they primarily reflect a drug-lysosome interaction rather than overt toxicity. Established functional evidence of nephrotoxicity (i.e., elevated serum creatinine, serum urea nitrogen) or histological lesions have not been observed after administration of single doses.

Numerous studies have characterized the histopathologic lesions and functional changes seen after repeated high and low dose aminoglycoside administration to Sprague Dawley[33,56,110,111,127,156,161,164,192,252,270] and Fischer 344[67,77-79,99,113,135,287] rats. Strain differences are restricted to differential sensitivity to AgTN, with the Fischer 344 being relatively susceptible. The earliest histological changes seen by light microscopy are confined to the proximal tubule and consist of vacuolar or granular degeneration and cloudy swelling, tubular dilatation, and disruption of brush borders. As dose or length of exposure increases, frank tubular necrosis is seen accompanied by peritubular inflammation and tubular atrophy. These changes are initially focal in distribution and become diffuse with increased dose of drug or length of exposure. Kidney weight increases as AgTN is induced. In animals that do not die of ARF, tubular regeneration is subsequently seen often in conjunction with necrosis and the other lesions outlined above. Necrosis is prominent 7 to 10 days after drug administration, with regeneration becoming prominent by 14 days. Animals examined a few weeks after the end of gentamicin administration have essentially normal-looking kidneys on histopathologic examination, with the only sign of prior AgTN being focal areas of fibrosis. Quantitative histopathologic scoring systems have been utilized to objectively evaluate these lesions in rats.[111,156]

Functional changes are fairly well correlated to the morphological lesions above. The earliest changes observed after 3 to 5 days of drug administration are polyuria, decreased urine osmolarity, and cylindruria, proteinuria, and enzymuria. This is followed by a decrease in GFR reflected by elevated serum creatinine and urea nitrogen concentrations by 7 to 10 days, peaking at approximately 14 days. This decrease in GFR parallels the onset of tubular necrosis described above. From this point on, renal function continues to improve regardless of whether drug administration is discontinued.[33,77,79,99,113,135,156,161,189,223,281] In vitro tissue slice studies demonstrate an enhanced rate of *p*-aminohippurate (PAH) uptake from medium as early as 1 day after drug administration which peaks at approximately 5 to 7 days. Subsequently, PAH uptake is depressed. In contrast, organic base uptake, *N*-methylnicotinamide (NMN) and tetraethylammonium (TEA), is depressed starting 3 to 4 days after the start of drug administration.[24,25,56,127,131,132,141,189] Other early changes in murine AgTN include an increased fractional excretion of glucose, sodium, and potassium and a decreased tubular reabsorption of filtered low molecular weight proteins such as lysozyme and beta microglobulin.[57,132,156,189]

The drug doses which produce the above changes are greater than those used clinically in man. If doses similar to those used in man on a milligram per kilogram basis are given to Sprague Dawley rats (4 to 5 mg/kg/day), histological but not functional evidence of AgTN was detected.[111] In a similar study,[164] the human dose only produced a decreased urine osmolarity with no evidence of histological or other functional damage. Based on the pharmacokinetic studies previously reviewed, the allometrically predicted (Equation 4) human equivalent rat dose of gentamicin would be 20 mg/kg/day. At this dose in Sprague Dawley rats, serum creatinine is minimally elevated, but no change in creatinine clearance was seen after 14 days of drug administration. Modest histopathologic evidence of AgTN was present.[164] When administered as a divided daily dose, peak and trough serum concentrations are similar to those seen in humans receiving therapeutic doses of gentamicin.[26] Nephrotoxicity was not detected in Sprague Dawley rats given gentamicin at 20 mg/kg as a single daily or divided daily doses.[7] This dose given for 28 days induced histopathologic but not functional (serum creatinine) evidence of AgTN.[110,111] These studies are difficult to directly compare since the length of drug administration is different and different stages of AgTN may be present (see discussion below). However, it is clear that doses in the above so-called low dose range do not induce a clinical toxic nephropathy and are probably more suitable for extrapolating to man than are higher doses which induce ARF.

For functional AgTN, reflected by elevated serum creatinine concentrations, to be induced in Sprague Dawley rats, gentamicin must be administered at a dose of at least 40 to 60 mg/kg for 7 to 12 days.[45,57,132,156,189] As dose is increased, functional and histological changes are more severe and occur earlier. Only 4 of 14 rats given gentamicin at a dose of 75 mg/kg/day for 28 days survived the study with deaths occurring from 17 to 23 days. Many mechanistic studies utilize doses of approximately 100 mg/kg for periods of up to a week or approximately 50 mg/kg for 10 to 14 days. In the Fischer 344 rat, 40 mg/kg/day produces frank renal failure marked by extensive necrosis and elevated serum urea nitrogen and creatinine concentrations by 7 days. At 20 mg/kg/day, increased serum urea nitrogen and creatinine concentrations are seen together with histological evidence of tubular necrosis by 2 weeks of treatment.[24,25,113,135,189] These doses are appropriate when overt ARF is desired.

Various strains of rats are appropriate to study AgTN, however, the proper doses must be utilized if a sensitive strain such as the Fischer 344 is selected. In all strains, results are similar if gentamicin is administered by s.c. or i.m. routes. Occasionally, the i.p. route is employed, but from a pharmacokinetic viewpoint, this would not be suggested because of the prolonged absorption half-life.

The enzymuria observed in AgTN is believed secondary to either loss of the integrity of the proximal tubule brush border and lysosomal degradation, or for low molecular weight filtered proteins, to a decreased reabsorption by the tubules. Excretion of the following enzymes have been noted to be increased in murine toxic nephropathies: muramidase (lysozyme), alanine aminopeptidase, alanine aminotransferase, alkaline phosphatase, alpha-fucosidase, aspartate aminotransferase, beta-galactosidase, beta-glucuronidase, beta-*N*-acetyl-hexosaminidase, gamma-glutamyl transferase, glutamate dehydrogenase, glutathione-*s*-transferase, lactate dehydrogenase, and *N*-acetyl-beta-D-glucosaminidase,[73,85,131,161,176,182,192,197,201,281] The major difference between enzymuria patterns in AgTN is the time of onset in relation to drug administration. Since some enzymes are found in the tubular brush border while others are intracellularly located, excretion of a specific enzyme is dependent on their origin and the degree of damage. In general, enzymuria is a very early, sensitive, but relatively nonspecific, indicator of AgTN since other renal diseases and altered pathophysiologic states can also produce it.[197,201] Optimally, urine enzymes should be monitored in conjunction with more specific renal function tests to accurately portray AgTN.

The phenomenon of tubular regeneration and recovery in the face of continued aminoglycoside administration merits further discussion. As noted above, histological evidence of tubular necrosis is followed by tubular regeneration and an ultimate resolution of AgTN. During the period when tubular regeneration is evident, the kidney appears refractory to further aminoglycoside-induced toxicity. However, if sufficient time elapses between episodes of aminoglycoside administration, susceptibility returns. This insensitivity is not absolute in that although GFR and in vitro PAH transport may be preserved and histological evidence of necrosis is not present, in vitro NMN transport is depressed and very high doses of drug will induce further renal injury. Previous necrosis is a necessary event since exposure to relatively nontoxic aminoglycosides such as netilmicin may actually potentiate AgTN from subsequent gentamicin administration. Acquired insensitivity to AgTN may also be conferred by nonaminoglycoside nephrotoxins such as mercuric chloride and potassium dichromate. In these cases, tubular injury may be excessive, but GFR is preserved. The precise mechanisms of this insensitivity are not known, but are probably related to the process of tubule cell regeneration. In the case of nonaminoglycoside-induced resistance, impaired uptake of drug may be responsible. In the case of gentamicin-induced insensitivity, another mechanism appears operative.[67,77,78,99,163,165,252] This cyclic behavior of resistance and susceptibility confounds chronic studies since the severity of toxicity must be evaluated at times when comparable stages of AgTN are occurring, otherwise a protocol producing early massive necrosis may be in a phase of regeneration while a protocol producing minimal toxicity may not have reached the regenerative stage. Sex differences may also be present because in Fischer 344 rats, females develop tubular necrosis and dysfunction with subsequent recovery earlier than do males given the same dose of drug.[77]

Other factors have also been found which modulate the toxic potential of aminoglycosides in rats. Sodium restriction, dehydration, and furosemide treatment preceding gentamicin dosing enhances the resulting AgTN, yet prior salt loading does not ameliorate it.[21,50,223] Metabolic acidosis induced by ammonium chloride in Sprague

Dawley rats likewise potentiated gentamicin AgTN.[114] Administering gentamicin as a divided daily dose rather than a single daily dose is more toxic.[23] Adult rats appear more sensitive to AgTN than do younger ones.[169] Diabetes[260] and dietary calcium loading[19] appear to confer some degree of protection against AgTN. In rats, concurrent cephalosporin administration appears to protect against AgTN.[72,153,162]

All of the work thus far reviewed in rats has concentrated on the aminoglycoside gentamicin. By consensus, neomycin appears to be the most and streptomycin the least toxic of all aminoglycosides studied. Within this range, the only conclusive comparisons based on firm experimental grounds would rank gentamicin as being more toxic than tobramycin followed by netilmicin.[118,156,189,252] However, such studies must be interpreted with caution because low- and high-dose investigations may yield conflicting results since the dose response curves of different aminoglycosides may not be parallel.[110,164]

The relationship of renal parenchymal concentrations of gentamicin to the occurrence of nephrotoxicity is at present an area of controversy. Aminoglycoside concentrations in the kidney are a function of total dose of drug administered, the dosing interval employed, length of administration, and at what stage in AgTN the tissue samples were collected for analysis. In all rat studies conducted to date, gentamicin accumulates in the renal cortex following multiple-dose administration to concentrations at least five times that found in serum. Cortical drug concentrations are always greater than medullary concentrations. At the relatively nontoxic gentamicin dose of 20 mg/kg/day in Sprague Dawley rats, twice daily dosing caused drug accumulation more rapidly and to a greater concentration than did single daily dose administration. In both situations, plateaus were reached in 4 to 7 days.[7] Similar results were obtained in Wistar rats given a single or divided daily dose of 10 mg/kg.[101] In rats, the cortical plateau gentamicin concentration appears to be approximately 500 to 700 μg/g.[7,26,160] The decline in renal gentamicin concentrations has been estimated to have a half-life of 109 hr in a renal homogenate,[267] 144 hr in cortex, and 133 hr in medulla[26] or, alternatively, to have a biphasic cortical half-life of 98 and 166 hr and a medullary half-life of 55 hr.[84] Different renal disease states alter intrarenal gentamicin disposition; bacterial pyelonephritis increases the cortical, medullary, and papillary concentrations compared to normals and alters the cortical to medullary concentration ratio[27] while ischemic injury reduces cortical drug concentrations.[51] When cortical gentamicin concentrations were monitored over the course of a chronic AgTN experiment in Fischer 344 rats, concentrations initially rose in a manner similar to that described above. However, as tubular necrosis became evident on histopathologic studies, concentrations declined. Concurrent with histopathologic evidence of tubular regeneration, gentamicin concentrations again rose.[77,78] Concentrations were not greater in rats receiving 80 rather than 40 mg/kg when tubular necrosis was present, a situation opposite to that seen in the above low-dose studies where AgTN was not induced. In a relatively low-dose study for 28 days utilizing Wistar rats in which AgTN was induced, serum creatinine and urea nitrogen concentrations increased in parallel to renal cortical concentrations.[83] In a high-dose study with Sprague Dawley rats, terminal renal drug concentrations were not correlated to the occurrence of renal impairment.[156]

As can be appreciated from these data, the correlation of cortical drug concentrations to the occurrence of nephrotoxicity is difficult to assess. Drug concentrations must be measured before toxicity ensues or at similar phases of the AgTN syndrome for meaningful comparisons to be made. If concentrations are measured before AgTN, they should have reached a plateau before comparisons are made. If concentrations are measured after AgTN ensues, the results obtained may be a result of rather than a cause of the nephrotoxicity. Tissue homogenates used to measure drug concentrations may contain a mixture of cortical and medullary tissue, and even when cortex is dissected out, samples will contain both necrotic and regenerating tubules making assessment difficult. Finally, different aminoglycosides accumulated to different degrees when administered according to the same dosage regimen making results obtained with one drug not readily transferable to others.

In conclusion, the rat has been extensively utilized as an animal model of AgTN. High-dose rat studies are appropriate for investigations designed to probe specific mechanisms of AgTN. When comparative studies are conducted, similar ages and strains of rats must be employed for results to be valid and dosage regimens employed must be the similar. When results are to be directly extrapolated to man, low-dose studies might be preferred.

Dogs

The second species which has been extensively utilized to study AgTN is the dog. The histopathologic appearance is essentially identical to that reviewed in rats and the reader is referred to two excellent studies in this area.[64,254] Briefly, by 5 days after high-dose gentamicin administration, brush border staining was decreased and hyaline droplet degeneration was present. Tubular necrosis was present by 9 days after drug administration. Cytosegresome formation on electron microscopy and increased fractional excretion of sodium have been detected after administration of a single 15 mg/kg dose of gentamicin to beagle dogs.[216] Light microscopic changes were not seen. The human equivalent therapeutic canine dose of gentamicin, determined from the allometric analysis presented earlier in this chapter, is 9 mg/kg/day. This dose (7 to 9 mg/kg/day for 7 to 14 days) when given to normal dogs in a laboratory setting does not produce functional AgTN, as evidenced by elevated serum creatinine or urea nitrogen concentrations.[64,211] In veterinary medicine, a dose of 6 to 9 mg/kg/day is a recommended canine clinical dose.[38,60,207] However, in a clinical study a dose of 6.6 mg/kg/day for 6 days caused an elevation in serum urea nitrogen in only 2 of 16 dogs for which post-treatment renal function was monitored.[148]

In a controlled study, 8 mg/kg/day, divided every 12 hr, produced a statistically significant decrease in urine osmolarity by day 7, an increase in fractional sodium excretion by day 8, an increase in serum urea nitrogen concentration by day 17, and an increase in serum creatinine concentration by day 18. The decrease in urine osmolarity, accompanied by polyuria, was preceded by an increase in the urinary excretion of prostaglandin E2 and followed by an increase in plasma renin activity. Urinary prostaglandin E2 activity decreased prior to azotemia, at which time the fractional excretion of sodium greatly increased.[172] The increase in urinary prostaglandin excretion was suggested to have induced a state of nephrogenic diabetes insipidus which caused the decreased urine osmolarity. This study documents a bimodal course of AgTN in the dog administered low-dose gentamicin therapy: an initial subclinical (subazotemic) phase marked by a urinary concentrating defect followed by a clinical (azotemic) phase. The primary difference with this study and the two previously referenced low-dose studies which did not produce clinical AgTN[64,211] is the longer length of drug administration in the present work.

High-dose studies of 30 mg/kg/day, divided every 8 hr, administered for 10 to 12 days, produced clinically significant increases in serum creatinine and serum urea nitrogen 9 to 12 days after the start of drug administration. Increases in urine enzyme excretion, beta-glucuronidase and *N*-acetylglucosaminidase, were noted as early as 2 days into the regimen. Glucosuria and urine osmolarity declined by day 7. Plasma potassium and calcium concentrations were significantly depressed.[3,64,254] Almost complete recovery occurred in dogs taken off of gentamicin for 28 days with only focal, tubulointerstitial nephritis being present on histopathologic examination.[64] Dogs dosed at 40 mg/kg/day i.m. for 15 days became moribund and were sacrificed, the necropsy revealing extensive proximal tubular necrosis.[31] In this study, dogs given 8 mg/kg/day for 50 days developed AgTN marked by elevated serum urea nitrogen concentrations and histologic evidence of tubular necrosis. In dogs given 5.6 mg/kg/day for 50 days, histologically only cloudy swelling of the renal tubule cells was present. In a recent study dosing gentamicin at 60 or 75 mg/kg/day for 10 days, dogs either died or became anuric.[200]

Based on limited comparative studies in dogs, the order of increasing nephrotoxic potential would be netilmicin, tobramycin, and gentamicin.[2,200,204,259,261] In these studies, netilmicin is essentially nontoxic in the dog. This is contrasted with one study in mongrel dogs which demonstrated significant netilmicin AgTN.[80] The pathophysiologic expression of AgTN in dogs treated with tobramycin is essentially identical to that of gentamicin.[204,279] Furosemide enhanced the nephrotoxicity of gentamicin and netilmicin, in the latter case from essentially none to a clinical state of renal failure.[2,3] Neonatal puppies are relatively resistant to AgTN, a fact attributed to an intrarenal distribution of blood flow in the newborn dog away from superficial cortical nephrons. In older puppies (30 days), sensitivity was similar to adults.[62] In 30-day-old puppies, histopathologic evidence of tubular necrosis was accompanied by a marginally decreased inulin clearance and a decreased tubular reabsorption of phosphate. In these studies, intrarenal gentamicin distribution paralleled that of radiolabeled microspheres, demonstrating the dependence of gentamicin distribution on renal blood flow in the first month of a puppy's life.

The increase in potassium excretion seen in canine AgTN[64] has been further studied using a 30 mg/kg twice-divided daily dose administered for 4 to 10 days.[35] These studies confirmed that gentamicin produces a potassium wasting evident by the 4th day. Further, prior potassium depletion increases the severity of the resulting AgTN evidenced by histopathology and by a fivefold greater increase in serum creatinine than seen in potassium supplemented dogs exhibiting AgTN. In this study, renal cortical gentamicin concentrations were also greater in the potassium depleted dogs than in the other AgTN dogs. An increase in fractional sodium excretion was not seen in either group although glucosuria was detected in both. In a second study,[65] this dose of drug reduced renal cortical concentrations of potassium, magnesium, and phosphorus. This was followed, in potassium depleted dogs developing AgTN, by increased cortical levels of calcium and sodium and reduced activity of Na/K ATPase activity. The possible significance of these findings to the pathogenesis of canine AgTN was discussed.

Finally, canine AgTN studies have helped to elucidate the effect of the dosage regimen design on the nephrotoxic potential of a constant daily dose of drug in normal[200,204,261] or subtotal nephrectomized dogs.[211] The normal dog studies employed a gentamicin dose of 45 mg/kg/day for 10 days administered by either constant i.v. infusion, divided into 6 injections every 4 hr, or given as a single daily injection. This dose of drug was selected as that which decreased pretreatment serum creatinine clearance by approximately 50%. A dose of 30 mg/kg/day produced minimal changes in GFR while 60 or 75 mg/kg/day was lethal or caused anuria. In the study employing subtotal nephrectomized dogs (three fourths to seven eighths renal mass surgically removed), the allometrically predicted normal dose of 9 mg/kg/day divided in 3 equal doses was employed and adjusted for the nephrectomy-induced decreased GFR by either increasing the dose interval (constant dose) or decreasing the dose (constant interval).[211,215] In all of these studies, the dose regimens characterized by prolonged intervals were less nephrotoxic than those which administered drug at short intervals. The high momentary peak serum drug concentrations produced by these regimens were not correlated to nephrotoxic potential. Rather, regimens which produced high trough concentrations were more nephrotoxic. In the high-dose studies in normal dogs, cortical gentamicin concentrations were not elevated in the animals with AgTN, while in the subtotal nephrectomized dogs given adjusted normal doses, cortical concentrations were greater in those which developed AgTN. Additional studies must be conducted to explain this discrepancy. However, a possible explanation is that in the high-dose work, cortical concentration was saturated and a dose-related AgTN was being induced, while in the low-dose work, cortical drug concentration was not saturated and a syndrome of AgTN similar to that seen clinically was occurring. This is supported in studies designed to describe the renal cortical deposition of gentamicin in normal dogs as a function of 4- to 6-hr i.v. infusions. Cortical uptake of drugs appears to saturate at high doses or high serum concentration values (6 to 8 μg/m*l*).[283] This phenomenon of a dose-related saturation of the renal cortex must be kept in mind when comparing the degree of AgTN in different animals when dosage regimens are not the same. It is probable, although not conclusively proven, that the so-called low-dose regimens induce toxicity only if this saturation is achieved, while in high-dose studies this point is attained much sooner. If this situation occurs, AgTN can be induced even when serum drug concentrations are in the so-called therapeutic range. Other pathophysiologic or pharmacologic factors which increase cortical drug accumulation may potentiate AgTN.

In the dog, care should be taken in interpreting serum creatinine and creatinine clearance as measures of GFR since males, and not females, actively secrete creatinine.[185,220] Therefore, as GFR decreases as a result of AgTN, tubular secretion will cause an overestimation of GFR if creatinine clearance is monitored in male dogs.

As can be appreciated from these canine studies, AgTN in dogs is very similar to that seen in rats and man. Dogs appear to be a good animal in which to study aminoglycoside toxicokinetics since drug disposition studies can be conducted in the same individual animals in which toxicologic and pathologic studies are also being performed.

Other Species

Rabbits and guinea pigs have been employed as animal models of AgTN. In a comparative study of amikacin, tobramycin, and gentamicin in rats, guinea pigs, and rabbits,[225] drugs were administered at a dose of 10 or 40 mg/kg/day for 5 days to rabbits and at 40 mg/kg to guinea pigs for 5 days without producing histological evidence of AgTN. The relative order of nephrotoxicity (lowest to greatest) was amikacin, tobramycin, and gentamicin. Renal function was not monitored. In a study designed to compare the nephrotoxic potential of a single or divided (8-hr intervals) regimen of 7.5 mg/kg/day in rabbits for 3 weeks or longer,[91] only the divided regimen resulted in significantly elevated serum creatinine concentrations. In this group, serum gentamicin half-life was prolonged compared to normal by day 7 of drug administration. Renal function did not improve for up to 28 days after dosing. A total of 6 of 11 rabbits receiving 30 mg/kg/8 hr gentamicin died in ARF with markedly elevated serum creatinine concentrations, prolonged gentamicin elimination half-life, and histological evidence of tubular necrosis and interstitial nephritis.[92] In another study assessing the nephrotoxicity of gentamicin doses of 8 or 16 mg/kg administered twice daily for 4 weeks, 4 of 7 rabbits treated with the low-dose regimen died after 22 days and 5 of 7 rabbits receiving the high dose died after 16 days of drug administration.[100] Serum creatinine concentrations were elevated in AgTN animals by 7 to 10 days and were accompanied by glucosuria and proteinuria. All animals which died had serum creatinine concentrations greater than 10 mg/d*l*. In contrast to the rat, concurrent cephalosporin administration (cefaclor, cefazolin, cephaloglycin) in rabbits potentiated rather than protected against AgTN induced by neomycin, tobramycin, or gentamicin administered for 5 to 7 days. The gentamicin dose used was 100 mg/kg/day which alone did not elevate serum creatinine concentration or produce histological evidence of tubular necrosis.[17] Cephaloridine did not potentiate AgTN.[74] This difference is attributed to the mechanism by which the different cephalosporins are secreted by the proximal tubule. In summary, rabbits appear to be the animal model selected to study cephalosporin-aminoglycoside interactions.

Gentamicin administration to mice at doses of 50 or 100 mg/kg/day for 7 to 10 days i.p. produces AgTN manifested by enzymuria (*N*-acetyl-beta-D-glucosaminidase) and ultrastructural evidence of cytosegresome formation, cytoplasmic edema, and mild mitochondrial and ribosomal changes. Lesions were not seen by light microscopy.[290] Concomitant administration of the loop diuretics furosemide, bumetamide, or pirelamide decreased the severity of ultrastructural changes, but did not influence urinary enzyme levels. The LD_{50} (SE) for gentamicin in the mouse was 45 (1) mg/kg intravenously and 274 (9) mg/kg subcutaneously.[279] In the same study, the tobramycin LD_{50} was 546 (2) mg/kg intravenously and 367 (3) mg/kg subcutaneously compared to 133 (4) mg/kg intravenously and 973 (75) mg/kg subcutaneously in rats. In a different study, the gentamicin LD_{50} was 75 mg/kg intravenously, 430 mg/kg intraperitoneally, and 485 mg/kg subcutaneously.[31] In both of these acute studies, the oral LD_{50} of gentamicin or tobramycin in mice was greater than 8000 to 9000 mg/kg, reflecting the lack of oral bioavailability of these polar compounds.

Cats administered 50 mg/kg/day of gentamicin for 65 days had markedly elevated serum urea nitrogen concentrations. In 6 cats given 35 mg/kg/day, serum urea nitrogen concentrations were 50 mg/d*ℓ* or greater after 6 to 13 days. Tubular necrosis was present on histologic examination.[279] Cats administered gentamicin at a daily dose of 60 mg/kg/day died at an average of 17 days of dosing from renal tubular necrosis. A total of 4 of 5 cats receiving 40 mg/kg/day died at 18 to 27 days into the dosing from ARF. A total of 3 of 4 cats given 20 mg/kg/day survived for 70 days. On necropsy, only mild or no tubular damage was present.[273] Cats are primarily employed as the animal model to study aminoglycoside ototoxicity and have not been utilized to study AgTN. It must be noted that in comparison to other species, cats appear to have a relatively more concentrated urine and retain the ability to produce concentrated urine even when GFR is significantly reduced.[221]

Aminoglycoside toxic nephropathy has also been studied in horses. Clinically, AgTN appears to be restricted to young animals with toxicity rarely reported in adults.[217,263] Daily doses of 8.8 mg/kg/day for 5 to 14 days induced AgTN marked by elevations in serum creatinine and serum urea nitrogen.[217] In an experimental study of AgTN in young horses, 1 of 3 horses given 8.8 mg/kg/day and 1 of 3 given 17.6 mg/kg/day for 14 days developed elevated serum urea nitrogen and creatinine concentrations and decreased urine osmolarity.[214] A total of 3 foals given 4.4 mg/kg/day for 14 days did not develop functional evidence of AgTN. All foals in this study had dose-related histopathologic evidence of proximal tubular nephrosis including tubular necrosis and regeneration. Ultrastructurally, cytosegresomes were present in gentamicin-treated animals. The clinical and pathologic expression of AgTN appeared similar to that in other species. Gentamicin trough serum concentrations increased in the foals exhibiting AgTN. These studies suggest that unlike other species, young foals are not resistant to AgTN although a mechanism for this phenomenon is not known. The severity of the AgTN seen even at the low dose above is not totally unexpected since the human-equivalent equine dose predicted using allometry would be only 2 to 3 mg/kg/day.

A few studies have been conducted in ruminants. Calves given 5 or 9 mg/kg/day neomycin intramuscularly for 12 to 13 days developed AgTN marked by increased serum creatinine and urea nitrogen concentrations, decreased creatinine clearance, decreased urine specific gravity, cylinduria, proteinuria, and enzymuria (gamma-glutamyltranspeptidase and alanine aminopeptidase).[66] On histopathologic examination, tubular hyaline droplet change, degeneration, and necrosis were present. Finally, neomycin concentrations in the renal cortex ranged from approximately 200 to 400 μg/g. Gentamicin-induced toxic nephropathy was not detected in adult sheep given 9 mg/kg/day for 7 days when serum creatinine and urea nitrogen concentrations were monitored and tissues were examined at necropsy by light microscopy.[36] Sheep have recently been advocated as a stable animal model for human renal diseases.[81] They may be a useful animal model of gentamicin toxic nephropathy because body weight is essentially identical to that seen in man and the pharmacokinetic parameters describing gentamicin disposition are also similar. Toxicological studies of AgTN in sheep have not been reported.

Gentamicin toxic nephropathy characterized by elevated levels of *N*-acetyl-beta-glucosaminidase in droppings was reported in two lanner falcons given the drug intramuscularly at a dose of 5 mg/day for 4 days.[88] Finally, rainbow trout have been suggested as being a sensitive animal model for AgTN,[258] and 8- to 10-day i.p. LD_{50}s for tobramycin and gentamicin were 53 and 55 mg/kg, respectively. Histopathologic examination of kidneys from these animals demonstrated a dose-dependent proximal tubular necrosis and glomerular degeneration. A single dose of either drug (15 to 30 mg/kg) caused a decrease in inulin clearance and a dose-dependent decrease in urine flow. These functional changes preceded histologic damage.

Recommendations

Based on the above animal studies of AgTN, a few general observations can be made concerning aminoglycoside comparative toxicology. First, if interspecies comparisons are to be made, the dose regimens selected must be comparable in the size of the dose, the interval between doses, and the time after dosing at which toxicity is being evaluated. This writer suggests that as a start, allometric relationships be utilized as a base on which doses are calculated so that the pharmacokinetic behavior of the compounds are accounted for. As an example, the following would be an equivalent set of daily therapeutic doses for rats, dogs, humans, and horses: 20, 9, 5, 3 mg/kg, respectively. If a severe AgTN reaction is desired, all doses could be multiplied by a constant, i.e. 27 mg/kg in a dog might be comparable to 60 mg/kg in a rat. An examination of the studies reviewed above would support this position. Dose intervals may also be adjusted to compensate for the differences in rates of elimination between species and differences in the time course of AgTN. Monitoring of serum drug concentrations can be helpful in this regard. Second, the stage of AgTN compared must be similar. Studies comparing a regimen demonstrating active tubular necrosis on histologic examination should not be compared with a study demonstrating primarily tubular regeneration. Third, the marker of AgTN utilized must have a similar sensitivity since enzymuria or decreased urine osmolarity, early indicators of AgTN, cannot be compared with equal weight to late indicator such as decreased GFR. In respect to this point, the marker of GFR selected must be appropriate for that species. Strain and sex differences must also be taken into account. Finally, the severity of AgTN is also a function of the specific drug selected and whether potentiating factors are present. These must be rigidly controlled in the experimental design. For example, pretreatment renal function should be balanced and all treatment groups should receive similar ancillary treatment (i.e., fluids, diuretics, etc.).

ADDITIONAL MODELS OF DRUG-INDUCED TOXIC NEPHROPATHY

Several other drugs and xenobiotics have also been associated with renal toxicity in animals [6,28,29,75,95,109,122,130,133,138,167,171,175,180,198,199,201,226,240,257,268,277a,296] (Table 5). Al-

Table 5
DRUGS AND XENOBIOTICS ASSOCIATED WITH TOXIC NEPHROPATHY

Allyl chloride
Aminoglycosides
Amphotericin B
Arsenic
Bacitracin
Beryllium
Bismuth
Bromobenzene
Cadmium
Carbon tetrachloride
Cephaloridine
Cephaloglycin
Cephalothin
Chloroform
Chloroprene
Chromium
Colistin
Cyclophosphamide
Cytembena
Daunorubicin
DDT
Demeclocycline
Dibromochloropropane
Dichloropropene
Diethylene glycol
Dinitrophenol
Doxorubicin
Ethylene glycol
Ethylene dibromide
Ethylene dichloride
Ethylenediaminetetraacetic acid (EDTA)
Folate
Furosemide
Gold
Glycerol
Hexachloro-1:3-butadiene
Lead
Lithium
Lomustine
Mercurial compounds (diuretics, HgCl, methyl-Hg, phenyl-Hg)
Methotrexate
Methylchloride
Methicillin
Methoxyflurane
Methycellulose
Mitomycin
Mycotoxins (citrinin, ochratoxin)
p-Nitrophenylarsonic
Nonsteroidal Anti-inflammatory drugs (NSAIDs, salicylates, indomethacin, phenacetin)
Osmotic agents
Oxalates
Paraquat
D-Penicillamine
Perchloroethylene
Polymyxins
Polyvinyl alcohol
Potassium dichromate
Puromycin aminonucleoside
Radiocontrast Media (iodinated)
Rifampin
Semustine
Streptozotocin
Sulfonamides
Tetrachlorodibenzo-*p*-dioxin (TCDD)
Tetracycline degradation product (epi-, anhydro-, epianhydro-)
Trichlorophenoxyacetic acid (2,4,5-T)
Trichloroethane
Uranyl nitrate
Vinyl chloride
Vinylidene chloride

though it is not the purpose of this chapter to provide a detailed review of the nephropathies induced by these compounds, it is worthwhile to briefly discuss a few of the more important syndromes encountered.

The predominant lesion induced by the nephrotoxins listed is tubular necrosis. Most current research has been directed at the subcellular mechanisms responsible. Cephalosporins have been extensively studied to determine the relationship of active tubular transport and intracellular biotransformation to nephrotoxicity.[226,268,296] Other nephrotoxins are used as animal models of ARF, the classic ones being mercuric chloride, uranyl nitrate, cis-platinum, and potassium dichromate. The histopathologic appearance of these tubular nephropathies is very similar to that seen with gentamicin, although the region of the nephron affected may be different. The functional changes induced are also similar, making the discussion of AgTN pathophysiology pertinent. However, a fulminating ARF marked by oliguria and anuria is often induced with these compounds. Early changes may also be different since varying mechanisms of toxicity may be operative, i.e., altered renal blood flow rather than tubule damage may be the earliest manifestation of the toxicoses.

A great deal of interest has recently focused on the ability of the kidney to metabolize foreign compounds which are presented for elimination, on the effects of metabolic inhibitors and inducers on this process, and finally on the nephrotoxic potential of the biotransformation products produced.[109,130,133] Those points will continue to be a major area of interest since many of these compounds are environmental contaminants whose nephrotoxic potential must be assessed. The nephrotoxicity of antineoplastic chemotherapeutic agents is also an area of increasing concern.[277a]

Analgesic nephropathy, an interstitial nephritis due to abuse of nonsteroidal anti-inflammatory agents, has also recently become an area of intensive research.[75,109,138,167,257] These compounds, which include phenacetin, salicylates, and indomethacin, induce in man renal papillary necrosis involving the Henle loops associated capillaries, and medullary interstitial cells. Multiple analgesics are often involved. As necrosis worsens, collecting ducts may be involved, with calcification occurring in advanced cases. Cortical interstitial nephritis may also occur. The initial functional changes are in the medullary interstitial cells which are responsible for prostaglandin production. This insult results in medullary vasoconstriction and subsequent ischemia. Hypertension and pyelonephritis are often seen in clinical cases. The prognosis of this nephropathy is poor and the sequelae is often end-stage renal disease. Work has also focused on the renal biotransformation of NSAIDs by microsomal enzymes to cytotoxic intermediate products. The acute laboratory animal model utilized in this syndrome is induced in most strains of rats by administering massive doses of drug (100 to 600 mg/kg) approaching the LD_{50}. The primary histopathologic change induced by these regimens is proximal tubular necrosis. Chronic dosing of rats with lower doses produces a nephropathy resembling that seen in chronic analgesic abuse in man with histopathologic changes involving the medulla and papilla. However, rats appear to be the animal species most sensitive to papillary necrosis and thus may not be appropriate for extrapolations to man. Acute studies are marked by oliguria or anuria while chronic administration induces polyuria.

Finally, some of the nephrotoxins listed are primary glomerular toxins which induce renal disease syndromes which differ significantly from tubular nephropathies. These include methycellulose, polyvinyl alcohol, gold, penicillamine, puromycin, and organic solvent inhalation.[105,106,188,213,256] Polyvinyl alcohol and methycellulose induce macromolecular hypertension in rats and dogs and a glomerulonephropathy characterized by mesangial hypertrophy and proliferation with minimal change in GFR. Solvent nephropathy is primarily an autoimmune glomerulonephritis. Whatever the cause, toxic glomerular disease usually presents as the nephrotic syndrome, the pathogenesis of which is beyond the scope of this article.

REFERENCES

1. **Adelman, M., Evans, E., and Schentag, J. J.,** Two compartment comparison of gentamicin and tobramycin in normal volunteers, *Antimicrob. Agents Chemother.*, 22, 800, 1982.
2. **Adelman, R. D., Spangler, W. L., Beasom, F., Ishizaki, G., and Conzelman, G.,** Furosemide enhancement of netilmicin nephrotoxicity in dogs, *J. Antimicrob. Chemother.*, 7, 431, 1981.
3. **Adelman, R. D., Spangler, W. L., Beasom, F., Ishizaki, G., and Conzelman, G.,** Furosemide enhancement of experimental gentamicin nephrotoxicity, comparison of functional-morphological changes with activities of urinary enzymes, *J. Infect. Dis.*, 140, 342, 1979.

4. Appel, G. B., Aminoglycoside nephrotoxicity. Physiologic studies of the sites of nephron damage, in *The Aminoglycosides, Microbiology, Clinical Use and Toxicology*, Whelton, A. and Neu, H. C., Eds., Marcel Dekker, New York, 1982, 269.
5. Appel, G. B. and Neu, H. C., Gentamicin in 1978, *Ann. Intern. Med.*, 89, 528, 1978.
6. Appel, G. B. and Neu, H. C., The nephrotoxicity of antimicrobial agents, *N. Engl. J. Med.*, 296, 772, 1977.
7. Aronoff, G. R., Pottratz, S. T., Brier, M. E., Walker, N. E., Fineberg, N. S., Glant, M. D., and Luft, F. C., Aminoglycoside accumulation kinetics in rat renal parenchyma, *Antimicrob. Agents Chemother.*, 23, 74, 1983.
8. Assael, B. M., Cavanna, G., Jusko, W. J., Marini, A., Parini, R., Sereni, F., and Vigano, A., Multiexponential elimination of gentamicin. A kinetic study during development, *Dev. Pharmacol. Ther.*, 1, 171, 1980.
9. Aubert-Tulkens, G., VanHoof, F., and Tulkens, P., Gentamicin-induced lysosomal phospholipidosis in cultured rat fibroblasts, *Lab. Invest.*, 40, 481, 1979.
10. Avasthi, P. S., Evan, A. P., Huser, J. W., and Luft, F. C., Effect of gentamicin on glomerular ultrastructure, *J. Lab. Clin. Med.*, 98, 444, 1981.
11. Baggot, J. D., *Principles of Drug Disposition in Domestic Animals*, W. B. Saunders, Philadelphia, 1977, 153.
12. Barza, M., Brown, R. B., Shen, D., Gibaldi, M., and Weinstein, L., Predictability of blood levels of gentamicin in man, *J. Infect. Dis.*, 132, 165, 1975.
13. Barza, M. and Lauermann, M., Why monitor serum levels of gentamicin?, *Clin. Pharmacokinet.*, 3, 202, 1978.
14. Barza, M., Murray, T., and Hamburger, R. J., Uptake of gentamicin by separated, viable renal tubules from rabbits, *J. Infect. Dis.*, 141, 510, 1980.
15. Baylis, C., Rennke, H. R., and Brenner, B. M., Mechanism of the defect in glomerular ultrafiltration with gentamicin administration, *Kidney Int.*, 12, 344, 1977.
16. Bendirdjian, J. P., Fillastre, J. P., and Foucher, B., Mitochondria modifications with the aminoglycosides, in *The Aminoglycosides, Microbiology, Clinical Use and Toxicology*, Whelton, A. and Neu, H., Eds., Marcel Dekker, New York, 1982, 325.
17. Bendirdjian, J. P., Prime, D. J., Browning, M. C., Hsu, C. Y., and Tune, B. M., Additive nephrotoxicity of cephalosporins and aminoglycosides in the rabbit, *J. Pharmacol. Exp. Ther.*, 218, 681, 1981.
18. Benet, L. Z. and Galeazzi, R., Noncompartmental determination of the steady state volume of distribution, *J. Pharm. Sci.*, 68, 1071, 1979.
19. Bennett, W. M., Elliott, W. C., Houghton, D. C., Gilbert, D. N., De Fehr, J., and McCarron, D. A., Reduction of experimental gentamicin nephrotoxicity in rats by dietary calcium loading, *Antimicrob. Agents Chemother.*, 22, 508, 1982.
20. Bennett, W. M., Hartnett, M. N., Craven, R., Gilbert, D. N., and Porter, G. A., Gentamicin concentrations in blood, urine, and renal tissue of patients with end stage renal disease, *J. Lab. Clin. Med.*, 90, 389, 1977.
21. Bennett, W. M., Hartnett, M. N., Gilbert, D., Houghton, D., and Porter, G. A., Effect of sodium intake on gentamicin nephrotoxicity in the rat, *Proc. Soc. Exp. Biol. Med.*, 151, 736, 1976.
22. Bennett, W. M., Plamp, C. E., Elliott, W. C., Parker, R. A., and Porter, G. A., Effect of basic amino acids and aminoglycosides on 3H gentamicin uptake in cortical slices of rat and human kidney, *J. Lab. Clin. Med.*, 99, 156, 1982.
23. Bennett, W. M., Plamp, W. M., Gilbert, D. N., Parker, R. A., and Porter, A. A., The influence of dosage regimen on experimental gentamicin nephrotoxicity, dissociation of peak serum levels from renal failure, *J. Infect. Dis.*, 140, 576, 1979.
24. Bennett, W. M., Plamp, W. M., Parker, R. A., Gilbert, D. N., Houghton, D. C., and Porter, G. A., Alterations in organic ion transport induced by gentamicin nephrotoxicity in the rat, *J. Lab. Clin. Med.*, 95, 32, 1980.
25. Bennett, W. M., Plamp, W. M., Parker, R. A., Gilbert, D. N., Houghton, D. C., and Porter, G. A., Renal transport of organic acids and bases in aminoglycoside nephrotoxicity, *Antimicrob. Agents Chemother.*, 16, 231, 1979.
26. Bergeron, M. G. and Trottier, S., Influence of single or multiple doses of gentamicin and netilmicin on their cortical, medullary and papillary distribution, *Antimicrob. Agents Chemother.*, 15, 635, 1979.
27. Bergeron, M. G., Trottier, S., Lessard, C., Beauchamp, D., and Gagnon, P. M., Disturbed intrarenal distribution of gentamicin in experimental pyelonephritis due to *Escherichia coli*, *J. Infect. Dis.*, 146, 436, 1982.
28. Berndt, W. O., Effects of toxic chemicals on renal transport processes, *Fed. Proc. Fed. Am. Soc. Exp. Biol.*, 38, 2226, 1979.

29. Berndt, W. O., Hayes, A. W., and Phillips, R. D., Effects of mycotoxins on renal function, mycotoxic nephropathy, *Kidney Int.*, 18, 656, 1980.
30. Bjornsson, T. O., Use of serum creatinine concentrations to determine renal function, *Clin. Pharmacokinet.*, 4, 200, 1979.
31. Black, J., Calesnick, B., Williams, D., and Weinstein, M. J., Pharmacology of gentamicin, a new broad spectrum antibiotic, *Antimicrob. Agents Chemother.*, 3, 138, 1963.
32. Blantz, R. C., The glomerulus, passive filter or regulatory organ, *Klin. Wochenschr.*, 58, 957, 1980.
33. Bowman, R., Silverblatt, F. J., and Kaloyanides, G. J., Comparison of the nephrotoxicity of netilmicin and gentamicin in rats, *Antimicrob. Agents Chemother.*, 12, 474, 1977.
34. Boxenbaum, H., Interspecies scaling, allometry, physiologic time, and the ground plan of pharmacokinetics, *J. Pharmacokinet. Biopharm.*, 10, 201, 1982.
35. Brinker, K. R., Bulger, R. E., Dobyan, D. C., Stacey, T. R., Southern, P. M., Henrich, W. L., and Cronin, R. E., Effect of potassium depletion on gentamicin nephrotoxicity, *J. Lab. Clin. Med.*, 98, 292, 1981.
36. Brown, S. A., Riviere, J. E., Coppoc, G. L., Hinsman, E. J., Carlton, W. W., and Steckle, R. R., Single intravenous and multiple intramuscular dose pharmacokinetics and tissue residue profile of gentamicin in sheep, *Am. J. Vet. Res.*, 47, 69, 1985.
37. Burkle, W. S., Comparative evaluation of the aminoglycoside antibiotics for systemic use, *Drug Intell. Clin. Pharm.*, 15, 847, 1981.
38. Burrows, G. E., Gentamicin, *J. Am. Vet. Med. Assoc.*, 175, 301, 1979.
39. Bush, M., Smeller, J. M., Charache, P., and Arthur, R., Biological half-life of gentamicin in gopher snakes, *Am. J. Vet. Res.*, 39, 171, 1978.
40. Cabrera, J., Arroyo, V., Ballesta, A. M., Rimola, A., Gual, J., Elena, M., and Rodes, J., Aminoglycoside nephrotoxicity in cirrhosis, *Gastroenterology*, 82, 97, 1982.
41. Carbon, C., Contrepois, A., Vigneron, A. M., and Lamotte-Barrilon, S., Effects of furosemide on extravascular diffusion, protein binding and urinary excretion of cephalosporins and aminoglycosides in rabbits, *J. Pharmacol. Exp. Ther.*, 213, 600, 1980.
42. Carbon, C., Contrepois, A., and Lamotte-Barrillon, S., Comparative distribution of gentamicin, tobramycin, sisomicin, netilmicin, and amikacin in interstitial fluid in rabbits, *Antimicrob. Agents Chemother.*, 13, 368, 1978.
42a. Carlier, M. B., Laurent, G., Claes, P. J., Vanderhaeghe, H. J., and Tulkens, P. M., Inhibition of lysosomal phospholipases by aminoglycoside antibiotics, in vitro comparative studies, *Antimicrob. Agents Chemother.*, 23, 440, 1983.
43. Carrie, B. J., Golbetz, A. B., Michaels, A. S., and Myers, B. D., Creatinine, an inadequate filtration marker in glomerular diseases, *Am. J. Med.*, 69, 177, 1980.
44. Carver, M. P., Coppoc, G. L., and Riviere, J. E., Gentamicin pharmacokinetics and tissue disposition in beagles, *Proc. Symp. Am. Acad. Vet. Pharmacol. Ther.*, 3, 209, 1982.
45. Carver, M. P., Shymodjeska, J. S., Brown, T. T., Rogers, R. A., and Riviere, J. E., Dose-response studies of gentamicin nephrotoxicity in rats with experimental renal dysfunction. I. Subtotal surgical nephrectomy, *Toxicol. Appl. Pharmacol.*, in press.
46. Chahwala, S. B. and Harpur, E. S., An investigation of the effects of aminoglycoside antibiotics on Na-K ATPase as a possible mechanism of toxicity, *Res. Commun. Chem. Pathol. Pharmacol.*, 35, 63, 1982.
47. Chiou, W. L., Extensive renal tubular reabsorption and secretion in man and its clinical significance. Creatinine. XI, *Res. Commun. Chem. Pathol. Pharmacol.*, 36, 349, 1982.
48. Chiou, W. L. and Hsu, F. H., Pharmacokinetics of creatinine in man and its implications in the monitoring of renal function and in dosage regimen modifications in patients, *J. Clin. Pharmacol.*, 15, 427, 1975.
49. Chiu, P. J. S., Brown, A., Miller, G., and Long, J. F., Renal extraction of gentamicin in anesthetized dogs, *Antimicrob. Agents Chemother.*, 10, 277, 1976.
50. Chiu, P. J. S. and Long, J. F., Effects of hydration on gentamicin excretion and renal accumulation in furosemide treated rats, *Antimicrob. Agents Chemother.*, 15, 214, 1978.
51. Chiu, P. J. S. and Long, J. F., Urinary excretion and tissue accumulation of gentamicin and para-aminohippurate in post ischemic rat kidneys, *Kidney Int.*, 15, 618, 1979.
52. Chung, M., Costello, R., and Symchowicz, S., Comparison of netilmicin and gentamicin pharmacokinetics in humans, *Antimicrob. Agents Chemother.*, 17, 184, 1980.
53. Chung, M., Parraircini, L., Assael, B. M., Cavanna, G., Radwanski, E., and Symchowicz, S., Comparative pharmacokinetics of aminoglycoside antibiotics in guinea pigs, *Antimicrob. Agents Chemother.*, 22, 1017, 1982.
54. Chung, M., Schrogie, J. J., and Symchowicz, S., Pharmacokinetic study of sisomicin in humans, *J. Pharmacokinet. Biopharm.*, 9, 535, 1981.
55. Cockcroft, D. W. and Gault, M. H., Prediction of creatinine clearance from serum creatinine, *Nephron*, 16, 31, 1976.

56. Cohen, L., Lapkin, R., and Kaloyanides, G. J., Effect of gentamicin on renal function in the rat, *J. Pharmacol. Exp. Ther.*, 193, 264, 1975.
57. Cojocel, C., Dociu, N., Maita, K., Sleight, S. D., and Hook, J. B., Effects of aminoglycosides on glomerular permeability, tubular reabsorption, and intracellular catabolism of the cationic low-molecular weight protein lysozyme, *Toxicol. Appl. Pharmacol.*, 68, 96, 1983.
58. Cojocel, C. and Hook, J. B., Aminoglycoside nephrotoxicity, *Trends Pharmacol. Sci.*, 4, 174, 1983.
59. Colburn, W. A., Schentag, J. J., Jusko, W. J., and Gibaldi, M., A model for the prospective identification of the prenephrotoxic state during gentamicin therapy, *J. Pharmacokinet. Biopharm.*, 6, 179, 1978.
59a. Collier, V. U., Lietman, P. S., and Mitch, W. E., Evidence for luminal uptake of gentamicin in the perfused rat kidney, *J. Pharmacol. Exp. Ther.*, 210, 247, 1979.
60. Conzelman, G. M., Pharmacotherapeutics of aminoglycoside antibiotics, *J. Am. Vet. Med. Assoc.*, 176, 1078, 1980.
61. Counts, G. W., Blair, A. D., Wagner, K. F., and Turck, M., Gentamicin and tobramycin kinetics, *Clin. Pharmacol. Ther.*, 31, 662, 1982.
62. Cowan, R. H., Jukkola, A. F., and Arant, B. S., Pathophysiologic evidence of gentamicin nephrotoxicity in neonatal puppies, *Pediatr. Res.*, 14, 1204, 1980.
63. Cronin, R. E., Aminoglycoside nephrotoxicity, pathogenesis and nephrotoxicity, *Clin. Nephrol.*, 11, 251, 1979.
64. Cronin, R. E., Bulger, R. E., Southern, P., and Henrich, W. L., Natural history of aminoglycoside nephrotoxicity in the dog, *J. Lab. Clin. Med.*, 95, 463, 1980.
65. Cronin, R., Nix, K., and Ferguson, E., Renal cortex ion composition and Na-K ATPase activity in early gentamicin nephrotoxicity, *Am. J. Physiol.*, 242F, 477, 1982.
66. Crowell, N. A., Divers, T. J., Byars, T. D., Marshall, A. E., Nusbaum, K. E., and Larsen, L., Neomycin toxicosis in calves, *Am. J. Vet. Res.*, 42, 29, 1981.
67. Cuppage, F. E., Setter, K., Sullivan, L. P., Reitzes, E. J., and Melnykovych, A. D., Gentamicin nephrotoxicity. II. Physiological, biochemical and morphological effects of prolonged administration to rats, *Virchows Arch. B*, 24, 121, 1977.
68. Custer, R. S., Bush, M., and Carpenter, J. W., Pharmacokinetics of gentamicin in blood plasma of quail, pheasants, and crane, *Am. J. Vet. Res.*, 40, 892, 1979.
69. Dahlager, J. I., The effect of netilmicin and other aminoglycosides on renal function, *Scand. J. Infect. Dis.*, 23, 96, 1980.
70. Dedrick, R. L. and Bischoff, K. B., Species similarities in pharmacokinetics, *Fed. Proc. Fed. Am. Soc. Exp. Biol.*, 39, 54, 1980.
71. Dedrick, R. L., Bischoff, K. B., and Zaharko, D. S., Interspecies correlation of plasma concentration history of methotrexate (NSC-740), *Cancer Chemother. Rep.*, 54, 95, 1970.
72. Dellinger, P., Murphy, T., Pinn, V., Barza, M., and Weinstein, L., The protective effect of cephalothin against gentamicin nephrotoxicity in rats, *Antimicrob. Agents Chemother.*, 9, 172, 1976.
73. Dierickx, P. J., Urinary gamma glutamyl transferase as an indicator of acute nephrotoxicity in rats, *Arch. Toxicol.*, 47, 209, 1981.
74. Dolislager, D., Fravert, D., and Tune, B. M., Interactions of aminoglycosides and cephaloridine in the rabbit kidney, *Res. Commun. Chem. Pathol. Pharmacol.*, 26, 13, 1979.
75. Dunn, M. J. and Zambraski, E. J., Renal effects of drugs that inhibit prostaglandin synthesis, *Kidney Int.*, 18, 609, 1980.
76. Elliott, W. C., Gilbert, D. N., DeFehr, J., Bennett, W. M., and McCarron, D. A., Protection from experimental gentamicin toxicity by dietary calcium loading, *Kidney Int.*, 21, 216, 1982.
77. Elliott, W. C., Houghton, D. C., Gilbert, D. W., Baines-Hunter, J., and Bennett, W. M., Gentamicin nephrotoxicity. I. Degree of permanence and acquired insensitivity, *J. Lab. Clin. Med.*, 100, 501, 1982.
78. Elliott, W. C., Houghton, D. C., Gilbert, D. W., Baines-Hunter, J., and Bennett, W. M., Gentamicin nephrotoxicity. II. Definition of conditions necessary to induce acquired insensitivity, *J. Lab. Clin. Med.*, 100, 513, 1982.
79. Elliott, W. C., Parker, R. A., Houghton, D. C., Gilbert, D. N., and Bennett, W. M., Comparative nephrotoxicity of dibekacin and gentamicin in rats, *Res. Commun. Chem. Pathol. Pharmacol.*, 33, 419, 1981.
80. Engle, J. E., Abt, A. B., Schenck, D. W., and Schoolwerth, C. A., Netilmicin and tobramycin, comparison of nephrotoxicity in dogs, *Invest. Urol.*, 17, 98, 1979.
81. Eschbach, J. W., Adamson, J. W., and Dennis, M. B., Physiologic studies in normal and uremic sheep. I. The experimental model, *Kidney Int.*, 18, 725, 1980.
82. Fabre, J. and Balant, L., Renal failure, drug pharmacokinetics and drug action, *Clin. Pharmacokinet.*, 1, 99, 1976.

83. Fabre, J., Fillastre, J. P., Morin, J. P., and Rudhardt, M., Nephrotoxicity of gentamicin. Action on subcellular organelles and pharmacokinetics in the kidney, *Contrib. Nephrol.*, 10, 53, 1978.
84. Fabre, J., Rudhardt, M., Blanchard, P., and Regamy, C., Persistence of sisomicin and gentamicin in renal cortex and medulla compared with other organs and serum of rats, *Kidney Int.*, 10, 444, 1979.
85. Feinfeld, D. A., Fleischner, G. M., and Arias, I. M., Urinary ligandin and glutathione-S-transferase in gentamicin-induced nephrotoxicity in the rat, *Clin. Sci.*, 61, 123, 1981.
86. Feldman, S., Josepovitz, C., Scott, M., Pastoriza, E., and Kaloyanides, G. J., Inhibition of gentamicin uptake in rat kidney by polycations, *Kidney Int.*, 19, 222, 1981.
87. Feldman, S., Wang, M. Y., and Kaloyanides, G. J., Aminoglycosides induce a phosphaolipidosis in the renal cortex of the rat, *J. Pharmacol. Exp. Ther.*, 220, 514, 1982.
88. Fernandez-Repollet, E., Rowley, J., and Schwartz, A., Renal damage in gentamicin treated lanner falcons, *J. Am. Vet. Med. Assoc.*, 181, 1392, 1982.
89. Follath, F., Vozeh, S., and Wenk, M., Pharmacokinetic models and dosage regimens, in *Pharmacokinetics*, Gladtke, H., Ed., Gustav Fischer Verlag, New York, 1980,109.
90. Forrey, A. W., Meijsen-Ludwick, B. T., O'Neill, M. A., Maxwell, A. D., and Cutler, R. E., Nephrotoxicity, a comparison in humans of gentamicin and gentamicin C1 administration, *Toxicol. Appl. Pharmacol.*, 44, 453, 1978.
91. Frame, P. T., Phair, J. P., Watanakunakorn, V. C., and Bannister, T. W. P., Pharmacologic factors associated with gentamicin nephrotoxicity in rabbits, *J. Infect. Dis.*, 135, 952, 1977.
92. Frame, P. T., Watanakunakorn, C., and Cluxton, R., Comparative nephrotoxicity of netilmicin and gentamicin in rabbits, *Curr. Ther. Res.*, 24, 344, 1978.
93. French, M. A., Cerra, F. B., Plaut, M. E., and Schentag, J. J., Amikacin and gentamicin accumulation pharmacokinetics and nephrotoxiciy in critically ill patients, *Antimicrob. Agents Chemother.*, 19, 147, 1981.
94. Garrett, E. R., Pharmacokinetics and clearances related to renal processes, *Int. J. Clin. Pharmacol. Biopharm.*, 16, 155, 1978.
95. Gartner, H. V., Drug associated nephropathy. I. Gomerular lesions, *Curr. Top. Pathol.*, 69, 143, 1980.
96. Gary, N. E., Buzzeo, L., Salaki, J., and Eisinger, R., Gentamicin associated acute renal failure, *Arch. Intern. Med.*, 136, 1101, 1976.
97. Gault, M. H. and Cockcroft, D. W., Creatinine clearance and age, *Lancet*, 2, 613, 1975.
98. Gibaldi, M. and Perrier, M., *Pharmacokinetics*, 2nd ed., Marcel Dekker, New York, 1982.
99. Gilbert, D. N., Houghton, D. C., Bennett, W. M., Plamp, C. E., Reger, K., and Porter, G. A., Reversibility of gentamicin nephrotoxicity in rats, recovery during continuous drug administration, *Proc. Soc. Exp. Biol. Med.*, 160, 99, 1979.
100. Ginsburg, D. S., Quintanilla, A. P., and Levin, M., Renal glycosuria due to gentamicin in rabbits, *J. Infect. Dis.*, 134, 119, 1976.
101. Guiliano, R. A., Pollet, D. E., Verpooten, G. A., Scharpe, S., Paulus, G. J., and DeBroe, M. E., Influence of dose regimen on renal accumulation of aminoglycosides, *Arch. Int. Pharmacodyn.*, 260, 277, 1982.
101a. Gordon, R. C., Regamey, C., and Kirby, W. M. M., Serum protein binding of the aminoglycoside antibiotics, *Antimicrob. Agents Chemother.*, 2, 214, 1972.
101b. Goodman, E. L., VanGelder, J., Holmes, R., Hull, A. R., and Sanford, J. P., Prospective comparative study of variable dosage and variable frequency regimens for administration of gentamicin, *Antimicrob. Agents Chemother.*, 8, 434, 1975.
102. Goto, M., Sugiyama, M., and Ishizaki, T., Pharmacokinetic studies with dibekacin, a new aminoglycoside, after intravenous and intramuscular administration to humans, *Antimicrob. Agents Chemother.*, 18, 372, 1980.
103. Gyselynck, A. M., Forrey, A., and Cutler, R., Pharmacokinetics of gentamicin, distribution and plasma and renal clearance, *J. Infect. Dis.*, 124S, 70, 1971.
104. Halkin, H., Lidje, M., and Rubinstein, E., The influence of endotoxin-induced pyrexia on the pharmacokinetics of gentamicin in the rabbit, *J. Pharmacol. Exp. Ther.*, 216, 415, 1981.
105. Hall, C. E. and Hall, O., Polyvinyl alcohol nephrosis, relationship of degree of polymerization to patholophysiologic effects, *Proc. Soc. Exp. Biol. Med.*, 112, 86, 1963.
106. Hall, C. E. and Hall, O., Comparison of macromolecular hypertension due to polyvinyl alcohol and methyl cellulose in respect to the role of sodium chloride, *Lab. Invest.*, 11, 826, 1962.
107. Hatala, M., Moravek, J., and Schuck, O., Gentamicin pharmacokinetics in rats with reduced renal parenchyma, *Int. J. Clin. Pharmacol.*, 11, 1, 1975.
108. Hewitt, W. L., Gentamicin toxicity in perspective, *Postgrad. Med. J.*, 50S, 55, 1974.

109. Hook, J. B., McCormack, M., and Kluwe, W. M., Biochemical mechanisms of nephrotoxicity, in *Reviews in Biochemical Toxicology*, Hodgson, E., Bend, J. R., and Philpot, J. R., Eds., Elsevier/North Holland, Amsterdam. 1979, 53.
110. Hottendorf, G. H., Barnett, D., Gordon, L. L., Christensen, E. F., and Madissoo, H., Nonparallel nephrotoxicity dose-response curves of aminoglycosides, *Antimicrob. Agents Chemother.*, 19, 1024, 1981.
111. Hottendorf, G. H. and Gordon, L. L., Comparative low-dose nephrotoxicities of gentamicin, tobramycin and amikacin, *Antimicrob. Agents Chemother.*, 18, 176, 1980.
112. Houghton, D. C., Campbell-Boswell, M. V., Bennett, W. M., Porter, G. A., and Brooks, R. E., Myeloid bodies in the renal tubules of humans, relationship to gentamicin therapy, *Clin. Nephrol.*, 10, 140, 1978.
113. Houghton, D. C., Hartnett, M., Campbell-Boswell, M., Porter, G., and Bennett, W., A light and electron microscopic analysis of gentamicin nephrotoxicity in rats, *Am. J. Pathol.*, 82, 590, 1976.
114. Hsu, C. H., Kurtz, T. W., Easterling, R. E., and Weller, J. M., Potentiation of gentamicin nephrotoxicity by metabolic acidosis, *Proc. Soc. Exp. Biol. Med.*, 146, 894, 1974.
114a. Hsu, C. H., Kurtz, T. W., and Weller, J. M., In vitro uptake of gentamicin by rat renal cortical tissue, *Antimicrob. Agents Chemother.*, 19, 192, 1977.
115. Huang, S. M., Huang, Y. C., and Chiou, W. L., Triexponential disposition pharmacokinetics of gentamicin in rabbits, *Res. Commun. Chem. Pathol. Pharmacol.*, 26, 115, 1979.
116. Hull, J. H., Hak, L. J., Koch, G. G., Wargin, W. A., Chi, S. L., and Mattocks, A. M., Influence of range of renal function and liver disease on the predictability of creatinine clearance, *Clin. Pharmacol. Ther.*, 29, 516, 1981.
117. Hull, J. H. and Sarubbi, F. A., Gentamicin serum concentrations, pharmacokinetic predictions, *Ann. Intern. Med.*, 85, 183, 1976.
118. Humes, H. D., Weinberg, J. M., and Knauss, T. C., Clinical and pathophysiologic aspects of aminoglycoside nephrotoxicity, *Am. J. Kidney Dis.*, 2, 5, 1982.
119. Humes, H. D. and Weinberg, J. M., Importance of membrane bound calcium on the hydroosmotic water flow response of ADH in toad urinary bladder, *Clin. Res.*, 28, 449, 1980.
120. Jusko, W. J., Guidelines for collection and pharmacokinetic analysis of drug disposition data, in *Applied Pharmacokinetics*, Evans, W. E., Schentag, J. J., and Jusko, W. J., Eds., Applied Therapeutics, San Francisco, 1980, 639.
121. Just, M. and Habermann, E., The renal handling of polybasic drugs. II. In vitro studies with brush border and lysosomal preparations, *Naunyn Schmiedebergs Arch. Pharmakol.*, 300, 67, 1977.
122. Kacew, S. and Hirsch, G. H., Evaluation of nephrotoxicity of various compounds by means of in vitro techniques and comparison to in vivo methods, in *Toxicology of the Kidney*, Hook, J. B., Ed., Raven Press, New York, 1981, 77.
123. Kahlmeter, G., Nephrotoxicity of gentamicin and tobramycin, *Scand. J. Infect. Dis.*, 19S, 15, 1979.
124. Kahlmeter, G. and Kamme, C., Prolonged excretion of gentamicin in a patient with unimpaired renal function, *Lancet*, 1, 286, 1975.
125. Kahlmeter, G., Netilmicin, clinical pharmacokinetics and aspects on dosage schedules, an overview, *Scand. J. Infect. Dis.*, 23, 74, 1980.
126. Kallay, Z., Trnovec, T., Kettner, M., Macickova, T., Navarova, J., and Greguskova, M., Kinetics of gentamicin accumulation in subcellular structures of the mouse kidney, *J. Pharm. Pharmacol.*, 34, 276, 1982.
127. Kaloyanides, G. J. and Pastoriza-Munoz, E., Aminoglycoside nephrotoxicity, *Kidney Int.*, 18, 571, 1980.
128. Kaye, D., Levison, M. E., and Labovitz, E. D., The unpredictability of serum concentrations of gentamicin. Pharmacokinetics of gentamicin in patients with normal and abnormal renal function, *J. Infect. Dis.*, 130, 150, 1974.
129. Keys, T. F., Kurtz, S. B., Jones, J. D., and Muller, S. M., Renal toxicity during therapy with gentamicin or tobramycin, *Mayo Clin. Proc.*, 56, 556, 1981.
130. Kluwe, W. M., The nephrotoxicity of low molecular weight halogenated alkane solvents, pesticides, and chemical intermediates, in *Toxicology of the Kidney*, Hook, J. B., Ed., Raven Press, New York, 1981, 179.
131. Kluwe, W. M. and Hook, J. B., Analysis of gentamicin uptake by rate renal cortical slices, *Toxicol. Appl. Pharmacol.*, 45, 531, 1978.
132. Kluwe, W. M. and Hook, J. B., Functional nephrotoxicity of gentamicin in the rat, *Toxicol. Appl. Pharmacol.*, 45, 163, 1978.
133. Kluwe, W. M. and Hook, J. B., Effects of environmental chemicals on kidney metabolism and function, *Kidney Int.*, 18, 648, 1980.
133a. Knauss, T. C., Weinberg, J. M., and Humes, H. D., Alterations in renal cortical phospholipid content induced by gentamicin, time course, specificity, and subcellular localization, *Am. J. Physiol.*, 244F, 535, 1983.

134. Kohlepp, S. J., Loveless, M. O., Kohnen, P. W., and Gilbert, D. N., Nephrotoxicity of the constituents of the gentamicin complex, *Proc. Intersci. Conf. Antimicrob. Agents Chemother.*, 22, 51, 1982.
135. Kosek, J. C., Mazze, R. I., and Cousins, M. J., Nephrotoxicity of gentamicin, *Lab. Invest.*, 30, 48, 1974.
136. Kuhar, M. J., Mak, L. I., and Lietman, P. S., Autoradiographic localization of 3H gentamicin in the proximal renal tubules of mice, *Antimicrob. Agents Chemother.*, 15, 131, 1979.
137. Kunin, C. M., Binding of antibiotics to tissue homogenates, *J. Infect. Dis.*, 121, 55, 1970.
138. Laberke, H. G., Drug associated nephropathy. II. Tubulo-interstitial lesions, *Curr. Top. Pathol.*, 69, 184, 1980.
139. Lancini, G. and Parenti, F., *Antibiotics, An Integrated View*, Springer-Verlag, New York, 1982.
140. Lane, A. Z., Wright, G. E., and Blair, D. C., Ototoxicity and nephrotoxicity of amikacin, *Am. J. Med.*, 62, 911, 1977.
141. Lapkin, R., Bowman, R., and Kaloyanides, G. J., Effect of gentamicin on *p*-aminohippurate metabolism and transport in rat kidney slices, *J. Pharmacol. Exp. Ther.*, 201, 233, 1977.
142. Lauermann, M. W. and Barza, M., Assays of aminoglycoside antibiotics. Purposes, procedures and pitfalls, in *The Aminoglycosides, Microbiology, Clinical Use and Toxicology*, Whelton, A. and Neu, H. C., Eds., Marcel Dekker, New York, 1982, 169.
143. Laurent, G., Carlier, M. B., Rollman, B., VanHoof, F., and Tulkens, P., Mechanisms of aminoglycoside-induced lysosomal phospholipidosis, in vitro and in vivo studies with gentamicin and amikacin, *Biochem. Pharmacol.*, 31, 3861, 1982.
144. Lecompte, J., Dumont, L., Hill, J., DuSouich, P., and LeDorier, J., Effect of water deprivation and rehydration on gentamicin disposition in the rat, *J. Pharmacol. Exp. Ther.*, 218, 231, 1981.
145. Lee, M. G., Chen, M. L., Huang, S. M., and Chiou, W. L., Pharmacokinetics of drugs in blood. I. Unusual distribution of gentamicin, *Biopharm. Drug Dispos.*, 2, 89, 1981.
146. Leitner, F. and Price, K. E., Aminoglycosides under development, in *The Aminoglycosides, Microbiology, Clinical Use and Toxicology*, Whelton, A. and Neu, H. C., Eds., Marcel Dekker, New York, 1982, 29.
147. Leroy, A., Humbert, G., Oksenhendler, G., and Fillastre, J. P., Pharmacokinetics of aminoglycosides in subjects with normal and impaired renal function, *Antibiot. Chemother.*, 25, 163, 1978.
148. Ling, G. V. and Ruby, A. L., Gentamicin for treatment of resistant urinary tract infections in dogs, *J. Am. Vet. Med. Assoc.*, 175, 480, 1979.
149. Lipsky, J. J., Cheng, L., Sacktor, B., and Lietman, P. S., Gentamicin uptake by renal tubule brush border membrane vesicles, *J. Pharmacol. Exp. Ther.*, 215, 390, 1980.
150. Lipsky, J. J. and Lietman, P. S., Aminoglycoside inhibition of a renal phosphatidylinositol phospholipase C, *J. Pharmacol. Exp. Ther.*, 220, 287, 1982.
151. Lipsky, J. J. and Lietman, P. S., Neomycin inhibition of adenosine triphosphatase, evidence for a neomycin-phospholipid interaction, *Antimicrob. Agents Chemother.*, 18, 532, 1980.
152. Lode, H., Kemmerich, B., and Koeppe, P., Comparative clinical pharmacology of gentamicin, sisomicin and tobramycin, *Antimicrob. Agents Chemother.*, 8, 396, 1975.
153. Luft, F. C., Cephalosporin and aminoglycoside interactions. Clinical and toxicologic implications, in *The Aminoglycosides, Microbiology, Clinical Use and Toxicology*, Whelton, A. and Neu, H. C., Eds., Marcel Dekker, New York, 1982, 387.
154. Luft, F. C., Aronoff, G. R., Evan, A. P., Connors, B. A., Weinberger, M. H., and Kleit, S. A., The renin-angiotensin system in aminoglycoside-induced acute renal failure, *J. Pharmacol. Exp. Ther.*, 222, 433, 1982.
155. Luft, F. C., Aronoff, G. R., Evan, A. P., and Connors, B. A., The effect of aminoglycosides on glomerular endothelium, a comparative study, *Res. Commun. Chem. Pathol. Pharmacol.*, 34, 89, 1981.
156. Luft, F. C., Bloch, R., Sloan, R. S., Yum, M. N., and Costello, R., Comparative nephrotoxicity of aminoglycoside antibiotics in rats, *J. Infect. Dis.*, 138, 541, 1978.
157. Luft, F. C., Brannon, D. R., Stropes, L. L., Costello, R. J., Sloan, R. S., and Maxwell, D. R., Pharmacokinetics of netilmicin in patients with renal impairment and in patients on dialysis, *Antimicrob. Agents Chemother.*, 14, 403, 1978.
158. Luft, F. C. and Evan, A. P., Glomerular filtration barrier in aminoglycoside induced nephrotoxic acute renal failure, *Renal Physiol.*, 3, 265, 1980.
159. Luft, F. C. and Evan, A. P., Comparative effects of tobramycin and gentamicin on glomerular ultrastructure, *J. Infect. Dis.*, 142, 910, 1980.
160. Luft, F. C. and Kleit, S. A., Renal parenchymal accumulation of aminoglycoside antibiotics in rats, *J. Infect. Dis.*, 130, 656, 1974.
161. Luft, F. C., Patel, V., Yum, M. N., Patel, B., and Kleit, S. A., Experimental aminoglycoside nephrotoxicity, *J. Lab. Clin. Med.*, 86, 213, 1975.
162. Luft, F. C., Patel, V., Yum, M. N., and Kleit, S. A., The nephrotoxicity of cephalosporin-gentamicin combinations in rats, *Antimicrob. Agents Chemother.*, 9, 831, 1976.

163. Luft, F. C., Rankin, L. I., Sloan, R. S., and Yum, M. N., Recovery from aminoglycoside nephrotoxicity with continued drug administration, *Antimicrob. Agents Chemother.*, 14, 284, 1978.
164. Luft, F. C., Rankin, L. I., Sloan, R. S., Fineberg, N. S., Yum, M. N., and Wong, L., Comparative low-dose nephrotoxicity of dibekacin, gentamicin and tobramycin, *J. Antimicrob. Chemother.*, 9, 297, 1982.
165. Luft, F. C., Yum, M. N., and Kleit, S. A., The effect of concomitant mercuric chloride and gentamicin on kidney function and structure in the rat, *J. Lab. Clin. Med.*, 89, 622, 1976.
166. Luft, F. C., Yum, M. N., Walker, P. D., and Kleit, S. A., Gentamicin gradient patterns and morphological changes in human kidneys, *Nephron*, 18, 167, 1977.
166a. Lullmann, H. and Vollmer, B., An interaction of aminoglycoside antibiotics with Ca binding to lipid monolayers and to biomembranes, *Biochem. Pharmacol.*, 31, 3769, 1982.
167. Maher, J. F., Toxic nephropathy, in *The Kidney*, Brenner, B. M. and Rector, F. C., Eds., W. B. Saunders, Philadelphia, 1976, 1355.
168. Mangione, A. and Schentag, J. J., Therapeutic monitoring of aminoglycoside antibiotics, an approach, in *Frontiers in Therapeutic Drug Monitoring*, Tognoni, G., Latini, R., and Jusko, W. J., Eds., Raven Press, New York, 1980, 143.
169. Marre, R., Tarara, N., and Louton, T., Age dependent nephrotoxicity and the pharmacokinetics of gentamicin in rats, *Eur. J. Pediatr.*, 133, 25, 1980.
170. Matzke, G. R., Gwizdala, C., Wery, J., Ferry, D., and Starnes, R., Evaluation of three gentamicin assay techniques, *Ther. Drug Monitor*, 4, 195, 1982.
171. Mazze, R. I., Methoxyflurane nephropathy, in *Toxicology of the Kidney*, Hook, J. B. Ed., Raven Press, New York, 1981, 135.
172. McNeil, J. S., Jackson, B., Nelson, L., and Butkas, D. E., The role of prostaglandins in gentamicin induced nephrotoxicity in the dog, *Nephron*, 33, 202, 1983.
173. Meisner, H., Effect of gentamicin on the subcellular distribution of renal beta-*N*-acetylglucosaminidase activity, *Biochem. Pharmacol.*, 30, 2949, 1981.
174. Meulemans, A., Manuel, C., and Trans Ba Huy, P., Radioimmunoassay of gentamicin in microliter and nanoliter samples of biological fluids, *Chemotherapy*, 27, 29, 1981.
174a. Mitchell, C. J., Bullock, S., and Ross, B. D., Renal handling of gentamicin and other antibiotics by the isolated perfused rat kidney, mechanisms of nephrotoxicity, *J. Antimicrob. Chemother.*, 3, 593, 1977.
175. Mitchell, J. R., McMurtry, R. J., Statham, C. N., and Nelson, S. D., Molecular basis of several drug-induced nephropathies, *Am. J. Med.*, 62, 518, 1977.
176. Mondorf, A. W., Urinary enzymatic markers of renal damage, in *The Aminoglycosides, Microbiology, Clinical Use and Toxicology*, Whelton, A., and Neu, H. C., Eds., Marcel Dekker, New York, 1982, 283.
177. Morin, J. P. and Fillastre, J. P., Aminoglycoside-induced lysosomal dysfunctions in the kidney, in *The Aminoglycosides, Microbiology, Clinical Use and Toxicology*, Whelton, A. and Neu, H. C., Eds., Marcel Dekker, New York, 1982, 269.
178. Morin, J. P., Viotte, G., Vandewalle, A., Van Hoof, F., Tulkens, P., and Fillastre, J. P., Gentamicin-induced nephrotoxicity, a cell biology approach, *Kidney Int.*, 18, 583, 1980.
179. Morin, J. P., Viotte, G., Van Hoof, F., Tulkens, P., Godin, M., and Fillastre, J. F., Functional, biochemical and morphological events related to gentamicin therapy in rats, *Drugs Exp. Clin. Res.*, 7, 345, 1981.
179a. Myers, D. R., DeFehr, J., Bennett, W. M., Porter, G. A., and Olsen, G. D., Gentamicin binding to serum and plasma proteins, *Clin. Pharmacol. Ther.*, 23, 356, 1978.
180. Mudge, G. H., Nephrotoxicity of urographic radiocontrast drugs, *Kidney Int.*, 18, 540, 1980.
181. Nagabhushan, T. L., Miller, G. H., and Weinstein, M. J., Structure-activity relationships in aminoglycoside-aminocyclitol antibiotics, in *The Aminoglycosides, Microbiology, Clinical Use and Toxicology*, Whelton, A. and Neu, H. C., Eds., Marcel Dekker, New York, 1982, 3.
182. Ngaha, E. D. and Madusolumuo, M. A., Renal effects of an aminoglycoside antibiotic in the rat, *Toxicol. Lett.*, 13, 23, 1982.
183. Neu, H. C., Pharmacology of aminoglycosides, in *The Aminoglycosides, Microbiology, Clinical Use and Toxicology*, Whelton, A. and Neu, H. C., Eds., Marcel Dekker, New York, 1982, 125.
184. Notari, R. E., *Biopharmaceutics and Pharmacokinetics, An Introduction*, 3rd ed., Marcel Dekker, New York, 1980.
185. O'Connell, J. M. B., Romeo, J. A., and Mudge, G. H., Renal tubular secretion of creatinine in the dog, *Am. J. Physiol.*, 203, 985, 1962.
186. Oeltgen, P. R., Hamann, S. R., and Blouin, R. A., Comparison of gentamicin assays, *Ther. Drug Monitor*, 2, 423, 1980.
187. Olsen, S. and Solez, K., Pathology of drug nephrotoxicity in humans, in *The Aminoglycosides, Microbiology, Clinical Use and Toxicology*, Whelton, A. and Neu, H. C., Eds., Marcel Dekker, New York, 1982, 355.

188. Pardo, V. and Shapiro, A. P., Ultrastructural glomerular lesions produced by synthetic polysaccharides, *Lab. Invest.*, 15, 617, 1966.
189. Parker, R. A., Bennett, W. M., and Porter, G. A., Animals models in the study of aminoglycoside nephrotoxicity, in *The Aminoglycosides, Microbiology, Clinical Use and Toxicology*, Whelton, A. and Neu, H. C., Eds., Marcel Dekker, New York, 1982, 235.
190. Pastoriza-Munoz, E., Bowman, R. L., and Kaloyanides, G. J., Renal tubular transport of gentamicin in the rat, *Kidney Int.*, 16, 440, 1979.
191. Pastoriza-Munoz, E., Timmerman, D., Feldman, S., and Kaloyanides, G. J., Ultrafiltration of gentamicin and netilmicin in vivo, *J. Pharmacol. Exp. Ther.*, 220, 604, 1982.
192. Patel, V., Luft, F. C., Yum, M. N., Patel, B., Zeman, W., and Kleit, S. A., Enzymuria in gentamicin-induced kidney damaged, *Antimicrob. Agents Chemother.*, 7, 364, 1975.
193. Pechere, J. C. and Dugal, R. D., Clinical pharmacokinetics of aminoglycoside antibiotics, *Clin. Pharmacokinet.*, 4, 170, 1979.
194. Pedersoli, V. M., Belmonte, A. A., Purohit, R. C., and Ravis, W. R., Pharmacokinetics of gentamicin in the horse, *Am. J. Vet. Res.*, 41, 351, 1980.
195. Pennington, J. E., Dale, D. C., Reynolds, H. Y., and MacLowry, J. D., Gentamicin sulfate pharmacokinetics, lower levels of gentamicin in blood during fever, *J. Infect. Dis.*, 132, 270, 1975.
196. Pien, F. D. and Ho, P. W. L., Antimicrobial spectrum, pharmacology, adverse effects, and therapeutic use of amikacin sulfate, *Am. J. Hosp. Pharm.*, 38, 981, 1981.
197. Piperno, E., Detection of drug induced nephrotoxicity with urinalysis and enzymuria assessment, *Toxicology of the Kidney*, Hook, J. B. Ed., Raven Press, New York, 1981, 31.
198. Porter, G. A. and Bennett, W. M., Nephrotoxic acute renal failure due to common drugs, *Am. J. Physiol.*, 241F, 1, 1981.
199. Porter, G. A. and Bennett, W. M., Toxic nephropathies, in *The Kidney*, Brenner, B. M. and Rector, F. C., Eds., W. B. Saunders, New York, 1981, 2045.
200. Powell, S. H., Thompson, W. L., Luthe, M. A., Stern, R. C., Grossniklaus, D. A., Bloxham, D. D., Groden, D. L., et al. Once-daily vs. continuous aminoglycoside dosing, efficacy and toxicity in animal and clinical studies of gentamicin, netilmicin and tobramycin, *J. Infect. Dis.*, 147, 918, 1983.
201. Price, R. G., Urinary enzymes, nephrotoxicity and renal disease, *Toxicology*, 23, 99, 1982.
202. Prince, R. A., Ling, M. H., Hepler, C. D., Rainville, E. C., Kealey, G. P., Donta, S. T., LeFrock, J. L., and Kowalsky, S. F., Factors associated with creatinine clearance changes following gentamicin therapy, *Am. J. Hosp. Pharm.*, 37, 1489, 1980.
203. Ratcliffe, R. M., Mirelli, C., Moran, E., O'Leary, D., and White, R., Comparison of five methods of the assay of serum gentamicin, *Antimicrob. Agents Chemother.*, 19, 508, 1981.
204. Reiner, N. E., Bloxham, D. D., and Thompson, W. L., Nephrotoxicity of gentamicin and tobramycin given once daily or continuously in dogs, *J. Antimicrob. Chemother.*, 4S, 85, 1978.
205. Regamey, G., Gordon, R. C., and Kirby, W. M. M., Comparative pharmacokinetics of tobramycin and gentamicin, *Clin. Pharmacol. Ther.*, 14, 396, 1973.
206. Ristuccia, A. M. and Cunha, B. A., The aminoglycosides, *Med. Clin. North Am.*, 66, 303, 1982.
207. Riviere, J. E., The aminoglycosides, in *The Bristol Veterinary Handbook of Antimicrobial Therapy*, Johnston, D. E., Ed., Veterinary Learning Systems, Princeton Junction, N.J., 1982, 186.
208. Riviere, J. E., A possible mechanism for increased susceptibility to aminoglycoside nephrotoxicity in chronic renal disease, *N. Engl. J. Med.*, 307, 252, 1982.
209. Riviere, J. E., Limitations on the physiologic interpretation of aminoglycoside body clearance derived from pharmacokinetic studies, *Res. Commun. Chem. Pathol. Pharmacol.*, 38, 31, 1982.
210. Riviere, J. E., Paradoxical increase in aminoglycoside body clearance in renal disease when volume of distribution increases, *J. Pharm. Sci.*, 71, 720, 1982.
211. Riviere, J. E., Carver, M. P., Coppoc, G. L., Carlton, W. W., Lantz, G. C., and Shy-Modjeska, J., Pharmacokinetics and comparative nephrotoxicity of fixed-dose versus fixed-interval reduction of gentamicin dosage in subtotal nephrectomized dogs, *Toxicol. Appl. Pharmacol.*, 75, 496, 1984.
212. Riviere, J. E. and Coppoc, G. L., Pharmacokinetics of gentamicin in the juvenile dog, *Am. J. Vet. Res.*, 42, 1621, 1981.
213. Riviere, J. E., Coppoc, G. L., Carlton, W. W., and Hinsman, E. J., Polyvinyl alcohol toxicosis as a model of glomerulonephritis in beagle dogs, *Am. J. Vet. Res.*, 41, 502, 1980.
214. Riviere, J. E., Coppoc, G. L., Hinsman, E. J., Carlton, W. W., and Traver, D. S., Species dependent gentamicin pharmacokinetics and nephrotoxicity in the young horse, *Fundam. Appl. Toxicol.*, 3, 448, 1983.
215. Riviere, J. E. and Davis, L. E., Renal handling of drugs in renal failure, in *Canine Nephrology*, Bovee, K. C., Ed., Harwal, Media, Pa., 1984, 643.
216. Riviere, J. E., Hinsman, E. J., Coppoc, G. L., and Carlton, W. W., Single dose gentamicin nephrotoxicity in the dog, early functional and ultrastructural changes, *Res. Commun. Chem. Pathol. Pharmacol.*, 33, 403, 1981.

217. Riviere, J. E., Traver, D. S., and Coppoc, G. L., Gentamicin toxic nephropathy in horses with disseminated bacterial infection, *J. Am. Vet. Med. Assoc.*, 180, 648, 1982.
218. Riviere, J. E. and Carver, M. P., Effects of familial hypothyroidism and subtotal surgical nephrectomy on gentamicin pharmacokinetics in beagle dogs, *Chemotherapy*, 30, 216, 1984.
219. Riviere, J. E., Shapiro, D. S., and Coppoc, G. L., Percutaneous absorption of gentamicin by the leopard frog, *Rana pipiens*, *J. Vet. Pharmacol. Ther.*, 2, 235, 1979.
220. Robinson, T., Harbison, M., and Bovee, K. C., Influence of reduced renal mass on tubular secretion of creatinine in the dog, *Am. J. Vet. Res.*, 35, 487, 1974.
221. Ross, L. A. and Finco, D. R., Relationship of selected clinical renal function tests to glomerular filtration rate and renal blood flow in cats, *Am. J. Vet. Res.*, 42, 1704, 1981.
222. Ross, M., Parker, R. A., and Elliot, W. C., Gentamicin induced resistance to ADH stimulated water flow in the toad bladder explained by drug induced pH changes, *Clin. Res.*, 28, 46, 1980.
223. Rougemont, D., Oeschges, A., Konrod, L., Thiel, G., Torhorst, J., Wenk, M., Wunderlich, P., and Brunner, F. P., Gentamicin induced acute renal failure in the rat, *Nephron*, 29, 176, 1981.
224. Sadee, W. and Beelen, G. C. M., *Drug Level Monitoring, Analytical Techniques, Metabolism, and Pharmacokinetics*, John Wiley & Sons, New York, 1980.
225. Sairio, E., Kasanen, A., Kangas, L., Nieminen, A. L., and Nieminen, L., The nephrotoxicity and renal accumulation of amikacin, tobramycin, and gentamicin in rats, rabbits and guinea pigs, *Exp. Pathol. Bd.*, 15S, 370, 1978.
226. Sanders, W. E. and Sanders, C. C., Toxicity of antibacterial agents. Mechanisms of action on mamalian cells, *Annu. Rev. Pharmacol. Toxicol.*, 19, 53, 1979.
227. Sastrasinh, M., Weinberg, J. M., and Humes, H. D., Effect of gentamicin on calcium uptake by renal mitochondria, *Life Sci.*, 26, 2309, 1982.
227a. Sastrasinh, M., Knauss, T. C., Weinberg, J. M., and Humes, H. D., Identification of the aminoglycoside binding site in rat renal brush border membranes, *J. Pharmacol. Exp. Ther.*, 222, 350, 1982.
228. Schacht, J., Isolation of an aminoglycoside receptor from guinea pig inner ear tissues and kidney, *Arch. Otorhinolaryngol.*, 224, 129, 1979.
229. Schacht, J., Purification of polyphosphoinositides by chromatography on immobilized neomycin, *J. Lipid Res.*, 19, 1063, 1978.
230. Schentag, J. J., Aminoglycosides, in *Applied Pharmacokinetics*, Evans, W. E., Schentag, J. J., and Juski, W. J., Eds., Applied Therapeutics, San Francisco, 1980, 174.
231. Schentag, J. J., Aminoglycoside pharmacokinetics as a guide to therapy and toxicology, in *The Aminoglycosides, Microbiology, Clinical Use, and Toxicology*, Whelton, A. and Neu, H. C., Eds., Marcel Dekker, New York, 1982, 143.
232. Schentag, J. J., Cerra, F. B., and Plaut, M. E., Clinical and pharmacokinetic characteristics of aminoglycoside nephrotoxicity in 201 critically ill patients, *Antimicrob. Agents Chemother.*, 21, 721, 1982.
233. Schentag, J. J. and Jusko, W. J., Renal clearance and tissue accumulation of gentamicin, *Clin. Pharmacol. Ther.*, 22, 364, 1977.
234. Schentag, J. J., Jusko, W. J., Vance, J. W., Cumbo, T. J., Abruytn, E., DeLattre, M., and Gerbracht, L. M., Gentamicin disposition and tissue accumulation on multiple dosing, *J. Pharmacokinet. Biopharm.*, 5, 559, 1977.
235. Schentag, J. J., Lasezkay, G., Cumbo, T. J., Plaut, M. E., and Jusko, W. J., Accumulation pharmacokinetics of tobramycin, *Antimicrob. Agents Chemother.*, 13, 649, 1978.
236. Schentag, J. J., Plaut, M. E., and Cerra, F. B., Comparative nephrotoxicity of gentamicin and tobramycin, pharmacokinetics and clinical studies in 201 patients, *Antimicrob. Agents Chemother.*, 19, 859, 1981.
237. Schentag, J. J., Plaut, M. E., Cerra, F. B., Wels, P. B., Walczak, P., and Buckley, R. J., Aminoglycoside nephrotoxicity in critically ill surgical patients, *J. Surg. Res.*, 26, 270, 1979.
238. Senekjian, H. O., Knight, T. F., and Weinman, E. J., Micropuncture study of the handling of gentamicin by the rat kidney, *Kidney Int.*, 19, 416, 1981.
239. Sheth, A. V., Senekjian, H. O., Babino, H., Knight, T. F., and Weinman, E. J., Renal handling of gentamicin by the Munich-Wistar rat, *Am. J. Physiol.*, 241F, 645, 1981.
240. Silverblatt, F. J., Antibiotic nephrotoxicity, a review of pathogenesis and prevention, *Urol. Clin. North Am.*, 2, 557, 1975.
241. Silverblatt, F. J., Autoradiographic studies of intracellular aminoglycoside disposition in the kidney, in *The Aminoglycosides, Microbiology, Clinical Use and Toxicology*, Whelton, A. and Neu, H. C., Eds., Marcel, Dekker, New York, 1982, 223.
242. Silverblatt, F. J., Pathogenesis of nephrotoxicity of cephalosporins and aminoglycosides, a review of current concepts, *Rev. Infect. Dis.*, 4S, 360, 1982.
243. Silverblatt, F. J. and Kuehn, C., Autoradiography of gentamicin uptake by the rat proximal tubule cell, *Kidney Int.*, 15, 335, 1979.

244. Silverman, M. and Mahon, W., Gentamicin interaction in vivo with luminal and antiluminal nephron surfaces of dog kidney, *Proc. Am. Soc. Nephrol.*, 12, 89, 1979.
245. Simmons, C. F., Jr., Bogusky, R. T., and Humes, H. D., Inhibitory effects of gentamicin on renal mitochondria oxidative phosphorylation, *J. Pharmacol. Exp. Ther.*, 214, 709, 1980.
246. Simon, V. K., Mosinger, E. V., and Malerczy, V., Pharmacokinetic studies of tobramycin and gentamicin, *Antimicrob. Agents Chemother.*, 3, 445, 1973.
247. Smith, C. R., Baughman, K. L., Edwards, C. Q., Rogers, J. F., and Lietman, P. S., Controlled comparison of amikacin and gentamicin, *N. Engl. J. Med.*, 296, 349, 1977.
248. Smith, C. R. and Lietman, P. S., Comparative clinical trials of aminoglycosides, in *The Aminoglycosides, Microbiology, Clinical Use and Toxicology*, Whelton, A. and Neu, H. C., Eds., Marcel Dekker, New York, 1982, 497.
249. Smith, C. R., Lipsky, J. J., Laskin, O. L., Hellmann, D. B., Mellits, E. D., and Longsteeth, J., Double-blind comparison of the nephrotoxicity and auditory toxicity of gentamicin and tobramycin, *N. Engl. J. Med.*, 302, 1106, 1980.
250. Smith, C. R., Maxwell, R. R., Edwards, C. Q., Rogers, J. F., and Lietman, P. S., Nephrotoxicity induced by gentamicin and amikacin, *Johns Hopkins Med. J.*, 142, 85, 1978.
251. Smith, T. R. and Smith, C. R., Risk factors for aminoglycoside nephrotoxicity, in *The Aminoglycosides, Microbiology, Clinical Use and Toxicology*, Whelton, A. and Neu, H. C., Eds., Marcel Dekker, New York, 1982, 401.
252. Soberon, L., Bowman, R. L., Pastoriza-Munoz, E., and Kaloyanides, G. J., Comparative nephrotoxicities of gentamicin, netilmicin, and tobramycin in the rat, *J. Pharmacol. Exp. Ther.*, 210, 334, 1979.
253. Souliere, C. R., Goodman, D. B. P., and Appel, G. B., Gentamicin selectively inhibits antidiuretic hormone induced water flow in the toad urinary bladder, *Kidney Int.*, 14, 733, 1978.
254. Spangler, W. L., Aldelman, R. D., Conzelman, G. M., and Ishizaki, G., Gentamicin nephrotoxicity in the dog, sequential light and electron microscopy, *Vet. Pathol.*, 17, 206, 1980.
255. Spreat, S. R. and Bickford, S. M., Pharmacodynamics of gentamicin in day old chicks, *W. Poult. Dis. Conf. Abstr.*, 26, 101, 1977.
256. Sterzel, R. B., Eisenbach, G. M., Seiler, M. W., and Hoyer, J. R., Uptake of polyvinyl alcohol by macrophages in the glomerular mesangium of rats, *Am. J. Pathol.*, 111, 247, 1983.
257. Stygles, V. G. and Iuliucci, J. D., Structural and functional alterations in the kidney following intake of nonsteroidal antiinflammatory agents, *Toxicology of the Kidney*, Raven Press, New York, 1981, 151.
258. Svec, A. L., Constantine, G. H., Larson, R. E., and Meyers, T. R., Single dose aminoglycoside nephrotoxicity and lethality in rainbow trout, *Toxicol. Appl. Pharmacol.*, 65, 153, 1982.
259. Szot, R. J., McCormick, G., Chung, M., Christie, B., Weinberg, E., and Schwartz, E., Comparative toxicity of netilmicin and tobramycin in dogs, *Toxicol. Appl. Pharmacol.*, 55, 169, 1980.
260. Teixiera, R. B., Kelley, J., Alpert, H., Pardo, V., and Vaamonde, C. A., Complete protection from gentamicin induced acute renal failure in the diabetes mellitus rat, *Kidney Int.*, 21, 600, 1982.
260a. Thompson, W. L., Aminoglycoside dosing regimens, once daily, thrice, or continuous administration?, in *The Aminoglycosides, Microbiology, Clinical Use and Toxicology*, Whelton, A. and Neu, H. C., Eds., Marcel Dekker, New York, 1982, 489.
261. Thompson, W. L. and Powell, S., Aminoglycoside nephrotoxicity in dogs, dependence on drug, dose, and regimen, *Scand. J. Infect. Dis.*, 23S, 91, 1980.
262. Thurau, K. and Boylan, J. W., Acute renal success, the unexplained logic of oliguria in acute renal failure, *Am. J. Med.*, 61, 308, 1976.
263. Tobin, T., Pharmacology review, streptomycin, gentamicin and the aminoglycoside antibiotics, *J. Equine Med. Surg.*, 4, 206, 1979.
264. Tofte, R. W., Canafax, D. M., Simmons, R. L., and Peterson, P. K., Aminoglycoside dosing in renal transplant patients, *Ann. Surg.*, 195, 287, 1982.
265. Trollfors, B., Alestig, K., Krantz, I., and Norrby, R., Quantitative nephrotoxicity of gentamicin in nontoxic doses, *J. Infect. Dis.*, 141, 366, 1980.
266. Tulkins, P. and Trouet, A., The uptake and intracellular accumulation of aminoglycoside antibiotics in lysosomes of cultured rat fibroblasts, *Biochem. Pharmacol.*, 27, 415, 1978.
267. Tulkens, P. and VanHoof, F., Comparative toxicity of aminoglycoside antibiotics towards lysosomes in a cell culture model, *Toxicology*, 17, 195, 1980.
268. Tune, B. M. and Fravert, D., Mechanism of cephalosporin nephrotoxicity, a comparison of cephaloridine and cephaloglycin, *Kidney Int.*, 18, 591, 1980.
269. Vandewalle, A., Farman, N., Morin, J. P., Fillastre, J. P., Hatt, P. Y., and Bonvalet, J. P., Gentamicin incorporation along the nephron, autoradiographic study on isolated tubules, *Kidney Int.*, 19, 529, 1981.
270. Vera-Roman, J., Krishnakantha, T. P., and Cuppage, F. E., Gentamicin nephrotoxicity in rats. I. Acute biochemical and ultrastructural effects, *Lab. Invest.*, 33, 412, 1975.

271. Wagner, J. G., *Fundamentals of Clinical Pharmacokinetics*, Drug Intelligence Publications, Hamilton, Ill., 1975.
272. Wagner, J. G., Linear pharmacokinetic equations allowing direct calculation of many needed pharmacokinetic parameters from coefficients and exponents of polyexponential equations which have been fitted to the data, *J. Pharmacokinet. Biopharm.*, 4, 443, 1976.
273. Waitz, J. A., Moss, E. L., and Weinstein, M. J., Aspects of the chronic toxicity of gentamicin sulfate in cats, *J. Infect. Dis.*, 124S, 125, 1971.
274. Wedeen, R. P., Batuman, V., Cheeks, C., Marquet, E., and Sobel, H., Transport of gentamicin in rat proximal tubule, *Lab. Invest.*, 48, 212, 1983.
275. Weinberg, J. M., Harding, P. G., and Humes, H. D., Mechanisms of gentamicin induced dysfunction of renal cortical mitochrondria. II. Effects on mitochondrial monovalent cation transport, *Arch. Biochem. Biophys.*, 205, 232, 1980.
276. Weinberg, J. M. and Humes, H. D., Mechanisms of gentamicin induced dysfunction of renal cortical mitochondria. I. Effects on mitochondrial respiration, *Arch. Biochem. Biophys.*, 205, 222, 1980.
277. Weinberg, J. M., Simmons, C. F., Jr., and Humes, H. D., Alterations of mitochondrial respiration induced by aminoglycoside antibiotics, *Res. Commun. Chem. Pathol. Pharmacol.*, 27, 521, 1980.
277a. Weiss, R. B. and Poster, D. S., The renal toxicity of cancer chemotherapeutic agents, *Cancer Treatment Rev.*, 9, 37, 1982.
278. Weiss, W., Sziegoleit, W., and Foster, W., Dependence of pharmacokinetic parameters on body weight, *Int. J. Clin. Pharmacol.*, 15, 572, 1977.
279. Welles, J. S., Emmerson, J. L., Gibson, W. R., Nickander, R., Owen, N. V., and Anderson, R. C., Preclinical toxicology studies with tobramycin, *Toxicol. Appl. Pharmacol.*, 25, 398, 1973.
280. Welling, P. G., Baumueller, A., Lau, C. C., and Madsen, P. O., Netilmicin pharmacokinetics after single intravenous doses to elderly male patients, *Antimicrob. Agents Chemother.*, 12, 328, 1977.
281. Wellwood, J. M., Lovell, D., Thompson, A. E., and Tighe, J. R., Renal damage caused by gentamicin, a study of the effects on renal morphology and urinary enzyme excretion, *J. Pathol.*, 118, 171, 1976.
282. Wheeler, L. A. and Sheiner, L. B., Clinical estimation of creatinine clearance, *Am. J. Clin. Pathol.*, 72, 27, 1979.
283. Whelton, A., Renal tubular transport and intrarenal aminoglycoside distribution, in *The Aminoglycosides, Microbiology, Clinical Use and Toxicology*, Whelton, A. and Neu, H. C., Eds., Marcel Dekker, New York, 1982, 191.
284. Whelton, A., Carter, G. G., Craig, T. J., Bryant, H. H., Herbst, D. V., and Walker, W. G., Comparison of the intrarenal disposition of tobramycin and gentamicin, therapeutic and toxicologic answers, *J. Antimicrob. Chemother.*, 4S, 13, 1978.
285. Whelton, A., Carter, G. G., Bryant, H. H., and Walker, W. G., Therapeutic implications of gentamicin accumulation in severely diseased kidneys, *Arch. Intern. Med.*, 136, 172, 1976.
286. Whelton, A. and Neu, H. C., Eds., *The Aminoglycosides, Microbiology, Clinical Use and Toxicology*, Marcel Dekker, New York, 1982.
287. Whelton, A. and Solez, K., Aminoglycoside nephrotoxicity, a tale of two transports, editorial, *J. Lab. Clin. Med.*, 99, 148, 1982.
288. Whelton, A. and Walker, W. G., Intrarenal antibiotic distribution in health and disease, *Kidney Int.*, 6, 131, 1974.
289. White, L. O., Scammel, L. M., and Reeves, D. S., Serum aminoglycoside assay by enzyme-mediated immunoassay (EMIT), Correlation with radioimmunoassay fluoroimmunoassay and acetyltransferase assays, *Antimicrob. Agents Chemother.*, 19, 1064, 1981.
290. White, P. H., Petersen, J., and Simpson, J. G., Gentamicin induced nephrotoxicity in mice. Protection by loop diuretics, *Br. J. Exp. Pathol.*, 62, 200, 1981.
291. Williams, P. D., Holohan, P. D., and Ross, C. R., Gentamicin nephrotoxicity. I. Acute biochemical correlates in rats, *Toxicol. Appl. Pharmacol.*, 61, 234, 1981.
292. Williams, P. D., Holohan, P. D., and Ross, C. R., Gentamicin nephrotoxicity. II. Plasma membrane changes, *Toxicol. Appl. Pharmacol.*, 61, 243, 1981.
293. Wilson, R. C., Whelan, S. C., Coulter, D. B., Mahaffey, E. A., Mahaffey, M. D., and Huber, T. L., Kinetics of gentamicin after intravenous, intramuscular, and intratracheal administration in sheep, *Am. J. Vet. Res.*, 42, 1901, 1981.
294. Wilson, T. W., Mahon, W. A., Inaba, T., Johnson, G. E., and Kadar, D., Elimination of trititated gentamicin in normal human subjects and in patients with severely impaired renal function, *Clin. Pharmacol. Ther.*, 14, 815, 1973.
295. Wise, R., Walker, J. M., and Mitchard, M., A comparison of the pharmacokinetics of amikacin and gentamicin, *J. Antimicrob. Chemother.*, 8S, 45, 1981.

296. Wold, J. S., Cephalosporin nephrotoxicity, in *Toxicology of the Kidney*, Hook, J. B., Ed., Raven Press, New York, 1981, 251.
297. Zaske, D. E., Aminoglycosides, in *Applied Pharmacokinetics*, Evans, W. E., Schentag, J. J., and Jusko, W. J., Eds., Applied Therapeutics, San Francisco, 1980, 210.
298. Zaske, D. E., Cipolle, R. J., Rotschafer, J. C., Solem, L. D., Mosier, N. R., and Strate, R. G., Gentamicin pharmacokinetics in 1,640 patients, method for control of serum concentrations, *Antimicrob. Agents Chemother.*, 21, 407, 1982.
299. Zaske, D. E., Cipolle, R. J., Strate, R. G., and Dickes, W. F., Increased genamicin dosage requirements, rapid elimination in 249 gynecology patients, *Am. J. Obstet. Gynecol.*, 8, 896, 1981.
300. Ziv, G. and Sulman, F. G., Distribution of aminoglycoside antibiotics in blood and milk, *Res. Vet. Sci.*, 17, 68, 1974.
301. Ziv, G., Nouws, J. F. M., and VanGinneken, C. A. M., The pharmacokinetics and tissue levels of polymyxin B, colistin and gentamicin in calves, *J. Vet. Pharmacol. Ther.*, 5, 45, 1982.

ANIMAL MODELS OF DIABETIC NEPHROPATHY

Elsa Janle-Swain

INTRODUCTION

Diabetic nephropathy in the human is one of the major secondary complications of diabetes. Diabetic nephropathy is the clinical term used to describe diabetic renal disease when it has progressed to the stage of a persistent proteinuria. However, long before there are any clinical manifestations of renal disfunction, structural and functional changes occur. These changes, beginning at the onset of diabetes, involve the entire kidney and probably all contribute to the eventual development of clinical nephropathy. Since the estimated number of diabetics in the U.S. is 5 to 20 million, diabetic nephropathy constitutes a major health problem.[1] Diabetic nephropathy occurs in 50% of children with type I, insulin-dependent diabetes mellitus (IDDM). This type of diabetes usually occurs before the age of 20 and is associated with beta cell destruction and insulin deficiency. These patients always require exogenous insulin. Onset of nephropathy occurs after an average duration of 20 years of diabetes,[2,3] and death usually occurs 5 to 6 years after the development of clinical proteinuria.[4] About 25% of the patients placed on hemodialysis are diabetic.[1] Thus, one fourth of the cost of dialysis therapy (over $1.5 billion in the U.S.) is devoted to diabetic patients. The prognosis for the diabetic patient placed on hemodialysis or receiving a kidney transplant is poor. There is only a 50% chance of survival after 3 years.[1] Renal failure is responsible for 50% of the deaths in patients with IDDM and 6% of the deaths in type II or noninsulin-dependent diabetics (NIDDM).[5]

Within the past several years the development of infusion pumps has made possible the continuous administration of insulin.[6,7] This method of treatment can provide a closer approach to normoglycemia in many patients than is possible with conventional treatment methods. However, this new technology requires considerable expense for equipment and supplies. It also requires a greater degree of participation and discipline on the part of the patient than older treatment methods. Whether or not there are advantages and benefits of these new treatment methods which justify the expense and effort remains to be determined. An important question, which can be addressed in animal models of diabetic renal pathology, is the effect of glycemic control on the development and progression of renal lesions. Some studies in humans[2,8-10] strongly indicate that there is a correlation between glycemic control and diabetic renal disease. However, other studies indicate that once nephropathy is established, improved glycemic control does not reverse it.[11,12] It may, however, reverse some of the early abnormalities, such as exercise-induced albuminuria[12] or slow the progression of the disease.[13]

Another question which might be addressed in animal models is the genetic component of nephropathy. Since both spontaneous and induced diabetic animal models are available, comparisons of the types of lesions and the rate of development can be made. In human type II or NIDDM diabetics there is a suggestion of a genetic component in the development of nephropathy. Those type II diabetics on chlorpropamide who develop a flush after drinking alcohol are less likely to develop nephropathy.[14]

RODENT MODELS OF DIABETES

A number of rodent models of diabetes have been developed. The main advantage of rodents as experimental models of disease is that they are less expensive to acquire and maintain than larger species. Diabetic rats develop kidney lesions relatively rapidly. There are also several disadvantages of rodent models. The renal lesions do not exactly parallel the renal lesions of human diabetics. This may be due to the short life in which there is not adequate time for the complete syndrome to develop. Also, because of the small size of the animals, repeated sampling on the same animals is not always possible. Also, since diabetes is induced this model provides a method of studying the consequences of the metabolic derangements of diabetes without confounding hereditary factors, which could influence the development of diabetic nephropathy.

Methods of Induction of Diabetes

The streptozotocin or alloxan diabetic rat is one of the best defined models of diabetic renal disease. The ease of diabetes induction and the rapidity with which renal lesions appear make it a convenient and cost effective model.

Diabetes is most frequently induced in rats with either alloxan or streptozotocin. Rats are commonly used between 80 and 250 g. Alloxan is administered intravenously in a caudal[15,16] or femoral vein[17] or the vena cava[18-20] in doses of 38 mg/kg[16,17] to 70 mg/kg[18] or intraperitoneally at 200 mg/kg.[18] Streptozotocin is administered intravenously in single doses of 40 to 90 mg/kg[21-29] or intraperitoneally at 40 mg/kg.[30-32]

Since alloxan is nephrotoxic, the kidneys may be protected during induction of diabetes with this drug by clamping the renal vessels during injection and for 5 min afterward[18,19,33,34] Alloxan does not cause glomerular changes,[18,33,34] but kidneys exposed to alloxan show more frequent tubular and interstitial damage than protected kidneys.[18,33] The ischemia caused by clamping a short time does not appear to have any detrimental effects on the kidney. Normal kidney morphology is maintained after 5 min of clamping in normal rats.[33,34] Rats in which the renal vessels have been clamped for 5 min after alloxan injection have normal laboratory values on the days following alloxan injection, and the kidneys appear normal in light microscopic sections.[18]

Rats can also be rendered diabetic by removal of 95% of the pancreas.[35,36] Onset of diabetes is slower than with chemical induction. Development of complete diabetes may require 3 months.[36] The procedure is only about 30% successful in female rats.[36] Castrated females, however, do develop diabetes, indicating that the ovaries may offer some protection. Criteria for successful induction of diabetes may be nonfasting blood glucose values between 300 and 400 mg/dℓ[15,37,38] or persistent glucosuria.[28,29,39]

Diabetic rats can usually be maintained on one to four units a day of subcutaneously injected intermediate-acting insulin such as lente or NPH[40,41] or long-acting insulin such as protamine zinc or ultralente insulin.[18,29] For short duration of insulin treatment (2 weeks), subcutaneously implanted osmotic minipumps, which deliver two to four units of insulin a day as a continuous dose, have been used.[26] To establish a severe diabetes, rats may be left untreated.[20]

Pathological and Physiological Changes of Kidneys

The structural changes in the kidney which occur early in the course of diabetes are very similar in the rat and the human.[42]

Kidney Size

Kidney size is increased in human diabetics.[2,43-45] It is often observed at the time of diagnosis of diabetes.[46-48] It has been demonstrated in both short-term [44,49,50] and long-term diabetic patients.[51]

In most studies on kidney size in diabetic rats, an increase has been found.[2,25,29,31,32,52-54] The increase begins within the first 2 to 4 days after induction of diabetes[47,32,52] and continues for at least 6 weeks.[52] After 4 weeks of diabetes the increase is 50% by weight[55] and by 6 weeks it has reached 90%.[52] Contrary to other studies, Weil et al.[56] did not find an increase in kidney size in rats after 9 months of diabetes. The difference in results from other studies may have been due to the greater duration of diabetes. The condition of the rats at the end of the experimental period and the level of glycemic control are not indicated. These factors as well as previous biopsies may have affected the final kidney weight.

Glomerular and Tubular Size

Various components of the nephron increase in size at different rates in diabetes.[32] During the first 7 days the increase in size is due to hypertrophy. This was determined by the protein/DNA ratio of the kidney homogenate.[31] Subsequent growth is a combination of hypertrophy plus hyperplasia. Quantitative stereological techniques are used to determine the various glomerular parameters from light and electron micrographs.[37,57] Glomerular volume[32,37,42] increases during the first few days of diabetes. The capillaries are widened and the tuft appears to be unfolded. Bowman's space becomes narrowed.[36] Quantitative studies of glomerular volume also indicate a significant age-related increase in glomerular volume in normal rats.[57] However, increases in glomerular volume are initially greater for diabetic rats. This accelerated rate of glomerular increase lasts from a few weeks[32] to 8 months.[57] The glomerular growth rate then decreases, and after 18 months the glomeruli of diabetic rats are the same size as those of age matched controls.[57] Gøtzsche et al.,[55] however, found that after 4 weeks of diabetes the glomerular volume of diabetic rats is 44% greater than controls.

Glomerular surface area also increases with age in control rats. In diabetics the glomerular surface area increases significantly at the onset of diabetes, but no subsequent increases are found.[37,47,57] Østerby and Gundersen[37] reported that tuft surface area increases 42% during the first 4 days of diabetes in rats, but no further increase is seen in the next 47 days. The surface area is related to changes in shape of the glomerular capillaries. After diabetes induction (4 days), the capillary length is increased, but the cross-sectional area remains normal. After 47 days, the length was found to be normal, but the cross-sectional area was increased.[37]

In human diabetics, glomerular volume[58,59] and capillary surface area are increased at the time of diagnosis.[42,45,53,59] The glomerular cells increase in size, but not in number.[59] After several years of diabetes, the glomerular volume is less than at diagnosis, but still greater than normal.[59] After many years of diabetes, there is obsolescence of some of the glomeruli. Hypertrophy of the remaining glomeruli may occur in a manner similar to the hypertrophy which occurs after unilateral nephrectomy. Thus, in long-term diabetics, glomerular volume is again increased.[60] Studies at autopsy show that the glomerular size is inversely related to the number of functional glomeruli. The increased glomerular size in long-term human diabetics also appears to be due to an increase in the number rather than the size of glomerular cells.[61]

In the diabetic rat, proximal tubule length and diameter also increase after onset of diabetes.[32,45,54] After 47 days, tubule length is 22% greater and diameter is 17% greater in diabetics than controls. The proximal tubules continue to grow for a longer time than the glomeruli.[32]

Effect of Glycemic Control on Kidney Size

The increase in kidney and glomerular size in the rat has been shown to be related to glycemic control.[19] The increase in kidney size can be minimized by initiation of insulin treatment within the first 8 days of diabetes.[52,53] If insulin treatment is initiated at the onset of glycosuria, cellular hypertrophy but not hyperplasia, as determined by kidney protein/DNA and DNA content, respectively, can be prevented.[52] All rats in a given group received the same insulin dose. Most but not all of the rats were free of glucosuria. If insulin treatment is delayed for 25 days, only a moderated reduction in kidney size occurs. This is due to a reduction in cell size, but not cell number.[52] Diabetic rats maintained in good glycemic control for 6 months with insulin have normal kidney weights and glomerular volumes, whereas rats maintained for the same period in poor glycemic control have increased kidney and glomerular sizes.[29,53]

In human diabetics, 8 days of insulin treatment and near normal blood glucose levels did not result in a reduction in kidney size in newly diagnosed diabetics.[48] However, after 3 months of insulin treatment,[43,46] kidney size was reduced to near normal levels, indicating that long-term control is a factor in kidney size.

Glomerular Basement Membrane

The glomerular basement membrane (GBM) is the central layer in the glomerular capillary wall between the epithelial and endothelial cells. The glomerular capillary forms the filtration barrier between the plasma and filtrate. Alterations in the GBM may affect the selective filtration function of the capillary.[62] The GBM is composed of glycoprotein and collagen. The carbohydrate component of the GBM stains distinctively with PAS (periodic Schiff stain). Semi-quantitative histological studies of the increase in GBM are done with PAS-stained light microscopic sections. Quantitative measurements are made with electron micrographs at 3200 to 23,000 magnification[20,57] using planimetry.[20]

Kidney tissue for most morphological studies is obtained fresh. When longitudinal studies are being conducted, biopsy samples can be obtained.[28,57,63] In terminal studies, rats are killed by decapitation, and the kidneys are quickly removed[15,30] or the rats are anesthetized with pentobarbital[52] or with diazepam and pentobarbital[29] before kidneys are removed.

The microscopic glomerular lesions seen in the diabetic rat resemble the lesions of early human diabetes.[2,56,64] Diffuse glomerular sclerosis, which can be easily recognized as an increase in glomerular PAS positive staining, consists of an increase in glomerular basement material and mesangial matrix material.

GBM thickening, similar to that which occurs early in human diabetes, develops very quickly in the diabetic rat. Quantitative measurements made on electron micrographs after 4 and 47 days of diabetes indicate rapid accumulation of GBM material.[37] The volume of GBM material increases within the first 4 days. There is no further accumulation in volume by 47 days due to the opposite changes in surface area and thickness.

GBM also thickens with age, but diabetic rats at any age have thicker GBM than age matched controls. In normal rats, males have a greater thickening than females, but diabetes removes this distinction.[19] Orskov et al.[18] found mild lobular fibrosis of the glomeruli in 26-month-old control rats. These lesions could be easily distinguished from diabetic glomerular lesions and were never found in 15-month-old control rats, an age at which diabetic rats showed severe glomerular lesions.

Quantitative studies of the rate of thickening in GBM in control and diabetic rats show a significant linear increase in GBM thickness with age and also a significant linear increase due to duration of diabetes.[57] The slope of the regression line is greater in diabetic than in normal rats. Extrapolation indicates that thickening began within days of diabetes induction.[57] Over the 18-month period of the study, the GBM thickening in diabetic rats resembled the increase in GBM thickening seen in the early and intermediate stages of human diabetic glomerulopathy.[57]

The initial rate of GBM thickening in rats is four times as great as the initial rate of GBM thickening seen in humans. However, rat GBM does not become as thick as that of human diabetics in the advanced stages of diabetic glomerulopathy.[57]

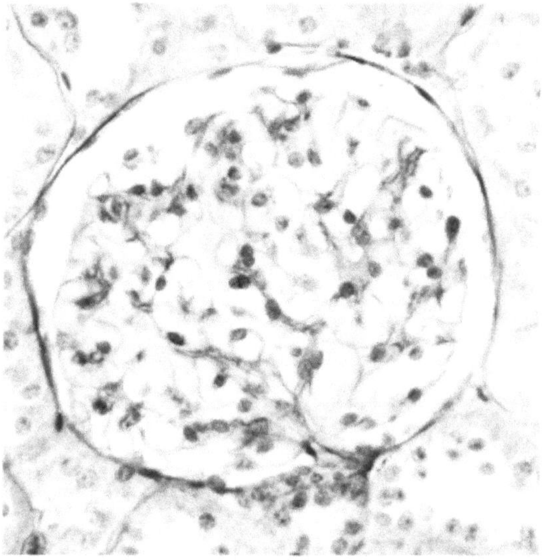

FIGURE 1. Light micrograph of a glomerulus of a control rat. (PAS stain.) (From Hägg, E., *Acta Pathol. Microbiol. Scand.*, 82, 211, 1974. With permission.)

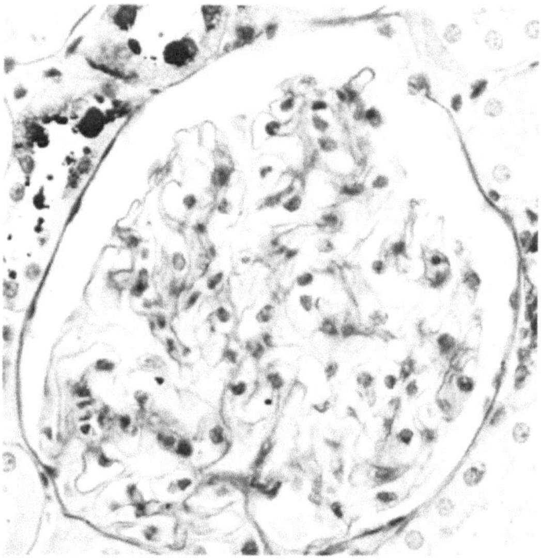

FIGURE 2. Light micrograph of a glomerulus from a diabetic rat, showing capillary basement membrane thickening. (PAS stain.) (From Hägg, E., *Acta Pathol. Microbiol. Scand.*, 82, 211, 1974. With permission.)

In the rat after 4 to 10 months of diabetes, the GBM becomes slightly thickened.[19,34,36,56,65] The increase in GBM material begins in the center of the tuft.[66] After 12 to 15 months the GBM has been found to be significantly thickened[18-20,33,67] (Figures 1 to 4). However, Anjo and Couturier[15] found no difference between the GBM of control rats and rats which had been diabetic for 12 months.

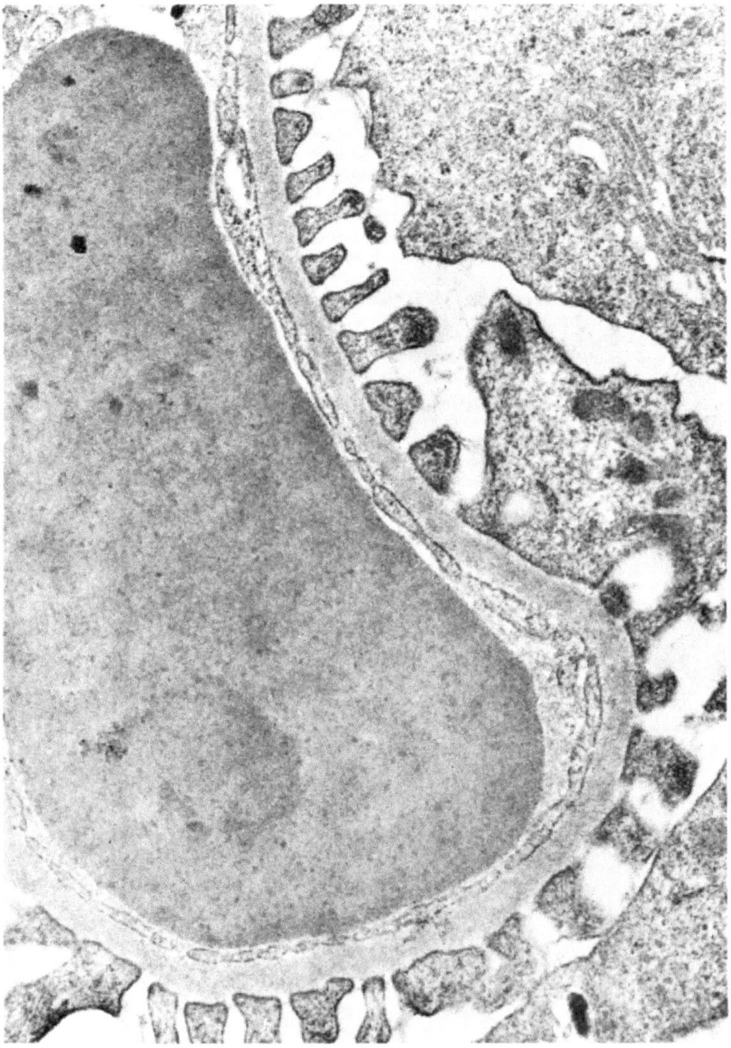

FIGURE 3. Electron micrograph of a capillary loop in a glomerulus of a control rat. (Magnification × 26,000.) (From Hägg, E., *Acta Pathol. Microbiol. Scand.*, 82, 211, 1974. With permission.)

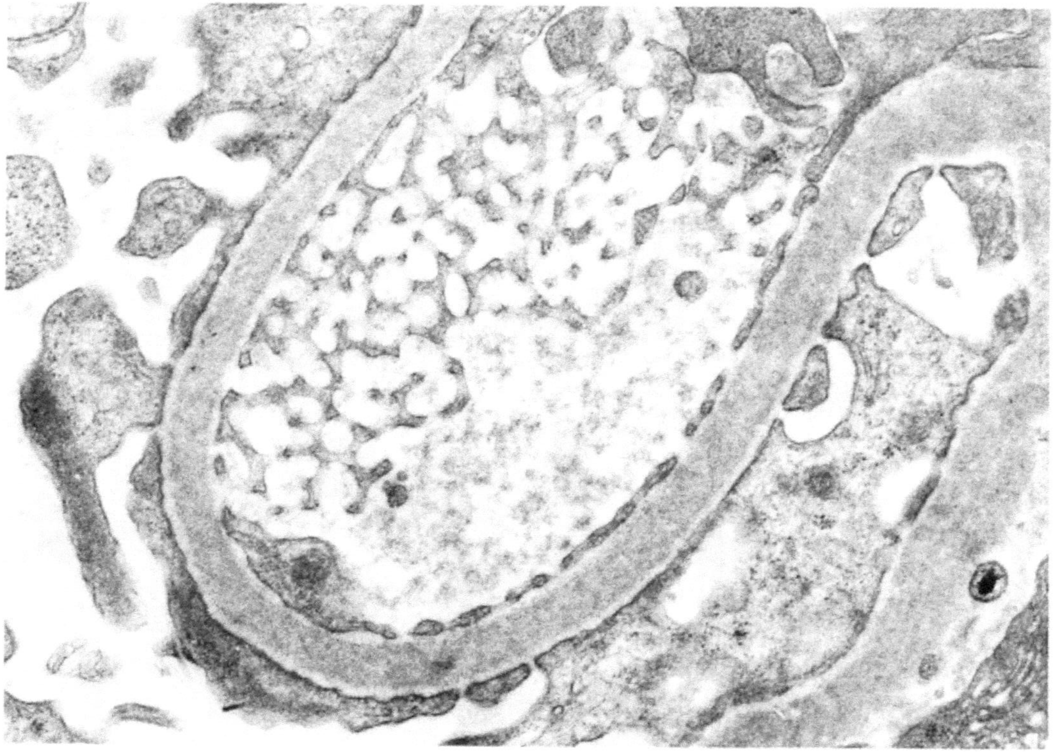

FIGURE 4. Electron micrograph of a glomerular capillary loop from a diabetic rat. Note thickened basement membrane and fused foot processes. (Magnification × 26,000.) (From Hägg, E., *Acta. Pathol. Microbiol. Scand.*, 82, 211, 1974. With permission.)

Hyperglycemia as well as duration of diabetes appears to be a significant factor in the GBM thickening. Rats maintained at different levels of hyperglycemia for 14 months show a significant positive correlation between GBM thickness and level of hyperglycemia.[40]

With light microscopy, deposits of increased density occurring both peripherally and axially have been observed.[18,33] These deposits occur most frequently on the endothelial side of the GBM.[18] With EM studies deposits of electron dense material on the endothelial side of the GBM have also been observed.[15] In one EM study localized thickenings of the GBM forming "spikes" into the epithelial cells have been observed.[33,34] These spikes often occur in pairs. In long-term diabetic rats (20 months), the electron dense deposits are observed in the areas between these spikes. These lesions are not uniformly distributed, occurring in some glomeruli, but not in others. Occasionally capillaries are completely occluded.[33] Labeling studies in normal rats indicate that the GBM is a product of the epithelial cells.[68] The PAS positive basement membrane (BM) material in diabetic rats is mostly homogeneous, but some areas of fibrillar structure and some vacuoles are observed.[18,33] The areas between the BM spikes contained thicker and more closely packed fibrils.[34] Occasionally well-defined caps similar to the fibrinoid caps of human glomerulosclerosis[67] were found on the epithelial side.[18] After 15 months of severe diabetes, lipid drops are occasionally observed in vacuoles of the glomerular capillaries.[67]

In human diabetics, GBM thickness is normal at diagnosis of IDDM.[69] However, the capillary surface area is increased in such diabetics,[58] and the total amount of GBM material is increased by 97%.[42] In the human, GBM thickening is detectable within 1.5 to 2.5 years of diagnosis of diabetes and is easily identified after 5 years.[70] After 10 to 30 years of diabetes, the GBM may reach 2 to 10 times normal thickness.[71] The fine structure of human GBM is usually normal in the early years of diabetes.[70] However, paired spike formation on the epithelial side of the GBM has also been observed in early human diabetes, and the area between the spikes contains whorled fibers.[70] The increase in GBM thickness in human diabetes appears to be the result of increased deposition of material on the epithelial side of the basement membrane.[71] Fibrin caps or crescents are exudative lesions which develop within the GBM [72] and project toward the urinary space.[33] They are homogeneous deposits, [33] but lipid deposits may appear as vacuoles.[72] Larger lesions stained with PAS are observed by light microscopy.[33] Smaller lesions can be seen in electron micrographs.[72]

Glomerular Epithelial and Endothelial Cells

Epithelial cells in alloxan diabetic rats show fusion of the foot processes in some studies,[33,34] but not in others.[15] Other abnormalities include the presence of large dense bodies of fibrillar or granular material, sometimes limited by a double membrane.[34] Some of the inclusions appear to be myelin in nature. Myelin inclusions are found in epithelial cells of both diabetic and normal rats, however, these are much larger in diabetic rats.[33,34]

In humans the foot processes of the epithelial cells remain normal in early diabetes, but become fused after the onset of proteinuria.[73] Early in the course of diabetes the cytoplasm, mitochondria, golgi apparatus,[73] and rough endoplasmic reticulum[70] increase in the epithelial cells. These changes are possibly related to an increased synthetic activity in GBM formation. A slight increase in the number of heterogeneous inclusion bodies has also been observed in human diabetics of 3.5 to 5 years duration.[70]. These are organelles limited by a double membrane and containing filamentous material as well as denser bodies.[70] It has been suggested that these may contain protein reabsorbed from the urinary space and therefore may be an indicator of the leakiness of the glomerular capillaries.

Bretzel et al.[74] found proliferation of endothelial cells in diabetic rats. Østerby-Hansen[69] found no endothelial cell abnormalities. Increases in endothelial cells have been reported in human diabetics, but it has been suggested that these may have been mesangial cells.[70]

Mesangial Changes

The mesangium consists of mesangial cells and the mesangial matrix material produced by these cells. Within the glomerulus the mesangium is bound by the capillary endothelial cell and the mesangial extension of the GBM.[75] The mesanguim extends through the hilum to the outside of the glomerulus.[2] The mesangial matrix material is similar to the GBM material[75] and in older literature is referred to as basement membrane-like material. The mesangium appears to perform a clearing function. Substances which enter the mesangium from the blood may be phagocytized by the mesangial cells or may be transported out of the glomerulus by the mesangium at the hilum.[75] It has also been suggested that the mesangium is involved in the maintenance and turnover of the GBM.[15,71]

Early mesangial changes in the rat resemble the changes in early human diabetes. In IDDM an increase in mesangial matrix material becomes detectable within 1.5 to 2.5 years and is easily identified by 5 years.[70] This histological abnormality is probably better correlated than any other with the functional abnormalities of diabetic nephropathy: proteinuria, hypertension, and decreased glomerular filtration rate (GFR).[76] As mesangial material increases, it invades the subendothelial space decreasing capillary filtration surface area.

In the rat there is a slight increase in mesangial matrix material in a predominantly diffuse pattern after 3 to 10 months of diabetes.[18,19,28,33,34,36,77,78] These deposits extend to the glomerular stalk and surround the intraglomerular afferent arteriole.[78] The accumulation of mesangial matrix material, together with the concurrent thickening of the GBM, resemble the early diffuse lesions of human diabetic glomerulopathy.[2,70]

The mesangial fine structure in the diabetic rat shows some abnormal electron dense fibrillar material.[33] Occasional irregular protrusions of mesangial material[56] and crescent-shaped or nodular hyaline deposits have been observed.[28,78] These hyaline deposits are found in the mesangium around the capillary loops or along the parietal layer of Bowman's capsule.[28]

With increasing duration of diabetes in the rat, the mesangial lesions become progressively more severe. After 10 to 12 months of diabetes, there is increased deposition of mesangial matrix material.[78] After 9 months, mesangial cell proliferation was observed by Weil et al.[56] Other investigators,[74,78] however, did not find any increase in the number of mesangial cells. In human diabetic glomerulosclerosis, an increase in the number of mesangial cells has been observed.[79] Anjo and Couturier[15] observed fibrillar and granular deposits surrounded by a smooth membrane within the cytoplasm of the mesangial cells of the diabetic rat. These inclusions resembled the BM of the capillaries. The authors suggest that these lesions imply a reduced ability of the mesangial cells to catabolize basement membrane material and that the resultant accumulation further impairs mesangial cell function.

Mesangial matrix material extending into the hilum causes a narrowing of the lumen of the intraglomerular arterioles.[36,78] Hyaline deposits in the mesangium partially surround capillary loops and are found along the parietal portion of the capsule and at the hilum. Weil et al.[56] noted prominent electron dense deposits and occasional focal segmental deposits of fibrinogen. After 12 to 15 months of diabetes, Hägg[19] observed focal deposits resembling fibrinoid caps in human diabetes.

In advanced stages (15 to 16 months of diabetes), there is extensive mesangial proliferation.[19,36,78] However, like GBM changes, the increases in mesangial matrix material

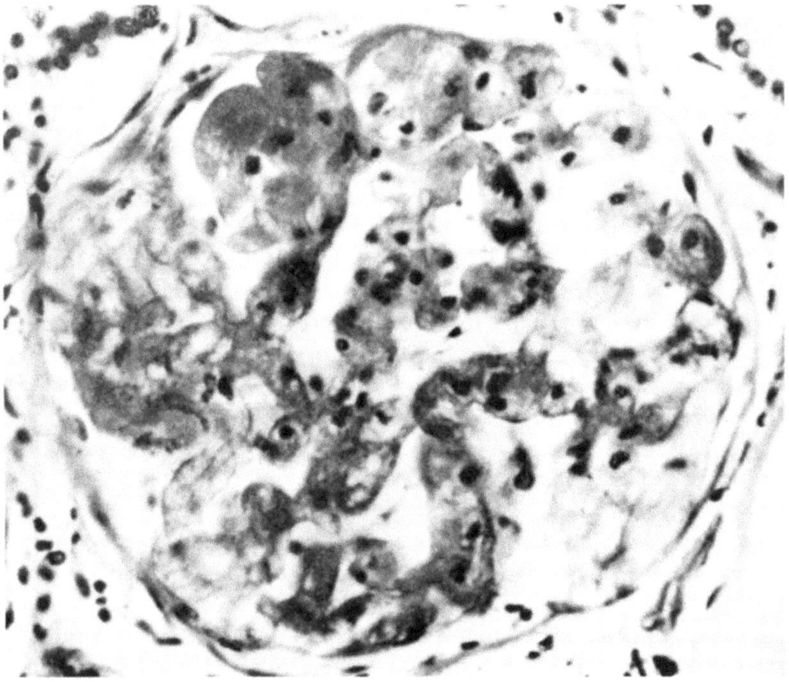

FIGURE 5. Glomerulus of a rat after 20 months of diabetes, showing PAS positive sausage-type lesions. (Magnification × 1000.) (From Olsen, S., *Int. Rev. Exp. Pathol.*, 7, 271, 1969. With permission.)

in the rat do not progress to the stage comparable with end-stage human glomerulopathy.[57] In the rat, sausage-like deposits are sometimes observed after 15 to 20 months of diabetes[18,33,67] (Figure 5). These lesions are not equivalent to the Kimmelsteil-Wilson lesion of advanced human diabetic glomerulosclerosis.[19,34,67] The sausage-like lesions in rats extend along the capillary lumen and an increase in size can result in the filling in of the capillary lumen.[18] These lesions are of two different types.[34] One type was highly cellular. The cells had increased amounts of endoplasmic reticulum and ribosomes. They also had large inclusions which were probably lipid in nature (Figure 6). The other type of lesion was predominantly mesangial matrix material.[33] In advanced human diabetes, the epithelial BM adjacent to the mesangium may develop a wrinkled appearance. Interruptions of this layer with mesangium protruding through the epithelium into the urinary space have also been observed.[71] Similar features have not been described in the diabetic rat.

The nodular, PAS positive (Kimmelsteil-Wilson) lesions are characteristic of advanced human diabetic glomerulosclerosis.[62,71] These lesions may be situated between the capillary loops and adherent to them [33] or may extend and adhere to the capsule.[72]

These lesions also have lipid inclusions.[72] It has been suggested that there are two types of nodular lesions.[72] The smaller nodules (about 40 μm) may result from extension of the process which led to diffuse glomerulosclerosis. These nodules consist of mesangial matrix material and mesangial cells. The larger nodules may be an organization of capillary microaneurysms. Microaneurysms are an additional consequence of advanced diabetic nephropathy.[72,80] They occur in patients with elevated blood urea nitrogen levels and are usually found in the same patients with Kimmelsteil-Wilson nodules.[72,80] Both capillary and arteriolar microaneurysms have been described, the former being more common.[80] As glomerulosclerosis progresses, nodules encroaching

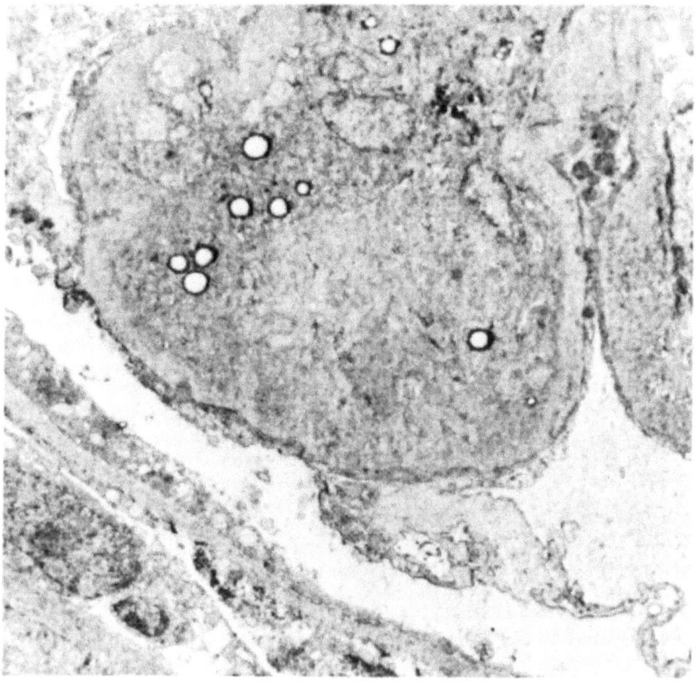

FIGURE 6. Electron micrograph of rat glomerulus after 20 months of diabetes, showing severe glomerulosclerosis. Note large accumulation of mesangial matrix material and mesangial cells containing lipid drops. (Magnification × 15,000.) (From Olsen, S., *Int. Rev. Exp. Pathol.*, 7, 271, 1969. With permission.)

on the capillary lumen cause many glomeruli to become obsolescent.[2,71,72] Microaneurysms, like Kimmelsteil-Wilson nodules, have not been described in the diabetic rat. These are both lesions of advanced diabetic nephropathy. Whether there is a difference in the pathological process in the rat kidney which leads to less severe renal complications or whether the rat simply does not live long enough to develop the complete picture of renal pathology is not known.

In rats mesangial matrix material also increases in controls with age, but increases are greater in diabetic rats.[19,57] Quantitative studies[57] showed a linear increase in mesangial volume of diabetic rats which paralleled the increase with age, but was positively displaced on the volume axis.

Functional impairment of the thickened areas of the mesangium in the diabetic rat has been demonstrated by the impaired ability to clear colloidal carbon.[2,38] The deposition of immunoglobulin and complement is a further indication of the impairment of the ability of the mesangium to clear the macromolecules.[78] IgG and C3 have been found in the mesangium after 2 to 4 months of diabetes.[56,74] After 4 to 6 months of diabetes, IgG and β1C [78] and to a lesser extent IgM[28] are present in the mesangium in a fibrillar and granular pattern radiating from the hilum to the periphery of the glomerulus.[78] Normal rats show either no deposits or trace amounts of deposits.[28] After 10 to 12 months of diabetes, these deposits become more diffusely distributed throughout the mesangium and have also been observed in the arteriolar walls and nodular masses. These immunofluorescent deposits correspond to the hyaline deposits seen with the light microscope. After 16 months of diabetes, IgG and β1C are also found in a faint linear pattern in the GBM. These changes, observed by immunohistochemical techniques, are similar to the changes observed in human diabetic glomeruli.[78] In humans, however, there is a more intense linear localization of IgG in the capillary walls.[81] In both the rat and human, these immunoglobulin proteins appear to be present in the kidney as a result of trapping from circulating blood rather than as a result of an immune stimulus within the kidney. Albumin, a globulin, and fibrinogen were demonstrated by immunofluorescent microscopy, in both diabetic and control rats.[78] Both had a diffuse pattern of distribution with no differences in intensity between diabetics and controls.

An additional similarity between diabetic glomerulopathy in the rat and the human is the accumulation of smooth muscle actomyosin in the mesangium.[82,83] In rats the actomyosin content is greater than controls after 4 months of diabetes and the difference is even more significant at 7 months.[83] In human diabetics who have received transplanted kidneys, increased mesangial actomyosin is observed 2 to 4 years after transplantation.[84] This increase in smooth muscle actomyosin may be the result of impaired mesangial clearance of macromolecules[82] or may result from altered hemodynamic parameters.[83]

Bowman's Capsule

In diabetic rats, Bowman's capsule may be moderately thickened[19,39] by glycoprotein deposits. In severe cases of diabetes, lipid deposits are also observed.[18] Adhesions between the visceral and parietal surfaces after 6 to 11 months of diabetes have been observed in some studies.[39,66] In other studies, no differences between diabetics and controls were observed after 12[15] or 20 months of diabetes.[34] In human glomerulosclerosis, there is thickening of the capsular basement membrane.[71] Adhesions are frequently found in areas of microaneurysms or Kimmelsteil-Wilson nodules.[72] Capsular drop lesions seen in long-term diabetic rats resemble the human exudative lesions.[19,74]

Arterioles

Several long-term studies indicate that diabetic rats do not develop the arteriolar hyalinosis seen in human diabetic nephropathy. In the human, the PAS positive thickening of the afferent and efferent arteriolar walls is usually observed after 5 to 10 years of diabetes.[72] This lesion has not been observed in the rat after 15 months of diabetes.[18,33,67] Mauer et al.[78] did observe occasional deposits in the arteriolar walls after 10 to 12 months of diabetes. Narrowing of the arteriolar lumen due to increased mesangial material around the intraglomerular portion of the arterioles was observed in the rat after 6 months of diabetes.[28] Medial hyperplasia and concentric fibrosis of the adventitia contributed to the thickening of the walls.[18]

Tubules

Tubular changes have been observed in diabetic rats and in humans. Within 6 days of diabetes induction in rats, there is an increase in DNA synthesis indicating a cellular proliferation.[85] After 3 months of diabetes, Weil et al.[56] observed nuclei of some of the proximal tubular cells extruded into the lumen. Tubular dilation has been observed after 4 to 6 months of diabetes in some studies[19,28] and after 10 to 15 months in another study.[33] Epithelial cells may appear flattened, and degenerated, desquamated epithelial cells may be observed in the tubular lumen.[18] After 6 months of diabetes, Lee et al.[28] observed cyst formation and focal areas of atrophy associated with some interstitial fibrosis in the rat. Bretzel et al.[74] observed protein cast within the tubules and lymphocytic infiltration of the interstitial space. Tubular dilation, atrophy, and interstitial fibrosis are more common in rats which have received alloxan without the protection of renal vessel clamping.[18] Vacuoles in the tubular epithelial cells may contain PAS positive cytoplasmic granules[28] or myelin figures.[56] Some thickening of the tubular basement membrane has been observed.[19,36,56] Tubular lesions are less frequent than glomerular lesions, but become progressively more frequent and severe with increased duration of diabetes. After 10 to 15 months of diabetes, flattening[33] or degeneration[18] of the tubular epithelium occurs. Lipid degeneration was observed occasionally after 10 months of diabetes and in all rats after 15 months of diabetes.[18] Glycogen deposits in the epithelium (Armani Ebstein lesions) have also been observed[15,18,74] after 10 to 12 months of diabetes. These lesions were more common in the distal and Henle's loop portion of the tubules.[18]

The degree of tubular changes appears to be correlated with the severity of the glomerular lesions.[18] The tubular lesions may be the result of the same pathogenic mechanism which causes the glomerular lesions or may be secondary to the glomerular lesions.[18]

The effect of aging on the development of tubular lesions has also been examined. Ørskov et al.[18] observed no tubular changes in control rats observed for up to 15 months. Hägg,[19] however, found tubular changes with age to be qualitatively similar in controls and diabetics, but lesions were more severe in old diabetic males.

Tubular and interstitial changes are also present in advanced human diabetic nephropathy. These are characterized by thickening of the tubular BM,[71] diffuse interstitial fibrosis, and tubular atrophy.[18]

Effect of Hyperinsulinemia on Renal Morphology

Since the treatment of diabetes with exogenous insulin often results in hyperinsulinemia, this factor has been considered as a possible contributing factor in diabetic renal pathology. However, Ørshov et al.[18] found that normal rats which received exogenous insulin did not show any of the morphological changes characteristic of diabetes.

Biochemical Alterations

One of the earliest detectable biochemical abnormalities in the streptozotocin diabetic rat is the rise in ornithine decarboxylase which is detectable within 24 hr of diabetes induction.[86] This enzyme also increases in other conditions which lead to kidney hypertrophy, such as unilateral nephrectomy. Immediate initiation of insulin therapy after the induction of diabetes prevents the rise in ornithine decarboxylase. However, initiation of insulin treatment after the rise has occurred does not restore control levels. After 1 week of diabetes, the levels fall below those of control. Whether or not the abnormalities in ornithine decarboxylase activity influence the development of renal pathology is unknown.

While histological studies indicate an increased amount of GBM material in diabetic rats, biochemical studies are contradictory about whether this is due to an increase in GBM synthesis. Some studies support an increased rate of synthesis while others do not. BM consists of a glycoprotein containing collagen (type IV) with a slightly different composition than connective tissue collagen. Like connective tissue collagen, it contains a high proportion of hydroxylysine and hydroxyproline, but it contains a greater number of half cystine residues and has a higher carbohydrate content.[62] The carbohydrate portion contains a polysaccharide and a disaccharide. The disaccharide unit is 2-*O*-a-D-glucosyl-D-galactose. It is formed by the glucosyltransferase catalyzed addition of glucose to the 2 position of galactose. The disaccharide is attached to 80% of the hydroxylysine residues via a β-glycosidic linkage. The polysaccharide is a branched chain structure containing sialic acid, fucose, galactose, mannose, and *N*-acetylglucosamine. This unit is attached to 6% of the protein asparagine residues.[62]

Biochemical studies on GBM have been performed on purified GBM,[62] isolated glomeruli,[35,87,88] or kidney homogenates.[89] Isolated glomeruli can be obtained by sieving and centrifugation techniques.[27,87,90] GBM can be isolated by ultrasonic disruption and differential centrifugation.[62]

The chemical composition of GBM in normal rats has been found to be the same as human GBM.[91] Studies of the biochemical changes in GBM protein in diabetes have concentrated on the analysis of the amino acids characteristic of the collagen component such as hydroxylysine and hydroxyproline.[92,93] The enzymes, lysine hydroxylase and proline hydroxylase, which catalyze their synthesis have also been studied. Changes of the glucose and galactose content of GBM as well as activities of the two glycosyltransferases necessary to form the disaccharide unit have also been used as indicators of changes in rate of GBM formation in diabetes. Glomeruli isolated from rat kidneys after 5 to 6 weeks of diabetes are larger than glomeruli from controls. Analysis of separated GBM showed that glomeruli from diabetic rats contained more BM material per gram of kidney cortex than controls, but no differences were found in the protein composition.[27]

Some studies have indicated that there is an early increase in synthesis of the collagen protein component. An increased incorporation of ^{3}H-proline into rat GBM has been observed after 9 days of alloxan diabetes,[16] and an increased incorporation of labeled lysine has been observed 7 weeks after 95% pancreatectomy.[35] An increase in the hydroxylation of proline[16] and lysine[35] has also been demonstrated by in vivo[16] and in vitro[35] use of tracers and chromatographic analysis of the hydrolyzed GBM protein. Another study in rats made diabetic with streptozotocin demonstrated an increase in hydroxylation of ^{14}C-proline and ^{14}C-lysine after 7 days of diabetes, but no increase in the total incorporation of label suggesting that in this study synthesis of total GBM protein was not significantly increased.[87] Another study by Beisswenger[91] failed to show either an increased incorporation of ^{3}H-lysine or conversion to hydroxylysine after 4 to 6 weeks of diabetes implying a normal rate of GBM synthesis.

Studies of hydroxylating enzyme activities have also given contradictory results. Grant et al.[87] found a threefold increase in the activities of both proline hydroxylase and lysine hydroxylase in 7-day diabetic rats. Cohen and Vogt[35] found an elevation in lysine hydroxylase but not proline hydroxylase after 4.5 weeks of streptozotocin-induced diabetes. Risteli et al.[89] found no changes in the activities of either proline or lysine hydroxylase after 4 weeks of diabetes. Significant increases were found in both enzymes at 12 weeks if the activities were expressed on a per kidney basis, but not if expressed on a per gram protein basis. Khalifa and Cohen[90] found an elevation in lysine hydroxylase in streptozotocin-induced diabetic rats. However, in rats which were 95% pancreatectomized and were therefore insulinopenic, but were not yet hyperglycemic, enzyme levels were normal. Thus, hyperglycemia and not lack of insulin was probably the cause of the elevation of lysine hydroxylase. Cohen and Khalifa[88] found that insulin treatment normalized lysine hydroxylase activity.

Glucosyltransferase was elevated in purified glomeruli isolated from streptozotocin diabetic rats[87] and in homogenates of kidney cortex from alloxan diabetic rats.[17] Risteli et al.[89] found levels of enzyme activity normal at 4 weeks in kidney homogenates from streptozotocin diabetic rats, but elevated at 12 weeks if activities were expressed on a per kidney basis. Levels of activity were normal if expressed on a per gram of protein basis. Duhault and Lonchampt[94] found no increase in glucosyltransferase in the renal cortex after 4 weeks of diabetes. Spiro and Spiro[17] found that the level increased with duration of diabetes. When insulin treatment was begun early, the enzyme level returned to normal. With diabetes of longer duration, insulin decreased the enzyme level, but did not return it to normal.[17] Islet transplantation also decreased the level of this enzyme.[74] Haft and Reddi[95] found an increase in glucosyltransferase per milligram of protein in streptozotocin diabetic rats in the cortical and medullary kidney fractions. The increase was greater in the medulla. No increase, however, was found in the isolated glomeruli.

Grant et al.[87] found galactosyl transferase elevated in renal cortical tissue. Risteli et al.[89] found this enzyme to be elevated after 12 weeks of diabetes if expressed on a per kidney basis, but normal if expressed on a per gram of protein basis. No elevation in enzyme level was found after 4 weeks of diabetes.[89]

Studies of GBM human diabetic kidneys have also yielded contradictory results. Beisswenger and Spiro[96-98] found increased quantities of BM material per kilogram of kidney cortex. Increases in glucosyl-galactosyl-hydroxy lysine were also found. There were lesser increases in hydroxyproline, hydroxylysine, and glycine. Significant decreases in lysine and lesser decreases in valine and tyrosine were also found. These differences in amino acid composition imply that diabetic GBM protein is more collagen-like.[12] Westburg and Michael[99] found a decrease in cystine to be the most significant difference between human diabetic and control GBM. He found no increase in hydroxylysine or glucosyl-galactosyl-hydroxylysine. In the carbohydrate analysis, a significant decrease in sialic acid was found. Klein et al.[60] found an increase in the amount of hydroxyproline per glomerulus in diabetics and inferred from the data that the amount of collagen was 40% greater in the diabetic than in the normal glomerulus. The age adjusted values of hydroxyproline per glomerulus were correlated with the duration of diabetes. Kefalides[100] also found no significant increase in hydroxylysine, hydroxyproline, or glycosyl-galactosyl-hydroxylysine.

Physiological Function in Diabetic Kidneys

Functional changes of kidneys occur early in the course of diabetes in the rat and are probably related to the physiological changes and also to the changes in kidney size. Glomerular filtration rate (GFR) is the rate at which fluid is filtered across the glomerular capillaries. It can be measured by the clearance of compounds such as inulin or iothalamate which are filtered, but neither secreted nor absorbed by the tubules. In humans whole kidney GFR is determined. In experimental animals it is possible to measure both whole kidney and single nephron GFR. In rats, for whole kidney measurements, timed urine collections are made via a ureteral catheter. Single nephron GFR is obtained by insertion of a fine pipette into the proximal tubule of a nephron.[25] Renal plasma flow (RPF) can be obtained from the clearance of compounds such as para-aminohippuric acid or hippuran, which are removed from the plasma by one pass through the kidney. Since filtration fraction (FF) = GFR/RPF, the RPF can also be obtained in experimental animals from the GFR and FF. The FF can be calculated by obtaining simultaneous concentration of inulin in the femoral artery (A), renal vein (V), and urine (U) by the formula $FF = [U]/([U]-[V])\times([A]-[V])/[A]$. This method has been used by Jensen et al.[25] to obtain FF and RPF in rats.

Whole kidney and single nephron GFRs are increased soon after diabetes induction.[2,25,54,101] Increased renal plasma flow,[2,25,54] increased transcapillary pressure (TCP),[2,54] as well as increased filtration surface area[34] are the probable contributors to the increased GFR. The fact that increased RPF and TCP are also found in partially pancreatectomized rats indicates that they are probably not side effects of the diabetes-inducing drugs.[2] Glomerular capillary filtration coefficient and filtration fraction were not found to be significantly different in diabetic and control rats.[25] Hostetter et al.[41,102] found that in moderately hyperglycemic rats both GFR and single nephron GFR were elevated, but in severely hyperglycemic rats, these values were lower than control. The decrease in GFR and single nephron GFR in severely hyperglycemic rats was due to a decrease in glomerular plasma flow. Also the increased TCP seen in moderately hyperglycemic rats was absent in severely hyperglycemic rats. The differences in GFR between moderately and severely hyperglycemic rats were not due to differences in blood volume, since relative blood volume in moderately and severely hyperglycemic rats was not significantly different from control rats. Mean arterial pressures have not been found to be significantly different in diabetic and control rats, but pressures in the efferent arteriole are greater in diabetics.[25]

The functional changes seen early in the course of diabetes in the rat parallel the early changes in human diabetes. In the human, there is also an increase in GFR[44,50] which has been correlated with the increase in kidney size[44,48,49] and glomerular surface area.[42] Insulin treatment has been shown to reverse this early increase in GFR,[46,48,103] and GFR is inversely related to glycemic control.[76] RPF was found in one study to be slightly but not significantly elevated in short-term diabetics,[44] and in other studies was found to be significantly elevated.[48-50] The RPF was decreased by insulin treatment.[48] In patients with diabetes of up to 10-year duration, RPF was found to be lower than in controls. RPF was also found to be correlated with kidney size.[44,49] When calculated on the basis of kidney weight, RPF per gram was found to be lower in recent onset diabetics than in controls.[50] Filtration fraction has been found to be significantly elevated in short-term diabetics[48,50] and was decreased by insulin treatment.[48]

A relationship between the altered hemodynamic parameters and the microscopic lesions of diabetic nephropathy has been suggested by several findings in human studies. Renal hypertension may be both a cause and a consequence of renal disfunction. Reduction of the number of functioning glomeruli or reduction of the filtration surface area in the functioning glomeruli by accumulation of mesangial material may contribute to the development of hypertension, which may in turn cause further nephron destruction. This becomes a vicious cycle, since increased nephron loss further increases hypertension.[76] An illustration of the contributory effect of hypertension to the development of diabetic nephropathy comes from the case of a diabetic patient with unilateral renal artery stenosis who had nodular glomerulosclerosis in the kidney without restricted blood flow, but only a slight increase in mesangial matrix material and no nodular lesions in the stenotic kidney.[104] Treatment of hypertension in diabetes reduces the rate of decline of kidney function.[2,105,106]

The contribution of hemodynamic alterations to structural lesions has been further investigated in the rat. Unilateral nephrectomy results in a rapid increase in RPF and capillary pressure.[64] In normal rats, unilateral nephrectomy results in an increase in kidney weight comparable to the increase seen after the induction of diabetes. However, the increase in glomerular volume is not as great.[32] In unilaterally nephrectomized diabetic rats, the increase in mesangial accumulation and IgG, C3,[22,64] and muscle actomyosin antigen[83] deposition are accelerated, thus supporting the passive accumulation of these substances. Similar changes were not observed in uninephrectomized control rats without diabetes. Hypertensive diabetic rats with a surgically created unilateral stenosis, which protects one kidney from hypertension, show a greater increase in mesangial matrix material and an increased deposition of IgG, IgM, and C3 in the unprotected kidney. Hypertensive, nondiabetic rats do not show these changes.[107] These results indicate that hypertension and renal loss can accelerate, but are not solely responsible for the characteristic lesion of diabetic glomerulopathy.

Proteinuria

Proteinuria develops quickly in rats. In 5 days after induction of diabetes, urinary protein losses are more than twice normal levels (13.9 mg/24 hr compared to 4.7 mg/24 hr for controls).[26] Weil et al.[56] found urinary protein loss significantly elevated in rats after 6 weeks of diabetes. Protein losses continue to increase with increased duration of diabetes.[26,56] Urinary protein losses also increase with age in control rats, but at 20 weeks protein losses in diabetic rats are 3 times greater than controls.[56] Urinary albumin losses also increase with age in rats,[19] but are greater for diabetics than controls.[19,108] Mauer et al.[78] also found an increase in urinary protein excretion with age in controls and diabetics, but found no significant difference between diabetic and control rats at any time during the 16-month study. Urinary albumin losses can be prevented for at least 6 months by treatment with insulin and maintenance of good glycemic control.[108,109] Initiation of insulin treatment after either 5 weeks or 3 months of diabetes can reduce existing proteinuria.[26] The relationship of proteinuria and insulin treatment suggests a strong influence of metabolic control on renal handling of proteins. Pennell et al.[26] found urinary protein losses short term in diabetic rats in the absence of increase GFR, implying that in diabetic rats, hemodynamic factors are not important contributing factors to proteinuria.

If alloxan is used to induce diabetes without protection of the kidney, an immediate but temporary rise in urinary albumin is seen.[19] Streptozotocin is also potentially toxic to renal tubules. Since lysozymuria may accompany tubular damage, urinary lysozyme has been used as an indicator of tubular damage.[26] In one study, tests 5 days after streptozotocin injection were negative,[26] indicating that early proteinuria is not a side effect of streptozotocin.

In human diabetics, urinary protein loss is used as a screening technique for the detection of diabetic nephropathy. Clinical proteinuria usually develops after about 15 to 20 years of diabetes. However, urinary excretion of albumin can be detected by the use of sensitive radioimmunoassays years before the development of clinical proteinuria.[12] Exercise increases this loss.[12] In the early stages, the albumin losses are due primarily to glomerular malfunction, since B2-microglobulin, an indicator of tubular function, is normal.[12] With the development of clinical proteinuria, protein losses of 10 to 30 g/day may occur. The GFR falls to below 40 mℓ/min/1.73 m^2, and the protein losses are due to both glomerular and tubular disfunction.[12]

Kidney Transplantation Studies in Diabetic Rats

Kidney transplantation between normal and diabetic rats has been used to elucidate the contribution of metabolic factors to the development of kidney lesions and has been useful in studying the time course of the development of renal lesions and the degree of reversibility of established lesions. The microsurgical techniques involved in kidney transplantation in the rat require the removal of the donor kidney and ureter with aortic, caval, and bladder cuffs. The kidney is transplanted into the recipient with aortic and caval end to side anastomoses, and the bladder cuff is sutured to the recipient's bladder.[28]

Kidneys transplanted from rats with diabetes of 6-month duration into normal rats show a decrease in mesangial matrix material in about 50% of the cases after 2 months in a normal recipient.[28,110] Interstitial fibrosis, round cell infiltration, and tubular atrophy are still present in some transplanted kidneys. IgG, IgM, and β1C either disappear completely or are greatly reduced. Kidneys from normal rats transplanted into diabetic rats quickly develop glomerulopathy.[2] After 2 months in a diabetic environment, 62% of transplanted kidneys had increased mesangial matrix material and hilar expansion with thickening of the arteriolar walls.[28] Interstitial fibrosis, tubular atrophy, tubular epithelial cell vacuolization, PAS positive cytoplasmic deposits, and tubular dilation were present in half of the kidneys. Deposition of IgG, IgM, and β1C occured in 75% of the transplanted kidneys. These lesions developed more rapidly in kidneys transplanted into diabetic rats than after the induction of diabetes.[28]

The development of these lesions in rats resembles the development of renal lesions in normal human kidneys transplanted into diabetic recipients. Within 2 years, hyaline arteriolar lesions develop[111] and deposition of IgG and albumin is seen.[112] By 4 years, mesangial thickening and nodular glomerulosclerosis have begun to appear.[2]

Islet Transplantation in Diabetic Rats

Islet transplantation in rats has also been used to study the degree of reversibility of established renal lesions and the contribution of such factors as hyperglycemia and altered hemodynamic parameters. Several methods of islet transplantation have been developed. When adult rats are used as donors, the islets can be isolated by collagenase digestion and ficoll separation.[113] Fetal or neonatal pancreases contain a high ratio of endocrine to acinar tissue and can be dissociated with collagenase and transplanted without prior islet separation.[63,114,115] Islet tissue can be implanted either into the peritoneal cavity[113,114] or the portal vein.[63,115-117] Transplantation of whole fetal pancreases under the kidney capsule is also possible.[118]

Islet transplantation restores normoglycemia, but often does not normalize insulin levels.[63,115] In 4 weeks after transplantation of islets into diabetic rats, kidney and glomerular size are decreased, but not to normal levels.[47] In 9 months after islet transplantation, glomerular volumes were not significantly different from controls.[23] Capillary surface area and GBM thickness did not change.[47] Intraportal transplantation of fetal pancreases after 4 weeks of diabetes also decreased kidney volume after 4 weeks of normoglycemia.[30,55] The increases of total protein DNA, RNA, and cell size caused by 4 weeks of diabetes were also reversed, but not normalized.[30] In these studies, transplantation did not tend to normalize glomerular volume, mesangial volume, filtration surface, or total capillary length.[55] Other studies,[74,110,119] however, have demonstrated a decrease in mesangial volume after islet transplantation. In the uninephrectomized rat, mesangial volume does not decrease after transplantation,[63,110] which supports the theory that hemodynamic parameters affect the development of microangiopathy. Bretzel et al.[74] found a proliferation of endothelial cells and narrowing of the capillary lumina with diabetes and normalization of these parameters with islet transplantation. If islet transplantation is performed after the GBM has become thickened by 7 to 13 months of diabetes, 2 to 6 months of normoglycemia does not decrease the GBM thickness.[110,115] Similar results were found after 4 weeks of normoglycemia following 4 weeks of diabetes.[30] The irreversibility of this lesion may be due to the slow rate of GBM turnover.[120] Neonatal pancreas transplantation does reverse urinary albumin excretion in rats.[110] This implies that the thickened GBM is not solely responsible for the increased capillary permeability to protein. Islet transplantation decreases the mesangial accumulation of IgG and C3.[74,110] In uninephrectomized diabetic rats, transplantation reduces IgG, but not C3.[63,110]

Islet transplantation also reverses some of the tubular lesions of diabetes. The tubular brush border enzyme, alanine amino peptidase, is elevated after 3 months of diabetes[119] and is reversed by 2 months of normoglycemia after islet transplantation. Islet transplantation early in the course of diabetes prevents the development of renal lesions,[2] indicating that if the metabolic abnormalities of diabetes can be prevented the renal lesions do not develop.

The BB Rat Model

The "BB" rat is a spontaneously diabetic rat first recognized in the Wistar colony of the Bio-Breeding Laboratories of Canada Ltd. in 1974. Diabetes in this model has many characteristics in common with human type I or IDDM.[121,122] Onset occurs in the young, nonobese rat between 60 and 120 days of age,[123,124] with varying degrees of severity[124] and with equal frequency in males and females.[123] The inheritance pattern of susceptibility is that of an autosomal recessive gene. However, only 50% of the susceptible animals develops diabetes.[123] The diabetic state is characterized by hyperglycemia, glucosuria, polyuria, polydypsia, hypoinsulinemia, and hyperketonemia.[121,123-126] Overt diabetes is usually preceded by abnormal glucose tolerance,[121,125] but some animals with abnormal glucose tolerance may remain normoglycemic in the fasting state and aglycosuric.[121,125] The severely diabetic animals are insulin dependent.[121]

A variety of hormonal and metabolic abnormalities have also been found in this strain. Glucagon[121,126] and somatostatin[121,127] are elevated in diabetic animals. Growth hormone secretion is initially normal, but decreases with metabolic decompensation.[121] Elevations in free fatty acid and branched chain amino acid levels are proportional to the severity of the diabetes.[124] There is increased hepatic gluconeogenesis and decreased glucose utilization as in human IDDM.[128]

The diabetes in this model appears to result from an autoimmune destruction of the islet B cells.[121-123,126,129-131] Factors such as hypersecretion of insulin and infectious agents have not yet been ruled out,[121] but seem unlikely.[122] Lymphocytic thyroiditis has also been found in this model and the incidence was significantly higher in diabetic animals, further implicating an autoimmune mechanism.[132] Alpha and D cells appear normal, but fewer in number.[122]

Extensive study of the secondary lesions of diabetes has not yet been done in this model. Neuropathy[121] and lesions of the testicles[121] and lymph nodes[121] have been identified.

Immunofluorescent studies of the renal glomeruli of normal, glucose-intolerant, and diabetic rats showed no differences in IgG and IgM within the glomeruli of normal and glucose-intolerant BB rats. In diabetic rats, IgM was similar in poorly controlled and well-controlled rats, but IgG deposition was greater in poorly controlled than in well-controlled rats. There were no differences between normal, glucose-intolerant, and diabetic rats in glomerular deposition of C3.[123,133]

The advantage of this model is its resemblance to human IDDM and the lack of need for surgery or administration of toxic drugs. A disadvantage is the susceptibility to infections and the requirement for a germ-free environment.[121,123]

Spontaneously Diabetic Mice

The C57/BLKsJ db/db mouse develops diabetes which resembles noninsulin dependent diabetes (type II) in humans. The disease is characterized by the hyperphagia, obesity, hyperinsulinemia, and down-regulation of insulin receptors.[134] Hyperglycemia develops at 4 to 8 weeks of age. Renal lesions include increased deposition of mesangial matrix material,[134] IgG, IgM, and C3.[135] Introduction of increased fiber into the diet of these mice at 5 to 6 weeks decreases the hyperglycemia and hyperinsulinemia. Renal lesions at 12 weeks are less than in conventionally fed diabetic mice.[134]

Like et al.[136] studied diabetic C57 BL/Ks-db/db at 12 to 22 months of age. In order to prolong the life of these mice beyond the normal 5- to 8-month duration, they were fed only 3 times per week between 7 and 11 weeks of age. At 5 to 6 months, glomeruli were larger and there was an increase in mesangial matrix material and mesangial cells. These lesions progressed with increasing duration of diabetes. At 18 to 20 months, thickening of the GBM and nodular subepithelial thickening were also observed.

Diabetes in the genetically diabetic KK mouse resembles type II diabetes.[137] In this model, both fasting glycemia and proteinuria increase with age. Diffuse glomerulosclerosis develops in some animals in as early as 2 to 4 months of age. By 19 to 22 months of age, 60% of the animals have developed this lesion. Nodular sclerosis also increases with age. The nodules seen in the glomeruli of these animals are not identical to the nodular lesion seen in human diabetics. At 2 months of age, there was an increase in the number of epithelial cells and an increase in glomerular diameter. Fibrin caps were frequently in 13- to 16-month-old animals. Occasional thickening of the capillary wall, glomerular aneurysms, and capsular adhesions were also found. Immunofluorescent studies demonstrated the presence of gamma globulin and complement in the mesangium and occasionally in the capillary wall. A correlation was found between blood glucose values and mesangial volume.

The Chinese Hamster

Certain inbred lines of Chinese hamsters (*Cricetulus griseus*) develop a spontaneous, insulin-deficient nonobese form of diabetes.[138-140] Several different strains have been developed. The severity of the diabetes varies from subline to subline. Some animals become glucosuric and ketonuric, while other animals develop glycosuria without ketonuria. Onset of hyperglycemia varies from strain to strain, but ranges from 2 to 10 weeks. Incidence of glucosuria ranges from 50 to 90%.[138] Diabetic females are more insulin deficient than males.[140] One subline shows a greater than normal glucagon response to arginine plus glucose.[140]

GBM thickness increases with age between 4 and 25 months in both controls and diabetics, but at any age is always greater in the diabetic.[141] Shirai et al.[142] found that the GBM is thickened in some areas and thin in others. Decreases in the number of both epithelial and endothelial cells have been observed.[143] Alpha- and beta- galactosidases are also reduced in most of the diabetic sublines, indicating an abnormal basement membrane metabolism,[141] and there is a correlation between the biochemical and morphological lesions. In vitro studies on kidney cells derived from diabetic sublines[144] have demonstrated abnormal glycosaminoglycan (GAG) production by the cells. Incorporation of ^{3}H into GAGs is also increased in diabetic cells and the ratios of the GAGs are also altered. Heparin sulfate is increased and chondroitin sulfate is decreased. The increase in heparin sulfate parallels the observation by Berenson et al.[145] in human kidneys. Kidneys from diabetic patients have a higher heparin sulfate content than normal control kidneys. The amount of heparin sulfate appears to be correlated with severity; kidneys from those patients having Kimmelstiel-Wilson nodules have the greatest amounts of heparin sulfate. Kidneys from patients with diffuse glomerulosclerosis show an increased hyaluronic acid content.

Increases in kidney size,[143] glomerular volume,[142] and capillary dilation have also been observed[138,142] in diabetic animals. Lawe,[143] however, did not find any increase in glomerular volume.

Mesangial matrix is increased,[138,142,146] and in the later stages, an increased number of mesangial cells have been seen.[142] In severe diabetes, mesangial cell changes including increased number of vesicles, multivesicular bodies, vacuoles, and mitochondrial disorganization are seen.[142] In one study, increases in mesangial matrix were also seen with aging, but not to the same extent as in diabetes.[142] In another study, no mesangial increases were observed in nondiabetics.[146]

Bowman's capsule increases slightly in thickness and occasional adhesions develop.[143] Fusion of foot processes of epithelial cells or nodules are not observed.[142] No arteriolar hyalinization or exudative lesions are observed.[143] Glycogen deposits are observed in the distal tubules of ketonuric animals.[146]

THE CANINE MODEL OF DIABETES

The dog is potentially a very useful model for the study of diabetic nephropathy. Its large size, long life span, tractable nature, and the ease of blood sampling make possible long-term studies with repeated sampling of blood and kidney tissue. Diabetic dogs have been maintained for periods of 5 to 8 years.[147,148] The diabetic dog is insulin dependent and can be maintained in either good or poor diabetic control as required by experimental design.

Methods of Induction of Diabetes

Diabetes can be induced in the dog by growth hormone, alloxan, streptozotocin, or pancreatectomy. Growth hormone given subcutaneously in doses of 2 to 16.5 mg/kg/day produces diabetes after 4 to 6 weeks of treatment.[147,149-151] Alloxan is given intravenously after an 18- to 24-hr fast in doses of 50 to 65 mg/kg.[148,152,153] Occlusion of the renal artery for 10 min following alloxan injection has also been used to protect against the nephrotoxic effects of alloxan.[153] If kidneys have not been protected during alloxan administration, death may occur during the first week from ARF.[152] Alloxan can also cause liver damage and the animal should be monitored for this.

The incidence of permanent diabetes following alloxan injection is 83 to 95%.[152,154] The most critical period is the first 48 hr after alloxan administration. Death can result from hypoglycemia during this period since large amounts of insulin are released from the destroyed β cells.[152] In order to avoid this, the dog should be fed 6 hr after alloxan treatment,[154] and supplemental feedings of dextrose twice a day should be continued for 2 to 3 days.[152] In spite of these precautions, it may be necessary to treat hypoglycemia with i.v. glucose. Usually 10 mℓ of 50% dextrose is sufficient for a 20-kg dog, but occasionally as much as 50 mℓ is necessary.[154] An alternate method of managing the dog in the postalloxanization period is by hyperalimentation.[155] A jugular catheter is implanted several days before the administration of alloxan, and 8 hr after alloxan administration hyperalimentation is begun. Since the fluids contain 25% glucose, hypoglycemia is usually not a problem. If the blood glucose does go below 70 mg/dℓ, 50% dextrose can be given via the catheter. Hyperalimentation during the postalloxan period has an additional advantage. During the 1st week to 10 days after alloxan treatment, many dogs are nauseated and anorexic. Many will refuse to eat, making stabilization difficult. With hyperalimentation, adequate nutrition is maintained during this period and the dogs appear to be less uncomfortable.

Intravenous streptozotocin can also be used to induce diabetes in the dog, but the required dose of 40 to 50 mg/kg is approximately the LD_{50}.[152,156] Diabetes can also be induced by a combination of alloxan and streptozotocin. The advantage of the combination is that a lower dose of each drug can be used, minimizing the toxic effects of each.[156,157]

Diabetes can also be induced by pancreatectomy. Total pancreatectomy produces an immediate diabetes, but since the exocrine function has also been eliminated, subsequent supplementation of diet with pancreatic enzymes is necessary.[152] Removal of 80 to 90% of the pancreas may cause a delayed diabetes, but without total loss of exocrine function.[152]

Diabetes occurs spontaneously in dogs with an incidence of 1 in 200 to 1 in 800. It often follows pancreatitis and occurs most frequently in older females with a history of obesity.[152,158] Two occurrences of a hereditary juvenile onset type of diabetes have been reported. Williams et al.[159,160] have reported diabetes in the golden retriever occurring at 2 to 5 months of age. Kramer et al.[152,161] have established a line of nonobese, keeshound dogs in which diabetes occurs between 2 and 6 months of age.

Pathologic Effects on the Kidney

Many of the morphological changes observed in the human diabetic kidney are also observed in the canine diabetic kidney. In the dog[149] as in the human,[44-50,162] there is an increase in the kidney size.

After 1 to 5 years of poorly controlled alloxan diabetes, the GBM is thickened.[147-149] Diabetic GBM thickness is two to four times greater than controls,[147] and there is both hypertrophy and hyperplasia of the endothelial cells.[147] Mesangial matrix material[147,148] and the number of mesangial cells are increased.[147] These structural changes are frequently associated with a decrease in the size of the capillary lumen.[147] An increase in mesangial matrix material has also been observed in the glomeruli of diabetic keeshound dogs[152] and other spontaneously diabetic dogs.[158] These changes resemble the increased GBM thickness[58,70,72,112,163] and increased mesangial matrix material[2,70,112] seen in human diabetes.

Glomerulosclerosis progresses to the formation of nodules both in alloxan diabetic dogs[147] and in spontaneously diabetic dogs.[158] Bloodworth,[72,147] who observed these lesions in alloxan diabetic dogs, felt they were the same as the Kimmelsteil-Wilson nodules found in human diabetes. Gepts and Toussaint,[158] who observed the lesions in spontaneously diabetic dogs, reported that the nodules were similar but not identical to the Kimmelsteil-Wilson nodules.

In the alloxan diabetic dogs, the glomeruli eventually fill up with mesangial matrix material, forming numerous large sclerotic glomeruli.[147] These resemble the large glomeruli of human obliterative glomerulosclerosis.[72] The small sclerotic glomeruli of ischemic obsolescence are also seen in the diabetic dog[147] as in the human.[72] Exudative lesions similar to those seen in human diabetic kidneys[72,73] are occasionally seen in the alloxan diabetic dogs.[147]

Thickening of Bowman's capsule is observed in long-term alloxan diabetic dogs.[147] Arteriolar sclerosis has been observed in a majority of alloxan diabetic dogs[147] as well as in some spontaneously diabetic dogs.[158] This lesion appears in the human diabetic after about 5 to 10 years of diabetes.[72]

Uninephrectomy also accelerates the course of diabetic glomerulosclerosis in dogs.[153] After a year of alloxan diabetes, uninephrectomized dogs have arteriolar hyalinosis and increased mesangial volume. In some dogs, there is early nodular glomerulosclerosis.

Tubular lesions have also been observed in the diabetic canine kidney. After 1 to 5 years of alloxan diabetes, thickening of the TBM is observed.[147] Armani Epstein lesions, which are glycogen deposits in the tubule epithelial cells, are not observed in these dogs. However, Black et al.[156] noted multifocal glycogen deposits in the renal tubular epithelial cells after 16 months of streptozotocin-alloxan diabetes and glycogen deposits have been observed in spontaneously diabetic dogs.[158] In the human diabetic, Armani Epstein lesions were observed before the discovery and use of insulin, but are rarely seen today.[73]

Very old nondiabetic dogs develop some of the same kidney lesions as diabetic dogs, but with lesser frequency. In 10- to 15-year-old dogs, GBM may be slightly thickened. Diffuse glomerulosclerosis with occasional nodules is seen. Some 10- to 15-year-old dogs have completely normal glomeruli, whereas no dogs with 5 years of diabetes are free of glomerulosclerosis.[147]

Some of the functional changes observed in human diabetic nephropathy have also been observed in the dog. GFR is increased in dogs made diabetic by pancreatectomy[53,164] as it is in human diabetics.[44,45,48-50,162,165] Proteinuria is a frequent occurrence in long-term diabetic dogs.[147,148] In humans proteinuria is seen in long-term diabetics.[53,166] In recent onset, diabetics urinary albumin excretion is often elevated.[45,153] In dogs the degree of proteinuria has been correlated with the degree of glomerulosclerosis.[147]

Effect of Glycemic Control on Renal Lesions

The degree of metabolic control has been shown to affect the rate of progression of glomerulosclerosis in the dog.[167] After 5 years of alloxan diabetes, dogs maintained in poor control have thickened GBM and increased mesangial matrix material. In dogs maintained in good control for the same length of time, these lesions are less severe or absent. The dog is also a good model to study the reversibility of renal lesions by improved glycemic control. A continuous insulin infusion pump can be carried in a pack attached to a harness (Figure 7). After development of glomerulopathy, improve-

FIGURE 7. The dog can be fitted with a harness and pack to carry an insulin infusion pump.

ment of glycemic control will improve some of the glomerular lesions, but not others. A total of 5 months of good glycemic control maintained by means of continuous s.c. insulin infusion does not decrease the amount of mesangial matrix material or the GBM thickness. It does, however, decrease the amount of fusion of the foot processes and the subendothelial deposits (Figures 8 and 9).[148]

In the human, reduction of subclinical albuminuria has been achieved by rigorous glycemic control.[45,166] In the dog, reduction of clinical proteinuria has been achieved by normalization of blood glucose using continuous i.p. insulin infusion.[148]

NONHUMAN PRIMATE MODELS OF DIABETES

The major advantage of using the nonhuman primate as a model of human diabetic nephropathy is the closeness of the relationship. As in the dog, the large size and longevity of the animals make possible prospective studies with repeated sampling. The major disadvantages of the model are the specialized facilities needed to house the animals, the expense involved in obtaining and maintaining the animals, and the long time period necessary for the development of renal lesions. However, this longer latency period may make this model closer to human renal pathology which also develops slowly over decades.

Diabetes can be induced in the monkey by pancreatectomy[168,169] or with alloxan [168,170] or streptozotocin.[168,169,171] Alloxan is administered intravenously at a dose of 100 mg/kg[170] and streptozotocin is administered at 30 to 55 mg/kg.[169,171] Strepto-

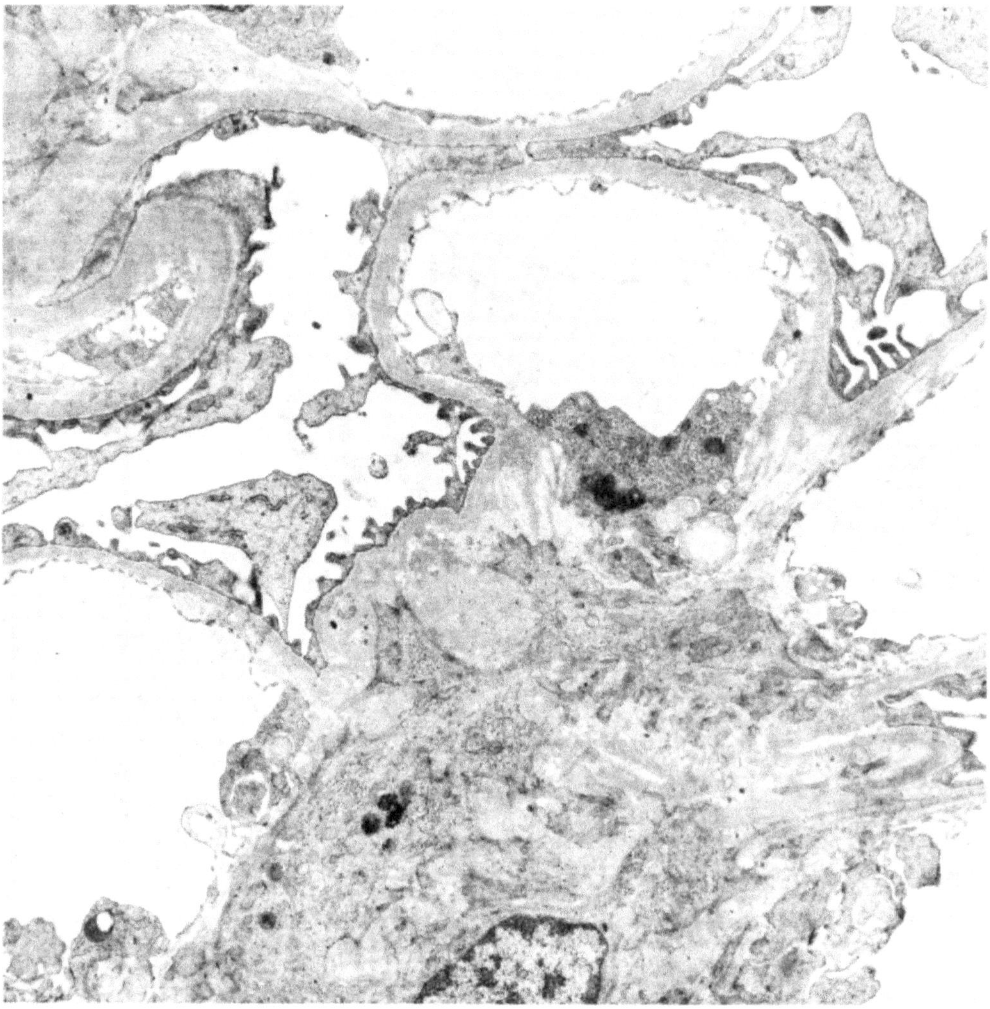

FIGURE 8. Electron micrograph from the kidney of a dog after 5 years of diabetes. Note basement membrane thickening, increased mesangial matrix material, fusion of foot processes, and subendothelial deposits. (Magnification × 6750.)

zotocin, which is now more commonly used, produces a model with diabetes varying in severity from carbohydrate intolerant to insulin dependent.[171] Fasting blood glucose values range from 240 to 493 mg/dℓ in pancreatectomized animals[169] and from 63 to 407 mg/dℓ in streptozotocin-treated animals.[169,171] In streptozotocin diabetes, glucagon and somatostatin are also elevated[171] as in human diabetes. Hemoglobin (Hb) $A1_c$ is elevated above control levels in some diabetic animals.[169] Unlike humans, nondiabetic monkeys do not have an $HbA1_c$ peak, and the usefulness and reliability of $HbA1_c$ as an indicator of severity of diabetes has not yet been demonstrated in this model. A number of spontaneously diabetic models have also been described. However, no renal lesions have yet been noted in these models.[168]

Thickened GBM has been observed in alloxan diabetic rhesus monkeys[172] and in an alloxan diabetic *Macaca mulatta* monkey after 41 months of diabetes.[170] One control *M mulatta* also showed some GBM thickening at 44 months of age.[170] Pancreatectomized baboons also develop thickened GBM.[173] Increased mesangial matrix material has

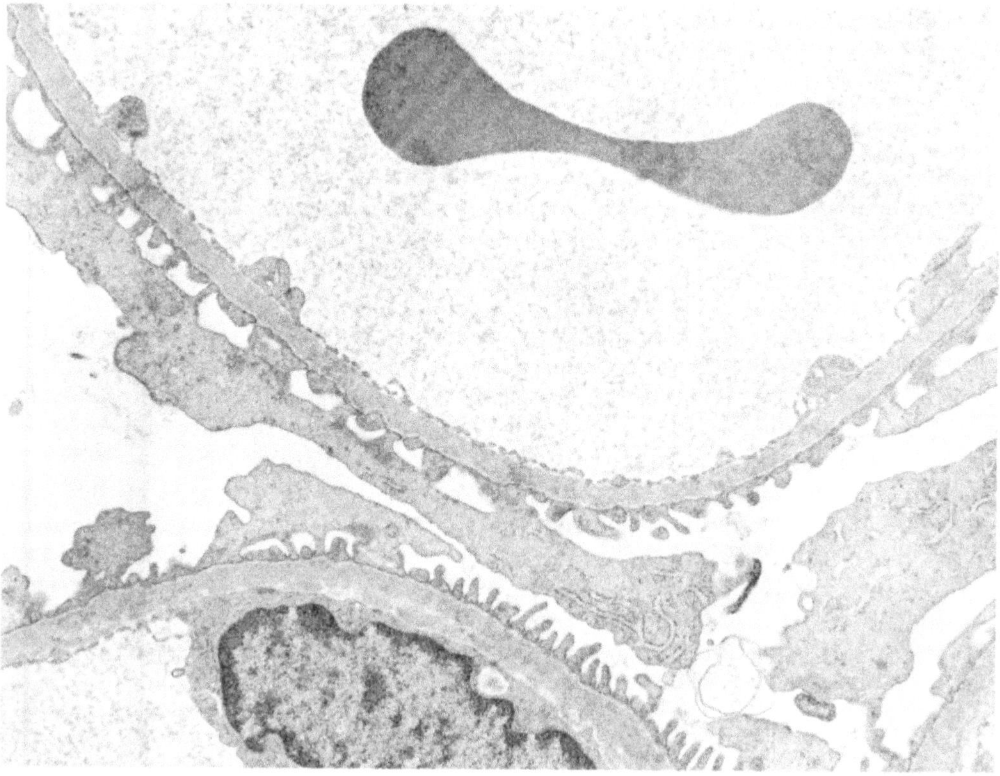

FIGURE 9. Electron micrograph from kidney of a dog with diabetes of 5-year duration after 5 months of continuous insulin infusion and improved glycemic control. Basement membrane remains thickened. Subendothelial deposits are decreased and there is less fusion of the foot processes. (Magnification × 7200.)

been observed in *M mulatta* monkeys maintained with a minimum amount of insulin after a duration of diabetes of 11 to 44 months.[170] In monkeys kept on a moderate insulin dosage, mesangial increase was noted in some at 15 to 73 months. In several other monkeys on moderate insulin doses, no glomerular changes were noted in biopsies taken between 19 and 31 months. Control monkeys showed no mesangial matrix accumulation up to 43 months of age.[170]

USES AND LIMITATIONS OF ANIMAL MODELS OF DIABETES

The rodent models of diabetes are the most easily studied. The rat with chemically or surgically induced diabetes is the most extensively studied model. Diabetes in these animals probably parallels the early stages of human diabetes fairly closely. The fact that the full picture of renal pathology is not attained may be due to the short life span of rodents. The main advantage of this model is that large numbers can be studied. Lesions develop rather quickly, and therefore answers to questions can be obtained in a reasonable amount of time. However, the danger is that since the pathology develops more rapidly than in humans, there may be basic differences. The fact that diabetes is induced rather than spontaneous has both advantages and disadvantages. The advantage is that the effect of physiological and metabolic abnormalities can be studied without the confounding factor of heredity. The disadvantages are (1) the influence of heredity on the development of lesions cannot be studied and (2) the possible pathology due to the method of diabetes induction.

A number of spontaneous rodent models have been developed, but further work needs to be done to characterize the diabetes in these animals and to determine how it is related to human diabetes. Only a small amount of work has been done on the effects of diabetes on the kidney in these models, and much more needs to be done.

The rodent models can be used to test some of the strategies of treatment of the renal complications of diabetes such as antihypertensive treatment and renal transplantation. However, the small size of these animals places a limit on the number of samples and the amount of data which can be collected.

The larger animal models of diabetic nephropathy, dogs and primates, have not been as well characterized as the rodent. This is due mainly to the greater expense of using these animals and to the longer time periods necessary for the lesion to develop. More needs to be done on both the spontaneous and induced models to determine similarities and differences to human diabetic nephropathy. These larger animals will probably be the best choice for future studies of strategies of treatment of diabetes to prevent or reverse renal complications. Because of their large size, repeated sampling of blood and kidneys is possible. The long-term consequences of the new intensive methods of diabetes treatment, such as use of insulin pumps, multiple insulin injections, and pancreas or B cell transplant, on the prevention or reversal of renal lesions have not yet been fully investigated. Since these treatments involve a great deal of expense and commitment on the part of the patient and in some cases danger of adverse consequences, the efficacy of these treatments in preventing the secondary complication of diabetes should first be established in animal models.

REFERENCES

1. Avram, M. M., Diabetic renal failure, *Nephron*, 31, 285, 1982.
2. Mauer, S. M., Steffes, M. W., and Brown, D. M., The kidney in diabetes, *Am. J. Med.*, 70, 603, 1981.
3. Marks, H. H., Longevity and mortality of diabetics, *Am. J. Public Health*, 55, 416, 1965.
4. Andersen, A. R., Andersen, J. K., Christiansen, J. S., and Deckert, T., Prognosis for juvenile diabetics with nephropathy and failing renal function, *Acta Med. Scand.*, 203, 131, 1978.
5. Diabetes Data, Publ. No. 78-1468, Department of Health, Education and Welfare, 1977.
6. Raskin, P., Treatment of insulin-dependent diabetes mellitus with portable insulin infusion devices, *Med. Clin. North Am.*, 66, 1269, 1982.
7. Sherwin, R. S., Tamborlane, W. V., Genel, M., and Felig, P., Treatment of juvenile-onset diabetes by subcutaneous infusion of insulin with a portable pump, *Diabetes Care*, 3, 301, 1980.
8. Takazakura, E., Nakamoto, Y., Hayakawa, H., Kawai, K., Muramoto, S., Yoshida, K., Shimizu, K., Shinoda, A., and Takeuchi, J., Onset and progression of diabetic glomerulosclerosis, *Diabetes*, 24, 1, 1975.
9. Ciavarella, A., Flammini, M., Stefoni, S., Borgnino, L. C., Forlani, G., Bacci, L., and Vannini, P., Kidney function after improved metabolic control in newly diagnosed diabetics and in diabetic patients with nephropathy, *Diabetes Care*, 5, 624, 1982.
10. Pirat, J., Glycemic control and development of diabetic nephropathy, *Acta Endocrinol.*, 97(Suppl. 242), 25, 1981.
11. Tamborlane, W. V., Puklin, J. E., Bergman, M., Verdonk, C., Rudolf, M. C., Felig, P., Genel, M., and Sherwin, R., Long-term improvement in metabolic control with the insulin pump does not reverse diabetic microangiopathy, *Diabetes Care*, 5(Suppl. 1), 58, 1982.
12. Viberti, G. C., Keen, H., Pickup, J. C., and Bilous, R. W., Proteinuria and diabetic control in insulin-dependent diabetes mellitus, *Acta Endocrinol.*, 97(Suppl. 242), 59, 1981.
13. Pitkänen, E. and Hulmi, S., Long-term subcutaneous insulin infusion therapy and the progression of nephropathy in type 1 (insulin-dependent) diabetes, *Diabetologia*, 22, 378, 1982.
14. Watkins, P. J., Predicting deterioration in renal function, *Acta Endocrinol.*, 97(Suppl. 242), 55, 1981.

15. Anjo, A. and Couturier, E., Mesangial changes of the renal glomerulus in long-term diabetic rats, *Pathol. Eur.*, 10, 21, 1975.
16. Brownlee, M. and Spiro, R. G., Glomerular basement membrane in the diabetic rat. In vivo studies, *Diabetes*, 28, 121, 1979.
17. Spiro, R. G. and Spiro, M. J., Effect of diabetes on the biosynthesis of the renal glomerular basement membrane. Studies on glycosyltransferase, *Diabetes*, 20, 641, 1971.
18. Ørskov, H., Olsen, T. S., Nielsen, K., Rafaelsen, O. J., and Lundbaek, K., Kidney lesions in rats with severe long-term alloxan diabetes. Influence of age, alloxan damage, and insulin administration, *Diabetologia*, 1, 172, 1965.
19. Hägg, E., Renal lesion in rats with long-term alloxan diabetes, *Acta Pathol. Microbiol. Scand.*, 82, 199, 1974.
20. Hägg, E., Glomerular basement membrane thickening in rats with long-term alloxan diabetes, *Acta Pathol. Microbiol. Scand.*, 82, 211, 1974.
21. Bretzel, R. G., Renal collagen glucosyltransferase activity following islet transplantation in streptozotocin-diabetic rats, *Diabetologia*, 21, 428, 1981.
22. Steffes, M. W., Brown, D. M., and Mauer, S. M., Diabetic glomerulopathy following unilateral nephrectomy in the rat, *Diabetes*, 27, 35, 1978.
23. Steffes, M. W., Brown, D. M., Basgen, J. M., and Mauer, S. M., Amelioration of mesangial volume and surface alterations following islet transplantation in diabetic rats, *Diabetes*, 29, 509, 1980.
24. Mauer, S. M., Sutherland, D. E. R., Steffes, M. W., Leonard, R. J., Najarian, J. S., Michael, A. F., and Brown, D. M., Pancreatic islet transplantation. Effect on the glomerular lesions of experimental diabetes in the rat, *Diabetes*, 23, 748, 1974.
25. Jensen, P. K., Christiansen, J. S., Steven, K., and Parving, H. -H., Renal function in streptozotocin diabetic rats, *Diabetologia*, 21, 409, 1981.
26. Pennell, J. P., Millard, M. M., and Ashby, M. H., Proteinuria in the acutely diabetic rat and its response to insulin treatment, *Diabetologia*, 21, 54, 1981.
27. Cohen, M. P. and Klein, C. V., Glomerulopathy in rats with streptozotocin diabetes, *J. Exp. Med.* 149, 623, 1979.
28. Lee, C. S., Mauer, S. M., Brown, D. M., Suterland, D. E. R., and Najarian, J. S., Renal transplantation in diabetes mellitus in rats, *J. Exp. Med.*, 139, 793, 1974.
29. Rasch, R., Prevention of diabetic glomerulopathy in streptozotocin diabetic rats by insulin treatment, *Diabetologia*, 16, 125, 1979.
30. Gøtzsche, O., Gundersen, H. J. G., and Østerby, R., Irreversibility of glomerular basement membrane accumulation despite reversibility of renal hypertrophy with islet transplantation in early experimental diabetes, *Diabetes*, 30, 481, 1981.
31. Seyer-Hansen, K., Renal hypertrophy in experimental diabetes, *Diabetologia*, 13, 141, 1977.
32. Seyer-Hansen, K., Hansen, J., and Gundersen, H. J. G., Renal hypertrophy in experimental diabetes: a morphometric study, *Diabetologia*, 18, 501, 1980.
33. Olsen, T. S., Diabetic glomerulosclerosis: a comparison between human and experimental lesion, *Int. Rev. Exp. Pathol.*, 7, 271, 1969.
34. Østerby-Hansen, R., Lundbaek, K., Olsen, T. S., and Ørskov, H., Kidney lesions in rats with severe long-term alloxan diabetes. III. Glomerular ultrastructure, *Lab. Invest.*, 17, 675, 1967.
35. Cohen, M. P. and Vogt, C., Evidence for enhanced basement membrane synthesis and lysine hydroxylation in renal glomerulus in experimental diabetes, *Biochem. Biophys. Res. Commun.*, 49, 1542, 1972.
36. Foglia, V. G., Mancini, R. E., and Cardeza, A. F., Glomerular lesions in the diabetic rat, *Arch. Pathol.*, 50, 75, 1950.
37. Østerby, R. and Gundersen, H. J. G., Fast accumulation of basement membrane material and the rate of morphological changes in cute experimental diabetic glomerular hypertrophy, *Diabetologia*, 18, 493, 1980.
38. Mauer, S. M., Steffes, M. W., Chern, M., and Brown, D. M., Mesangial uptake and processing of macromolecules in rats with diabetes mellitus, *Lab. Invest.*, 41, 401, 1979.
39. Beveridge, J. M. R. and Johnson, S. E., Studies on diabetic rats: the production of cardiovascular and renal disease in diabetic rats, *Br. J. Exp. Pathol.*, 31, 285, 1950.
40. Fox, C. J., Darby, S. C., Ireland, J. T., and Sönksen, P. H., Blood glucose control and glomerular capillary basement membrane thickening in experimental diabetes, *Br. Med. J.*, 2, 605, 1977.
41. Hostetter, T. H., Troy, J. L., and Brenner, B. M., Glomerular hemodynamics in experimental diabetes mellitus, *Kidney Int.*, 19, 410, 1981.
42. Gundersen, H. J. G., Gøtzsche, O., Hirose, K., Kroustrup, J. P., Mogensen, C. E., Seyer-Hansen, K., and Østerby, R., Early structural changes in glomerular capillaries and their relationship to long-term diabetic nephropathy, *Acta Endocrinol.*, 97(Suppl. 242), 19, 1981.
43. Mogensen, C. E., Østerby, R., and Gundersen, G., Early functional and morphological vascular renal consequences of the diabetic state, *Diabetologia*, 17, 71, 1979.

44. Mogensen, C. E. and Andersen, M. J. F., Increased kidney size and glomerular filtration rate in early juvenile diabetes, *Diabetes*, 22, 706, 1973.
45. Mogensen, C. E., Steffes, M. W., Deckert, T., and Christiansen, J. S., Functional and morphological renal manifestations in diabetes mellitus, *Diabetologia*, 21, 89, 1981.
46. Mogensen, C. E. and Andersen, M. J. F., Increased kidney size and glomerular filtration rate in untreated juvenile diabetes: normalization by insulin treatment, *Diabetologia*, 11, 221, 1975.
47. Østerby, R., Götzsche, O., Gundersen, H. J. G., Hirose, K., Lundbaek, K., and Seyer-Hansen, K., Progression of glomerular structural kidney alterations in diabetic rat and man, *Acta Endcrinol.*, 97(Suppl. 242), 63, 1981.
48. Christiansen, J. S., Gammelgaard, J., Tronier, B., Svendsen, P. A., and Parving, H. -H., Kidney function and size in diabetics before and during initial insulin treatment, *Kidney Int.*, 21, 683, 1982.
49. Christiansen, J. S., Gammelgaard, J., Frandsen, M., and Parving, H. -H., Increased kidney size, glomerular filtration rate and renal plasma flow in short-term insulin-dependent diabetics, *Diabetologia*, 20, 451, 1981.
50. Puig, J. G., Antón, F. M., Grande, C., Pallardo, L. F., Arnalich, F., Gill, A., Vázquez, J. J., and Garcia, A. M., Relation of kidney size to kidney function in early insulin-dependent diabetes, *Diabetologia*, 21, 363, 1981.
51. Kahn, C. B., Raman, P. G., and Zic, Z., Kidney size in diabetes mellitus, *Diabetes*, 23, 788, 1974.
52. Seyer-Hansen, K., Renal hypertrophy in streptozotocin diabetic rats, *Clin. Sci. Mol. Med.*, 51, 551, 1976.
53. Viberti, G. C., Early functional and morphological changes in diabetic nephropathy, *Clin. Nephrol.*, 12, 47, 1979.
54. Jensen, P. K., Christiansen, J. S., Steven, K., and Parving, H. -H., Renal function in diabetic rats, *Acta Endocrinol.*, 97(Suppl. 242), 25, 1981.
55. Gøtzsche, O., Gundersen, H. J. G., and Østerby, R., Reversibility and non-reversibility of renal morphology changes in early experimental diabetes after pancreatic transplantation, *Acta Endocrinol.*, 91(Suppl. 227), 30, 1979.
56. Weil, R., Nozawa, M., Koss, M., Weber, C., Reemtsma, K., and McIntosh, R., The kidney in streptozotocin diabetic rats, *Arch. Pathol. Lab. Med.*, 100, 37, 1976.
57. Hirose, K., Østerby, R., Nozawa, M., and Gundersen, J. G., Development of glomerular lesions in experimental diabetes in the rat, *Kidney Int.*, 21, 689, 1982.
58. Østerby, R. and Gundersen, H. J. G., The glomerular basement membrane in early diabetes, *Front. Matrix Biol.*, 7, 72, 1979.
59. Østerby, R. and Gundersen, H. J. G., Glomerular size and structure in diabetes mellitus. I. Early abnormalities, *Diabetologia*, 11, 225, 1975.
60. Klein, L., Butcher, D. L., Sudilowsky, O., Kikkawa, R., and Miller, M., Quantification of collagen in renal glomeruli isolated from human nondiabetic and diabetic kidneys, *Diabetes*, 24, 1075, 1975.
61. Gundersen, H. J. G. and Østerby, R., Glomerular size and structure in diabetes mellitus. II. Late abnormalities, *Diabetologia*, 13, 43, 1977.
62. Spiro, R. G., Search for a biochemical basis of diabetic microangiopathy, *Diabetologia*, 12, 1, 1976.
63. Steffes, M. W., Vernier, R. L., Brown, D. M., Basgen, J. M., and Mauer, S. M., Diabetic glomerulopathy in uninephrectomized rats resists amelioration following islet transplant, *Diabetologia*, 23, 347, 1982.
64. Mauer, S. M., Steffes, M. W., and Brown, D. M., Animal models of diabetic nephropathy, *Adv. Nephrol.*, 8, 23, 79.
65. Rasch, R., Prevention of diabetic glomerulopathy in streptozotocin diabetic rats by insulin treatment: glomerular basement membrane thickness, *Diabetologia*, 16, 319, 1979.
66. Greenberg, S. R., Glomerular changes in chronic alloxan diabetes, *Arch. Pathol.*, 73, 263, 62.
67. Olsen, T. S., Ørskov, H., and Lundbaek, K., Kidney lesions in rats with severe long-term alloxan diabetes, *Acta Pathol. Microbiol. Scand.*, 66, 1, 1966.
68. Kurtz, S. M. and Feldman, J. D., Experimental studies on the formation of the glomerular basement membrane, *J. Ultrastruct. Res.*, 6, 19, 1962.
69. Østerby-Hansen, R., A quantitative estimate of the peripheral glomerular basement membrane in recent juvenile diabetics, *Diabetologia*, 1, 97, 1965.
70. Østerby, R., Early phases in the development of diabetic glomerulopathy, *Acta Med. Scand.*, 574 (Suppl. 1), 1, 1975.
71. Huang, T. W., The nature of basal lamina alterations in human diabetic glomerulosclerosis, *Am. J. Pathol.*, 100, 225, 1980.
72. Bloodworth, J. M. B., A reevaluation of diabetic glomerulosclerosis 50 years after the discovery of insulin, *Hum. Pathol.*, 9, 439, 1978.
73. Arieff, A. I. and Myers, B. D., Diabetic nephropathy, in *Kidney*, Brenner, B. M. and Rector, F. C., Eds., W. B. Saunders, Philadelphia, 1976, chap. 36.

74. Bretzel, R. G., Breidenbach, C., Hofmann, J., and Federlin, K., Islet transplantation in experimental diabetes of the rat. IV. Rate of regression of diabetic kidney lesions after isogeneic islet transplantation: quantitative measurement, *Horm. Metab. Res.*, 11, 200, 1979.
75. Michael, A. F., Keane, W. F., Raij, L., Vernier, R. L., and Mauer, S. M., The glomerular mesangium, *Kidney Int.*, 17, 141, 1980.
76. Mauer, S. M., Steffes, M. W., Goetz, F. C., Sutherland, D. E. R., and Brown, D. M., Diabetic nephropathy: a perspective, *Diabetes*, 32(Suppl. 2), 52, 1983.
77. Rasch, R., Prevention of diabetic glomerulopathy in streptozotocin diabetic rats by insulin treatment: the mesangial regions, *Diabetologia*, 17, 243, 1979.
78. Mauer, S. M., Michael, A. F., Fish, A. J., and Brown, D. M., Spontaneous immunoglobulin and complement deposition in glomeruli of diabetic rats, *Lab. Invest.*, 27, 488, 1972.
79. Iidaka, K., McCoy, J., and Kimmelstiel, P., The glomerular mesangium. A quantitative analysis, *Lab. Invest.*, 19, 573, 1968.
80. Nakamoto, Y., Takazakura, E., Hayakawa, H., Kawai, K., Doni, K., Fujioka, M., Kida, H., Hattori, N., and Takeuchi, J., Intrarenal microaneurysms in diabetic nephropathy, *Lab. Invest.*, 42, 433, 1980.
81. Gallo, G., Elution studies in kidneys with linear deposition of immunoglobulin in glomeruli, *Am. J. Pathol.*, 61, 377, 1970.
82. Scheinman, J. I., Fish, A. J., and Michael, A. F., The immunohistopathology of glomerular antigens. II. The glomerular basement membrane, collagen and actomyosin antigens in normal and diseased kidneys, *J. Clin. Invest.*, 54, 1144, 1974.
83. Scheinman, J. I., Steffes, M. W., Brown, D. M., and Mauer, S. M., The immunohistopathology of glomerular antigens. III. Increased mesangial actomyosin in experimental diabetes in the rat, *Diabetes*, 27, 632, 1978.
84. Scheinman, J. I., Fish, A. J., Matas, A. J., and Michael, A. F., The immunohistopathology of glomerular antigens. II. The glomerular basement membrane, actomyosin and fibroblast surface antigens in normal, diseased and transplanted human kidneys, *Am. J. Pathol.*, 90, 71, 1978.
85. Romen, W. and Takahashi, A., Autoradiographic studies on the proliferation of glomerular and tubular cells of the rat kidney in early diabetes, *Virchows Arch. (Cell Pathol.)*, 40, 339, 1982.
86. Levine, J. H., Buse, M. G., Leaming, A. B., and Raskin, P., Effect of streptozotocin induced diabetes on renal ornithine decarboxylase activity, *Diabetes*, 29, 532, 1980.
87. Grant, M. E., Harwood, R., and Williams, I. F., Increased synthesis of glomerular basement membrane collagen in streptozotocin diabetes, *J. Physiol.*, 257, 56p, 1976.
88. Cohen, M. P. and Khalifa, A., Effect of diabetes and insulin on rat renal glomerular protocollagen hydroxylase activities, *Biochim. Biophys. Acta*, 496, 88, 1977.
89. Risteli, J., Koivisto, V. A., Akeblom, H. K., and Kivirikko, K. I., Intracellular enzymes of collagen biosynthesis in rat kidney in streptozotocin diabetes, *Diabetes*, 25, 1066, 1976.
90. Khalifa, A. and Cohen, M. P., Glomerular proto-collagen lysylhydroxylase activity in streptozotocin diabetes, *Biochim. Biophys. Acta*, 386, 332, 1975.
91. Beisswenger, P. J., Glomerular basement membrane. Biosynthesis and chemical composition in the streptozotocin diabetic rat, *J. Clin. Invest.*, 58, 844, 1976.
92. Spiro, R. G., Studies on the renal glomerular basement membrane, *J. Biol. Chem.*, 242, 1915, 1967.
93. Kefalides, N. A. and Denduchis, B., Structural components of epithelial and endothelial basement membranes, *Biochemistry*, 8, 4613, 1969.
94. Duhault, J. and Lonchampt, M., Glucosyltransferase activity and diabetic microangiopathy, *Biomedicine*, 27, 127, 1977.
95. Haft, D. E. and Reddi, A. S., Glucosyltransferase activity in kidney fraction of normal and streptozotocin diabetic rats, *Biochim. Biophys. Acta*, 584, 1, 1979.
96. Beisswenger, P. J. and Spiro, R. G., Human glomerular basement membrane: chemical alteration in diabetes mellitus, *Science*, 168, 596, 1970.
97. Beisswenger, P. J. and Spiro, R. G., Studies on the human glomerular basement membrane. Composition, nature of the carbohydrate units and chemical changes in diabetes mellitus, *Diabetes*, 22, 180, 1973.
98. Beisswenger, P. J., Specificity of the chemical alteration in the diabetic glomerular basement membrane, *Diabetes*, 22, 744, 1973.
99. Westberg, N. G. and Michael, A. E., Human glomerular basement membrane: chemical composition in diabetes mellitus, *Acta Med. Scand.*, 194, 37, 1973.
100. Kefalides, N. A., Biochemical properties of human glomerular basement membrane in normal and diabetic kidneys, *J. Clin. Invest.*, 53, 403, 1974.
101. Hostetter, T. H., Troy, J. L., and Brenner, B. M., Glomerular dynamics in rats with diabetes mellitus, *Kidney Int.*, 14, 725, 1978.

102. Hostetter, T. H., Renal microcirculation in diabetes mellitus, *Acta Endocrinol.*, 97(Suppl. 242), 22, 1981.
103. Mogensen, C. E., Renal functional changes in diabetes, *Diabetes*, 25(Suppl. 2), 872, 1976.
104. Berkman, J. and Rifkin, H., Unilateral nodular diabetic glomerulosclerosis (Kimmelstiel-Wilson): report of a case, *Metabolism*, 22, 715, 1973.
105. Mogensen, C. E., Progression in nephropathy in long term diabetes with proteinuria and effect of initial hypertensive treatment, *Scand. J. Clin. Invest.*, 36, 383, 1976.
106. Mogensen, C. E., Long-term antihypertensive treatment (over six years) inhibiting the progression of diabetic nephropathy, *Acta Endocrinol.*, 97(Suppl. 242), 31, 1981.
107. Mauer, S. M., Steffes, M. W., Azar, S., Sandberg, S. K., and Brown, D. M., The effects of Goldblatt hypertension of development of glomerular lesions of diabetes mellitus in the rat, *Diabetes*, 27, 738, 1978.
108. Rasch, R., Prevention of diabetic glomerulopathy in streptozotocin diabetic rats by insulin treatment: albumin excretion, *Diabetologia*, 18, 413, 1980.
109. Rasch, R., Studies on the prevention of glomerulopathy in diabetic rats, *Acta Endocrinol.*, 97(Suppl. 242), 43, 1981.
110. Mauer, S. M., Brown, D. M., and Steffes, M. W., Studies on the reversibility of kidney changes in experimental diabetes in the rat, *Acta Endocrinol.*, 97(Suppl. 242), 29, 1981.
111. Mauer, S. M., Barbosa, J., Vernier, R. L., Kjellstrand, C. M., Baselmeyer, T. J., Simmons, R. L., Najarian, J. S., and Goetz, F. C., Development of diabetic vascular lesions in normal kidneys transplanted into patients with diabetes mellitus, *N. Engl. J. Med.*, 295, 916, 1976.
112. Mauer, S. M., Miller, K., Goetz, F. C., Barbosa, J., Simmons, R. L., Najarian, J. S., and Michael, A. F., Immunopathology of renal extracellular membranes in kidneys transplanted into patients with diabetes mellitus, *Diabetes*, 25, 709, 1976.
113. Ballinger, W. F. and Lacy, P. E., Transplantation of intact pancreatic islets in rats, *Surgery*, 72, 175, 1972.
114. Leonard, R. J., Lazarow, A., and Hegre, O. D., Pancreatic islet transplantation in the rat, *Diabetes*, 22, 413, 1973.
115. Steffes, M. W., Brown, D. M., Basgen, J. M., Matas, A. J., and Mauer, M., Glomerular basement membrane thickness following islet transplantation in the diabetic rat, *Lab. Invest.*, 41, 116, 1977.
116. Kemp, C. B., Knight, M. J., Scharp, D. W., Lacy, P. E., and Ballinger, W. F., Transplantation of isolated pancreatic islets into the portal vein of diabetic rats, *Nature (London)*, 244, 447, 1973.
117. Kemp, C. B., Knight, M. J., Scharp, D. W., Ballinger, W. F., and Lacy, P. E., Effect of transplantation site on results of pancreatic islet isografts in diabetic rats, *Diabetologia*, 9, 486, 1973.
118. Brown, J., Molnar, I. G., Clark, W., and Mullen, Y., Control of experimental diabetes mellitus in rats by transplantation of fetal pancreas, *Science*, 184, 1377, 1974.
119. Bretzel, R. G., Schneider, J., Draeger, K. E., Weise, M., and Federlin, K., III, Reversibility of diabetic glomerular and tubular changes by islet transplantation, *Diabetologia*, 19, 260, 1980.
120. Price, R. G. and Spiro, R. G., Studies on the metabolism of renal glomerular basement membrane, *J. Biol. Chem.*, 252, 8597, 1977.
121. Marliss, E. B., Nakhooda, A. F., Poussier, P., and Sima, A. A. F., The diabetic syndrome of the "BB" Wistar rat: possible relevance to type I (insulin dependent diabetes) in man, *Diabetologia*, 22, 225, 1982.
122. Seemayer, T. A., Oligny, L. L., Tannenbaum, G. S., Goldman, H., and Colle, E., Dynamic time course studies of the spontaneously diabetic BB Wistar rat, *Am. J. Pathol.*, 106, 237, 1982.
123. Like, A. A., Butler, W., Williams, R. M., Appel, M. C., Weringer, E. J., and Rossini, A. A., Spontaneous autoimmune diabetes mellitus in the BB rat, *Diabetes*, 31(Suppl. 1), 7, 1982.
124. Nakhooda, A. F., Like, A. A., Chappel, C. I., Murray, F. T., and Marliss, E. B., The spontaneous diabetic Wistar rat. Metabolic and morphologic studies, *Diabetes*, 26, 100, 1077.
125. Nakhooda, A. F., Like, A. A., Chappel, C. I., Wei, C. -N., and Marliss, E. B., The spontaneously diabetic Wistar rat (the BB rat). Studies prior to and during development of the overt syndrome, *Diabetologia*, 14, 199, 1978.
126. Seemayer, T. A., Oligny, L. L., Tannenbaum, G. S., Goldman, H., and Colle, E., Spontaneous diabetes mellitus in the BB Wistar rat, *Am. J. Pathol.*, 101, 485, 1980.
127. Patel, Y. C., Wheatley, T., Malaisse-Lagae, F., and Orci, L., Elevated portal and peripheral blood concentration of immunoreactive somatostatin in spontaneously diabetic (BBL) Wistar rats. Suppression with insulin, *Diabetes*, 29, 757, 1980.
128. Appel, M. C., Like, A. A., Rossini, A. A., Carp, D. B., and Miller, T. B., Hepatic carbohydrate metabolism in the spontaneously diabetic Bio-Breeding Worcester rat, *Am. J. Physiol.*, 240, E83, 1981.
129. Like, A. A., Rossini, A. A., Guberski, D. L., and Appel, M. C., Spontaneous diabetes mellitus: reversal and prevention in the BB/W rat with antiserum to rat lymphocytes, *Science*, 206, 1421, 1979.

130. Seemayer, T. A., Schurch, W., and Kalant, N., B cell lymphoproliferation in spontaneously diabetic BB Wistar rats, *Diabetologia*, 23, 261, 1982.
131. Dryberg, T., Nakhooda, A. F., Backkeskov, S., Lernmark, A., Poussier, P., and Marliss, E. B., Islet cell surface antibodies in the spontaneously diabetic BB Wistar rat, *Diabetes*, 31, 278, 1982.
132. Sternthal, W., Like, A. A., Sarantis, K., and Braverman, L. E., Lymphocytic thyroiditis and diabetes in the BB/Wistar rat: a new model of autoimmune endocrinopathy, *Diabetes*, 30, 1058, 1981.
133. Like, A. A., Appel, M. C., and Ericson, S., Renal glomerular studies in the Bio Breeding/Worcester (BB/W) diabetic rats, *Diabetes*, 29, 80A, 1980.
134. Lee, S. M., The effect of a high fiber diet on diabetic nephropathy in the db/db mouse, *Diabetologia*, 22, 349, 1982.
135. Mauer, S. M., Steffes, M. W., Michael, A. F., and Brown, D. M., Studies of diabetic nephropathy in animals and man, *Diabetes*, 25(Suppl. 2), 850, 1976.
136. Like, A. A., Lavine, R. L., Poffenbarger, P. L., and Chick, W. L., Studies in the diabetic mouse. VI. Evolution of glomerular lesions and associated proteinuria, *Am. J. Pathol.*, 66, 193, 1972.
137. Wehner, H., Höhn, D., Faix-Schade, U., Huber, H., and Walzer, P., Glomerular changes in mice with spontaneous hereditary diabetes, *Lab. Invest.*, 27, 331, 1972.
138. Gerritsen, G. C., The Chinese hamster as a model for the study of diabetes mellitus, *Diabetes*, 31(Suppl. 1), 14, 1982.
139. Gerritsen, G. C. and Blanks, M. C., Characterization of Chinese hamsters by metabolic balance, glucose tolerance and insulin secretion, *Diabetologia*, Suppl. 10, 493, 1974.
140. Frankel, B. J., Heldt, A. M., and Grodsky, G. M., Insulin and glucagon release in the diabetic Chinese hamster: differences among inbred sublines, *Diabetologia*, 22, 292, 1982.
141. Soret, M. G., Peterson, T., Block, E. M., and Chang, A. Y., Glomerular capillary basement membrane thickening in spontaneously diabetic Chinese hamsters, *Diabetes*, 27(Suppl. 2), 446, 1978.
142. Shirai, T., Welsh, G. W., and Sims, E. A. H., Diabetes mellitus in the Chinese hamster. II. Evolution of renal glomerulopathy, *Diabetologia*, 3, 266, 1967.
143. Lawe, J. E., Renal changes in hamster with hereditary diabetes mellitus, *Arch. Pathol.*, 79, 166, 1962.
144. Ginsberg, L. C., Wyse, B. M., and Chang, A., Analysis of glycosaminoglycan from normal and Chinese hamster cells, *Diabetes*, 30, 393, 1981.
145. Berenson, G. S., Ruiz, B. S., Dalferes, E. R., Dugan, F. A., and Radhakrishnamurthy, B., Acid mucopolysaccharide changes in diabetic kidneys, *Diabetes*, 19, 161, 1970.
146. Soret, M. G., Dulin, W. E., Mathews, J., and Gerritsen, G. C., Morphologic abnormalities observed in retinal, pancreas and kidney of diabetic Chinese hamsters, *Diabetologia*, 10, 567, 1974.
147. Bloodworth, J. M. B., Experimental diabetic glomerulosclerosis. II. The dog, *Arch. Pathol.*, 79, 113, 1965.
148. Janle-Swain, E., Thornhill, J. A., Carter, J. M., Hinsman, E., Jackson, H. D., and Ash, S. R., Case study of a diabetic dog with chronic membranous glomerulopathy treated with continuous intraperitoneal insulin infusion, *Am. J. Vet. Res.*, 43, 2044, 1982.
149. Bloodworth, J. M. B., Engerman, R. L., and Powers, K. L., Experimental diabetic microangiopathy. Basement membrane statistics in the dog, *Diabetes*, 18, 455, 1969.
150. Pierluissi, J. and Campbell, J., Metasomatotrophic diabetes and its induction: basal insulin secretion and insulin release responses to glucose, glucagon, arginine and meals, *Diabetologia*, 18, 223, 1980.
151. Hausler, H. R., Sibay, T. M., and Campbell, J., Diabetes in a dog following diabetes induced by growth hormone, *Diabetes*, 13, 122, 1964.
152. Engerman, R. L. and Kramer, J. W., Dogs with induced or spontaneous diabetes as models for the study of human diabetes mellitus as models for the study of human diabetes mellitus, *Diabetes*, 31(Suppl. 1), 26, 1982.
153. Steffes, M. W., Buchwald, H., Wigness, B. D., Groppoli, T. J., Rupp, W. M., Rhode, T. D., Blackshear, P. J., and Mauer, S. M., Diabetic nephropathy in the uninephrectomized dog: microscopic lesions after one year, *Kidney Int.*, 21, 721, 1982.
154. Janle-Swain, E., unpublished data.
155. Janle-Swain, E., Carter, J. M., and Doyon, C., Administration of insulin in TPN fluids during the post-alloxanization period of diabetes induction in dogs, *Conference of Research Workers in Animal Diseases*, Chicago, 73, 1981.
156. Black, H. E., Rosenblum, I. Y., and Caper, C. C., Chemically induced (streptozotocin-alloxan) diabetes mellitus in the dog, *Am. J. Pathol.*, 98, 295, 1980.
157. Issekutz, B., Issekutz, T. B., Elahi, D., and Borkow, I., Effect of insulin infusion on the glucose kinetics in alloxan-streptozotocin diabetic dogs, *Diabetologia*, 10, 323, 1974.
158. Gepts, W. and Toussaint, D., Spontaneous diabetes in dogs and cats. A pathological study, *Diabetologia*, 3, 249, 1967.
159. Williams, M., Gregory, R., Schall, W., Rovner, D., and Padgett, G., Diabetes mellitus in a colony of golden retrievers, *Fed. Proc. Fed. Am. Soc. Exp. Biol.*, 39, 637, 1980.

160. Williams, M., Gregory, R., Schall, W., Gossain, V., Bull, R., and Padgett, G., Characterization of naturally occurring diabetes in a colony of golden retrievers, *Fed. Proc. Fed. Am. Soc. Exp. Biol.*, 40, 740, 1981.
161. Kramer, J. W., Nottingham, S., Robinette, J., Lenz, G., Sylvester, S., and Dessouky, M. I., Inherited, early onset, insulin-requiring diabetes mellitus of keeshound dogs, *Diabetes*, 29, 558, 1980.
162. Christiansen, J. S., Frandsen, M., Swendsen, P. A., Gammelgaard, J., and Parving, H. -H., Rapid changes in kidney function in diabetics and normal man, *Acta Endocrinol.*, 97(Suppl. 242), 11, 1981.
163. Williamson, J. R. and Kilo, C., Basement-membrane thickening and diabetic microangiopathy, *Diabetes*, 25(Suppl. 2), 925, 1976.
164. Del Castillo, E., Fuenzalida, R., and Uranga, J., Increased glomerular filtration rate and glomerulopressin activity in diabetic dogs, *Horm. Metab. Res.*, 9, 46, 1977.
165. Ditzel, J. and Junker, K., Abnormal glomerular filtration rate, renal plasma flow and renal protein excretion in recent and short-term diabetics, *Br. Med. J.*, 2, 13, 1972.
166. Viberti, G. C., Pickup, J. C., Jarrett, R. J., and Keen, H., Effect of control of blood glucose on urinary excretion of albumin and β2 microglobulin in insulin dependent diabetes, *N. Engl. J. Med.*, 300, 368, 1979.
167. Engerman, R., Bloodworth, J. M. B., and Nelson, S., Relationship of microvascular disease in diabetes to metabolic control, *Diabetes*, 26, 760, 1977.
168. Howard, C. F., Nonhuman primates as models for the study of human diabetes mellitus, *Diabetes*, 31(Suppl. 1), 37, 1982.
169. Widness, J. A., Schwartz, R., Thompson, D., Tsuboi, K. K., Reynolds, W. A., Chez, R. A., and Schwartz, H. C., Hemoglobin A1c in the glucose-intolerant streptozotocin-treated or pancreatectomized macaque monkey, *Diabetes*, 27, 1182, 1978.
170. Gibbs, G. E., Wilson, R. B., and Gifford, H., Glomerulosclerosis in the long-term alloxan diabetic monkey, *Diabetes*, 15, 258, 1966.
171. Jones, C. W., Reynolds, W. A., and Hoganson, G. E., Streptozotocin diabetes in the monkey: plasma levels of glucose, insulin, glucagon and somatostatin, with corresponding morphometric analysis of islet endocrine cells, *Diabetes*, 29, 536, 1980.
172. Bloodworth, J. M. B., Engerman, R. L., and Anderson P. J., Microangiopathy in the experimentally diabetic animal, in *Vascular and Neurological Changes in Early Diabetes*, Camerine-Davalos, R. A. and Cole, H. S., Eds., Academic Press, New York, 1973, 245.
173. Stout, C., Folse, D., Weeford, A., Whorton, E., Davis, E., Beathard, G., Crosby, W. M., Kling, O. R., Williams, G. R., Price, W. E., Geyer, J. R., Padala, R., and Kimmelstiel, P., Diffuse glomerulosclerosis in pancreatectomized baboons, *Fed. Proc. Fed. Am. Soc. Exp. Biol.*, 39, 638, 1980.

INDEX

A

B

C

N

O

P

Q

R

S

T

U